Durkheim's Philosophy of Science and the Sociology of Knowledge

Science and Its Conceptual Foundations,
David L. Hull, editor

Durkheim's Philosophy of Science and the Sociology of Knowledge

Creating an Intellectual Niche

Warren Schmaus

The University of Chicago Press
Chicago and London

Warren Schmaus is Associate Professor of Philosophy in the Department of
Humanities and Director of the Program for Science and Technology in Context
at Illinois Institute of Technology.

The University of Chicago Press, Chicago 60637
The University of Chicago Press, Ltd., London
© 1994 by The University of Chicago
All rights reserved. Published 1994
Printed in the United States of America

03 02 01 00   99 98 97 96 95 94   5 4 3 2 1

ISBN (cloth): 0-226-74251-2
ISBN (paper): 0-226-74252-0

Library of Congress Cataloging-in-Publication Data

Schmaus, Warren.
    Durkheim's philosophy of science and the sociology of knowledge :
creating an intellectual niche / Warren Schmaus.
        p.   cm. — (Science and its conceptual foundations)
    Includes bibliographical references and index.
    ISBN 0-226-74251-2. — ISBN 0-226-74252-0 (pbk.)
    1. Science—Philosophy.   2. Science—History.   3. Science—Social
aspects.   4. Durkheim, Emile, 1858–1917.   I. Title.   II. Series.
Q175.S356   1994   93-28623
                                                           CIP

To Tekla and Alexander

Contents

Preface and Acknowledgments

My interest in Durkheim began when I was looking for some interesting ways in which I could use what I had learned by doing my dissertation on Auguste Comte. To the extent that Comte's views on scientific method were not well understood, I thought at the time, neither would the connections between him and subsequent thinkers. In France at least, Durkheim was clearly the most important thinker to be stimulated by Comte's suggestion that philosophical reflection upon the methods of science was an important first step toward establishing a science of society. A paper I wrote comparing the philosophies of science of Durkheim and Comte (Schmaus 1985) was kindly accepted for publication by *Studies in History and Philosophy of Science*. Many of the arguments and interpretations in part 2 of this book made their first appearance in that paper, versions of which were presented in 1983 at the annual meeting of Cheiron and, thanks to an invitation from Robert Richards, at one of the monthly meetings of Chicago Group in the History of the Social Sciences, organized by George Stocking, at the Morris Fishbein Center for the Study of the History of Science and Medicine at the University of Chicago.

An invitation from Larry Laudan to participate in a conference on "Testing Theories of Scientific Change" in 1986 led me to investigate the extent to which Durkheim actually followed his philosophy of science in one of his substantive sociological works, *Suicide*. I was struck by just how much this work conformed to his explicit methodology and I conceived the idea of a book-length study in which I would include similar analyses of *The Division of Labor in Society* and *The Elementary Forms of the Religious Life*.

A sabbatical during the academic year 1987–88 made it possible to devote my full time to this project. I am grateful to the philosophy department at the University of Chicago for inviting me to be a visiting scholar that year, a position that provided me with privileges at the Regenstein Library that I deeply appreciated. I also would like to thank the Illinois Institute of Technology for granting me this sabbatical and in particular the Department of Humanities for providing a generous faculty development grant.

During this sabbatical I developed further some thoughts on Durkheim's sociology of knowledge first presented in another Cheiron paper and became intrigued with his ideas on the division of intellectual labor. I presented papers on these topics at the annual meeting of the Society for Social Studies of Science in 1987 and 1990 and at the Center for the Interdisciplinary Study of Science and Technology at Northwestern University in 1988. The latter presentation resulted from an invitation from David Hull, who has probably given me more intellectual help and encouragement than any other individual since I first arrived in Chicago in 1980. An earlier version of my account in chapter 2 of Durkheim on specialization also appeared in *Social Epistemology* (Schmaus 1992) as the result of an invitation from Steve Fuller.

In addition to David Hull, the editor of this series, I would like to thank Susan Abrams at the University of Chicago Press and the anonymous reviewers of this manuscript for their useful suggestions and advice. Thanks are also due to Stephen P. Turner for helpful discussions and suggestions and for sending me papers and other materials and to Robert Strikwerda for sending me a copy of his dissertation on Durkheim. I would also like to thank those who have participated in Illinois Institute of Technology's fortnightly philosophy colloquium over the years who have read and commented on more papers on Durkheim than they would care to count. The participants in this group who have criticized my work on Durkheim include Michael Davis, Jill Gordon, Douglas Jesseph, Robert Ladenson, Stephen Menn, Rachel Rue, Fay Horton Sawyier, John Snapper, David Weberman, and Vivian Weil. My father-in-law, Richard Thome, deserves my gratitude for going above and beyond the duties incumbent upon that relationship and reading and commenting on a draft of the entire manuscript. Special thanks go to the librarians at the Galvin Library at the Illinois Institute of Technology. Above all, however, I owe a profound debt of gratitude to my wife, Constance, and to my children, Tekla and Alexander, for their patience and forbearance during this project.

Finally, I would like to thank the following publishers for kindly allowing me to use material from two of my articles: Pergamon Press, Inc., for material from "Hypotheses and Historical Analysis in Durkheim's Sociological Methodology: A Comtean Tradition," *Studies in History and Philosophy of Science* 16 (1985):1–30; Taylor and Frances, Ltd., for material from "Research Programs as Intellectual Niches," *Social Epistemology* 6 (1992):13–22.

I

Creating a Niche

1 *Interpreting Durkheim*

Émile Durkheim (1858–1917) is considered one of the founders and patron saints of sociology. Although he could hardly be said to have invented the idea of a science of society, he played an important role in establishing sociology as a tradition of scientific research. In the words of his collaborator and nephew Marcel Mauss, Durkheim provided sociology with "a method and a subject matter" (Mauss 1928:v). This conception of the methods and domain of sociological inquiry was not, however, to be left for others to draw out from some paradigm scientific achievement, as Kuhn (1962) might suggest. Durkheim explicitly defined the explanatory goals of sociology—both in the sense of the kind of facts it is supposed to explain and in the sense of the type of explanation it must provide—and the methods appropriate for pursuing these goals. He wrote extensively on such topics, drawing heavily on the philosophical tradition in which he was educated.

Durkheim not only worked out a philosophy of the social sciences but carried out a lifelong program of sociological research characterized by these philosophical conceptions. To the extent that philosophers now tend to ignore Durkheim, we are missing a favorable opportunity to analyze the relationship between the development of a new science and the philosophy of science that inspired it. His career also affords an example of the way in which the philosophy of science develops through empirical scientific work. Historians and sociologists, as well as philosophers of science, will learn from this study how Durkheim expanded and refined John Stuart Mill's methods of eliminative induction and Francis Bacon's method of crucial test through actual use of these methods. Explaining Durkheim's thought in the context of nineteenth-century philosophy of science, this book illustrates the value of intellectual history for understanding the history of science and bucks the trend toward social histories of science. Sociologists will find evidence in this account of Durkheim's career that intellectual factors such as prevailing conceptions of method have the same right to be considered social facts as do things like social structure or social interests.

This book should prove to be of more than just historical interest for sociologists and anthropologists who consider Durkheim one of the patron saints of their disciplines. They will discover that Durkheim's thought is much richer than the positivism and empiricism with which he has been labeled since Talcott Parsons. Social scientists, as well as philosophers concerned with the social sciences, will be interested in Durkheim's attempt to reconcile a humanistic concern for the meanings of social facts with a desire to follow the methods of the natural sciences. Durkheim's thought, I will argue, provides a model for resolving the current crisis in sociology in which theoretical and empirical studies are going their separate ways and even empirical investigations are divided between interpretative and more positivist approaches. This crisis is exacerbated by the antitheoretical assumptions shared by the prevailing positivist, nominalist, and phenomenological methodologies in the social sciences today.

In my study of Durkheim, sociologists will find an alternative hypothetico-deductive methodology that links the theoretical with the empirical and the meanings of social facts with their explanations. Durkheim's hypothetico-deductive methodology allows for the postulation of what he calls "collective representations," or mental representations shared among members of a social group, as a kind of theoretical entity modeled on the unobservable entities of the natural sciences. By associating the meanings of social facts with these representations, one can provide a way for these meanings to play a role in scientific explanations. Such an approach is only now being rediscovered by philosophers and psychologists who have become dissatisfied with behaviorist, reductive materialist, and artificial intelligence models of the mind. Recognizing the need to take meanings into account to explain and even to classify human actions, these thinkers attempt to provide a naturalistic account of the way in which meanings are represented in the brains of the agents responsible for these actions. Of course, Durkheim was able to provide an explanatory role for meanings just to the extent that he did not distinguish explanation from definition in the first place. Nevertheless, we can retain from Durkheim the insights that types of social actions cannot be individuated without recourse to their shared meanings and that, for these meanings to play an explanatory role, representations of these meanings must be physically realized in the brains of social actors.

Above all, however, my purpose in this book is to explain the direction taken by Durkheim's research program in terms of its core conceptions of the goals and methods of sociology. In the beginning of his career, when he wrote *The Division of Labor in Society* (1893b) and *Suicide* (1897a), Durkheim seemed to be concerned with the sociology of law and morality in contemporary European societies. The first book draws on comparative law to argue that a new form of social solidarity is evolving in modern societies. The second argues for the efficacy of social causes by demonstrating that they account for variations in statistical rates of suicide. In an article published at the turn of the century, he characterized his sociological research as the study of moral and legal rules, with respect to both their causes, or, as he says, their "development and genesis," and their effects or functions (1900b:126). Fifteen years later, however, he designated his research as the sociology of religion and knowledge, citing *The Elementary Forms of the Religious Life* (1912a) and an article on primitive classifications written with Marcel Mauss and published in 1903 (1915a:10).[1] In these works he drew on ethnographic evidence to explain the formation of collective representations in so-called primitive societies.

Only a superficial interpretation of Durkheim's works, however, would characterize this shift in empirical focus as a fundamental change in his conception of sociology. Durkheim always conceived the goal of sociology to be the explanation of the causes and functions of collective representations. Even in his earliest works, he thought of moral and legal rules as consisting ultimately in collective representations. The religious, moral, legal, and other collective representations characteristic of a society constitute what he called its "collective consciousness." As W. Paul Vogt (1979:102) argues, in a broad sense all of Durkheim's sociology can be considered a sociology of knowledge or at least a sociology of belief or mental phenomena. In fact, as early as the 1890s, Durkheim had characterized his course on the history of socialist thought in this way, that is, as an investigation into the causes that can explain why people hold socialist ideas (1928a:2–3). Durkheim shifted his attention to the study of primitives because he thought that this was the best way to achieve his goals and not because his goals had changed.

Durkheim's sociological research program is defined by its explanatory goals in two senses of the term, one having to do with the content and the other with the form of a sociological explanation, as well as by its methods. Part 2 of this book provides an analysis of these goals and methods. In chapter 3 I argue that Durkheim identified the content, or what Mauss called the "subject matter," of sociology with collective representations, which he regarded as real but unobservable entities. The second sense of explanatory goal concerns the type of explanation he wanted to provide. In chapter 4, I show that he adhered to an essentialist, causal model of explanation. Durkheim's essentialism, in which the same kind of fact must always proceed from the same kind of causal entity, entails a separation of sociological and psychological causes into two distinct kinds. I then argue in chapter 5 that he subscribed to a hypothetico-deductive methodology appropriate for theorizing about unobservable causal entities. His methodology also provided criteria for the comparative evaluation of theories. Many of these criteria, however, are only implicit in his concept of the explanatory goals of sociology. That is, his judgments concerning his rivals' theories are grounded in his philosophical assumptions about models of causation, the status of theoretical entities, the relations among the sciences, and the role of intentionality.

In his empirical works Durkheim employed several modes of theory evaluation for which he did not explicitly provide in *The Rules of Sociological Method* (1895a). Specifically, he eliminated the theories of his rivals not only because of their empirical inadequacies but for their conceptual problems as well, including ambiguity, question begging, inconsistency, methodological weaknesses, and so on.[2] However, Durkheim's rejecting theories for such reasons does not necessarily constitute a departure from his explicit methodology. There are at least two hypotheses regarding the interpretation of Durkheim. The first would be that Durkheim's methodology is only that which he made explicit in *The Rules* and nothing more. On this hypothesis, any mode of inference Durkheim used that is not provided for by the very letter of *The Rules,* such as rejecting a rival's theory because of its internal inconsistencies or ambiguities, would count as a "deviation" from his methodology. The alternative hypothesis would be that in *The Rules* Durkheim discussed only those aspects of his methodology that he felt were not generally accepted and needed to be defended, including his concept of the explanatory goals of sociology. On this assumption, we could say that Durkheim neither articulated nor defended many kinds of arguments in *The Rules* that he nevertheless employed elsewhere because he felt that there was nothing particularly sociological or novel about these kinds of arguments. Rather, he would have thought, these patterns of persuasion are common to all intellectual endeavors and are what make the pursuit of knowledge possible in the first place.

In this work I have chosen the latter interpretative hypothesis. Durkheim did not feel a need in *The Rules of Sociological Method* to defend all of his methods of criticizing alternative points of view because he regarded some of these methods as accepted by the scientific community. In a work on the history of educational thought, Durkheim argued that in spite of the advent of empirical methods in the sciences, there will always be a place for the art of persuasion even in the physical sciences:

If science, to the extent that it progresses and arrives at more precise and better demonstrated results, makes controversy retreat, it cannot however drive it away from this world. There is thus always a place for discussion in intellectual life and for the art of discussion in education.[3]

Dialectics, he said, is the art of discussion and includes methods of analogy, comparison, generalization, and supposition. An exhaustive account of these methods would belong not in a treatise on sociological method so much as in a textbook on logic or rhetoric, which, I believe, Durkheim had no interest in writing.

One cannot learn all the rules that govern a research program simply by reading its methodological manifestos if these are concerned with only those aspects of its methodology that are not generally accepted. To the extent to which scientists want to succeed in persuading the relevant research community to accept their ideas, they must conform to patterns of argument that are already accepted by their intended audience. In order to get a sense of what are considered acceptable types of argument in the relevant scientific or academic community, one must analyze the arguments scientists actually make. Much of the thinking of scientists is directed toward producing arguments that conform to what they believe to be the rules governing persuasion within that community. Scientists sometimes will even abandon those of their own ideas that they believe they cannot defend in admissible ways. The social rules governing acceptable patterns of argument must then be included as an important part of a scientist's methodology whenever one attempts to account for a scientist's beliefs or actions in terms of his or her methods.

For my purposes, therefore, it will not suffice simply to offer an interpretation of *The Rules* and his other writings on method. I will also need to analyze the kinds of arguments Durkheim made in his evaluations of the sociological theories of his competitors and in defense of his own theories. Part 3 of this book provides this analysis, discussing *The Division of Labor, Suicide,* and *The Elementary Forms of the Religious Life,* in chapters 6, 7, and 8, respectively. I show not only how the forms of argument used in these works are grounded in Mill's methods of elimination and Bacon's notion of a crucial test but how Durkheim developed these methods further. In Durkheim's works, especially *Suicide* and *The Elementary Forms of the Religious Life,* these methods became ways not just to eliminate causal hypotheses but to criticize the research programs of his rivals. For instance, he expanded the concept of a crucial test beyond that of an experiment that logically refutes one of two competing causal hypotheses that offer opposite predictions. For Durkheim the notion of a crucial test also included the situation in which one hypothesis fails to make any prediction at all or explains the experimental results in some ad hoc, question begging, or other unacceptable manner. That is, a crucial test can be a confrontation between competing research programs in which one of these programs fails to live up to social rules governing scientific argument. I will summarize Durkheim's methodological development in a ninth, concluding chapter in part 4.

According to Durkheim rules of method and argument can be considered social rules that govern the competition among research programs. For Durkheim science is an "eminently social thing" (1910a(iii)(2):44), a cooperative enterprise with a strong tradition of methodological rules that are "veritable social institutions" (1903a(i):395). The rules that govern scientific research, Durkheim thought, are not necessarily characteristic of science as a whole. Each scientific discipline, he said, must have "its own object, its method, its spirit" (1902b:2). "In order for sociology to be possible it must above all have an object that belongs only to it," he wrote (1897a:ix). The methods of each discipline "coordinate the steps [*démarche*]" of the scientists who cultivate it (1902b:359). According to Durkheim these rules of method are invested with an authority analogous to that of law and morality (1910a(iii)(2):44; cf. 1902b:359). Durkheim thought that both science and morality are characterized by rules that carry the same sort of authority as religion. Science differs from religion as a social institution, he said, only to the extent that science allows a greater degree of freedom of opinion with respect to the content of beliefs (1899a(ii):157; 1901h:425).

On my interpretative hypothesis, Durkheim's working methods can be regarded as sophisticated refinements of his explicit methodology that reveal the germs of a sociology of science. Other scholars, however, have chosen the alternative interpretative hypothesis and regarded his actual procedures as illegitimate departures from those he prescribed in *The Rules of Sociological Method*. These interpreters have read this work as advocating some sort of crude empiricism and seem to have thought that his methods are—or perhaps should be—concerned only with the manipulation of data and not with comparative theory evaluation. Ernest Wallwork, for instance, has argued that *The Rules* prescribes a "positivist" methodology, while Durkheim's substantive works proceed by a "dialectical" method of argument (1972:5–7). Sometimes the objection that Durkheim deviated from his explicit methodology has been made only with regard to his later works, especially *The Elementary Forms of the Religious Life*. Roger Lacombe, for example, has argued that this work and Durkheim's article on incest are concerned only with the interpretation of primitive cultures and not with the establishing of laws.[4] Similar charges regarding a deviation from his expressed methodology, however, have also been raised against Durkheim's earlier empirical works, especially *Suicide*. Maurice Halbwachs (1930:3) and Steve Taylor (1982:35) have argued that *Suicide* persuades us not so much through an appeal to facts and statistics as through a use of dialectic and eliminative arguments. Jack Douglas (1967:153) dismissed *Suicide* as an unscientific piece of "casuistry." More recently, Mike Gane (1988:50–51) considers the "deductive" method of *Suicide* to be the very reverse of the "inductive" method endorsed in *The Rules*. This unfortunately still rather prevalent interpretation of Durkheim's methodology makes him out to be a rather uninteresting and inconsistent thinker who has little to offer as a model for contemporary sociology.

One might object that not all of the rules governing a scientific community can be ascertained through analyzing the texts they produce because some of these rules are not directly relevant to written persuasion but concern professional relationships among colleagues (for example, Hull 1988). I have tried to discover Durkheim's understanding of such relationships through an analysis of the way in which he and his collaborators organized their journal, *L'Année sociologique*. In the following chapter I pursue this analysis to the extent that I am able, drawing on book reviews, Mauss's autobiography, posthumously published letters, and other sources. Admittedly, more could be learned about this aspect of Durkheim's research program had his personal papers not been destroyed in World War II.[5] Nevertheless, there remains sufficient evidence that Durkheim tolerated considerable disagreement with his collaborators. Some, for example, rejected his realist conception of the goals of sociology and his appeal to unobservable entities. As I will show in chapter 3, François Simiand's review of *Suicide,* which attacked Durkheim's concepts of social forces and suicidogenic currents, provides an instance of this antirealist attitude among some of Durkheim's collaborators (Simiand 1898:649–50). These were not necessarily followers or even those with whom he found himself in intellectual agreement. For such purposes as founding a journal or gaining a place on the curriculum, he seems to have understood, one may form professional allegiances with those with whom one disagrees intellectually. That intellectual commitments do not reduce to professional interests or group membership is also one of my core assumptions. It is simply a mistake to assume that all of Durkheim's collaborators on this journal pursued the same explanatory goals and employed the same methods. His conceptions of the goals and methods of sociology were shared by at best a small number of these collaborators, including Marcel Mauss, Henri Hubert, Paul Fauconnet, and to some extent Célestin Bouglé.

To say that Durkheim pursued a coherent plan of research and that he remained consistent in his methods and explanatory goals is not to say that his thinking never changed. In fact, he underwent several changes in his explanatory theories. He replaced his early concepts of moral or social density and mechanical and organic solidarity with the concept of social integration and regulation. These explanatory concepts gave way in turn to more detailed analyses of the collective representations that hold a society together. As I argue in part 3, much of Durkheim's theoretical development can be interpreted as a struggle to solve conceptual problems in his theories and to bring his theories more in line with his explanatory goals. I will also show how this struggle contributed to his decision to turn to the study of primitives.

Durkheim's sociological theories reflect the philosophical tradition in which he was educated. As Martin Hollis (1985:233) suggests, Durkheim may be considered a "way of ideas" philosopher. This philosophical tradition, which goes back to Plato and Aristotle, takes its name from the fact that it identifies knowledge, thought, and the meanings of our concepts with mental representations or "ideas" in the head. For concepts that referred to concrete objects, these representations were quite literally supposed to be mental pictures. John Locke, in his *Essay concerning Human Understanding* (1690), had tried to explain how more abstract notions are synthesized from these concrete ideas in the minds of individuals. Analogously, Durkheim had originally sought to explain how the collective representations that constitute our moral and legal rules and other abstract concepts had been synthesized from simpler, more concrete representations. Just as individual psychological processes were thought to be governed by the laws of the association of ideas, Durkheim had hoped that sociology would discover the "laws of collective ideation" that govern this synthesis of collective representations (1898b:47 n. 1; 1901c:xviii). For Durkheim, mental representations were real entities governed by natural laws not unlike the laws governing entities in physics and chemistry. He was seeking a naturalistic account of the formation of the shared mental states in which the meanings of a society's beliefs and practices are represented.

However, the way of ideas tradition proved inadequate for Durkheim's purposes. As Durkheim realized, Locke's notion of an abstract general idea had been criticized by philosophers beginning with George Berkeley. In a lecture on Jean-Jacques Rousseau first given in the academic year 1901–2 we find Durkheim attributing to Étienne Bonnot de Condillac the insight that there can be no abstract general ideas (*connaissances*) without language.[6] Thus in order to provide an account of the meaning of the moral rules and other general concepts shared by societies, Durkheim needed to find an alternative to the traditional notion of abstract general ideas. Furthermore, Durkheim did not believe that the formation of collective representations, especially those that characterize contemporary societies, could be explained solely in terms of the sort of individual psychological processes that Locke had invoked in his account of abstract general ideas.

As I explain in chapter 5, Durkheim had always believed that the way to analyze our collective representations was to trace their development back through the historical record. Around the time when he began reading in the comparative study of religions, however, it seems that he began to think that one could push this analysis back even further through the study of ethnographies. Specifically, he thought that the answer to the problem of the formation of collective representations could be found in the study of the formation of religious ideas in so-called primitive societies. Durkheim, that is, shifted the focus of his empirical research to the sociology of religion and knowledge in order to solve certain conceptual problems that he encountered in the sociology of morality and law. He characterized this period, around 1894–95, as marking a "line of demarcation in the development of my thought" (1907b:613). From this period we find him increasingly emphasizing the role of linguistic and other symbols in the process of ideation.[7] In *The Elementary Forms of the Religious Life* he sketched out a theory of the formation of collective representations in which arbitrary and conventional symbols played a significant role. In the concluding chapter I will evaluate Durkheim's success in explaining shared meanings and solving the conceptual problems with his theory of collective representations. I will argue that although it may be necessary to associate the meanings of social facts with mental representations for purposes of explaining social action, it is unnecessary and even disadvantageous to *identify* these meanings with mental representations.

Durkheim's shift to the study of primitive religions also had to do, at least in part, with his perceptions of competing research programs. He recognized that one research program can compete successfully with other research programs for intellectual recognition only as long as it pursues a clear, coherent set of goals, with methods that are conducive to those goals, in subject areas where its proponents have some reasonable expectations to achieve these goals. By solving conceptual problems in his research program through the study of primitive beliefs he believed he could then return to compete more effectively with other research programs studying law and morality. Thus, having at least begun in *The Elementary Forms of the Religious Life* to work out certain difficulties with collective representations, we find him at the end of his career turning once again to the sociology of morality (for example, 1920a). That is, he understood the idea of making a strategic retreat in order better to prepare for the next battle.

Durkheim's choice of research speciality, it turns out, was not just a personal decision but in part the result of intense negotiation with his professional colleagues. He saw his research program as engaged in a struggle for intellectual existence with other research programs both within and outside his circle of collaborators on *L'Année sociologique*. As we shall see in the following chapter, the way in which Durkheim and his collaborators divided up the work on this journal illustrates Durkheim's general views about how the division of intellectual labor provides a solution to this struggle. For Durkheim, for a group of collaborators to avoid competing with one another can be a successful survival strategy that benefits all concerned.

I describe Durkheim's shifting of research specialties in terms of the creation of intellectual "niches" that permit the continued survival of his research program. These were intellectual and not institutional niches he was creating, as he was competing not merely for an academic post in France but for international intellectual recognition for himself and for his sociological research program. Different intellectual niches within the same research program are created as one brings to new empirical domains the same intellectual commitments to conceptions of method and explanation. Durkheim consciously sought over the course of his career to create new intellectual niches for his program that were distinct from other fields of social inquiry, such as "moral" or social statistics, ethnography, and comparative law. Because an intellectual niche is individuated by an empirical subject matter in conjunction with conceptions of goals and methods, two research programs can study the same empirical subject yet occupy distinct intellectual niches just to the extent that they are doing entirely different things with the same facts. Durkheim felt that the methodological and conceptual resources of his research program allowed him to create a niche distinct from that occupied by other academics and social scientists working with ethnographic material. In creating this niche Durkheim did not necessarily abandon older ones that he had created around social statistics and comparative law. Rather, he put these niches aside until he felt ready to return to them.

Change and Continuity in Durkheim

Durkheim's shift in empirical concerns is often taken as evidence of a significant change in his most basic conceptions of sociology. In characterizing his change of research specialty in terms of his desire to solve certain kinds of problems within a more general research program, however, I am arguing for a fundamental and underlying continuity in his intellectual career. This continuity becomes most apparent when his intellectual career is understood in the context of his philosophical presuppositions concerning the goals and methods of sociology. Until now, however, there seems to have been little interest in interpreting Durkheim in this philosophical context. Social scientists appear to be more interested either in appealing to Durkheim in order to legitimize their own ideas or in treating him as a scapegoat for what they perceive as wrong with sociology. In either case, writers end up forcing Durkheim into one or another of the Procrustean beds of contemporary "isms" in the social sciences. In fact, Robert Alun Jones stirred up considerable controversy when he suggested that Durkheim should instead by understood in some sort of intellectual and historical context.[8]

Talcott Parsons's *The Structure of Social Action* (1937) provides something of a paradigm example of both the legitimizing and criticizing strategies. Parsons interpreted Durkheim's intellectual development as one of a gradual movement first toward and then "beyond" a model of sociological explanation to which Parsons himself subscribes. He saw Durkheim as having adhered to a so-called positivist model of explanation in his early works, especially *The Division of Labor in Society.*[9] In this work Durkheim supposedly explained social actions in terms of the actor's rational responses to objectively perceived conditions in both the physical and social environments, including the presence of social sanctions. That is, Parsons interpreted Durkheim as having assumed that the social actor's relationship to reality is that of a scientific investigator: hence the term *positivist*. Between this "positivist" and Durkheim's ultimate "idealist" stages, Parsons posited two intermediate phases. In the first of these Durkheim supposedly came to recognize that the content of the collective consciousness is as important as its strength. In the second or "voluntarist" phase, Durkheim purportedly "discovered" Parsons's notion of the internalization of norms and values and explained the behavior of social actors in these terms. This phase, then, corresponds to Parsons's own model of explaining social action (Parsons 1937:61, 67, 438, 304–7; cf. Pope 1973:400). Instead of resting with this voluntarist model, however, when Durkheim turned to the study of religion he moved to what Parsons regarded as a final, idealist stage. As Parsons saw it, Durkheim's concern with religious ideas represented a complete break with his positivism just to the extent that religious ideas have no referent in empirical reality and thus no meaning for the positivist (Parsons 1937:410, 420–27).

Whitney Pope has debated Parsons's way of interpreting Durkheim (Pope 1973, 1975; Parsons 1975). Pope challenged Parsons's claim that Durkheim's goal was to explain the behavior of individual social actors in the first place (1973:400). He also argued that to the extent that Durkheim had a conception of internalization, he held it from the beginning of his career (Pope 1973:405). Here Pope cited Ernest Wallwork, who had compared Parsons's notion of internalization to Durkheim's notion that all the members of a society participate in the collective consciousness (Wallwork 1972:40 n. 44). More recently, however, Stephen Turner has questioned whether Durkheim's notion of the collective consciousness and Parsons's notion of the internalization of norms and values are all that similar. He has argued that Parsons's distinction between internalized and externally imposed social constraints assumes the very thing that Durkheim would deny, that is, that these internalized norms guide the individual through being available to introspection or reflection (1986:126–28). The collective representations that constitute the collective consciousness impose themselves unconsciously and even mechanically on the minds of individual members of a society and cannot be identified with Parsons's notion of internalized norms and values. Indeed, Jerrold Seigel (1987:486 n. 8, 505) has suggested that to the extent that Durkheim appealed to something like societal norms and values in his later work, he was becoming more of a determinist than a "voluntarist" in Parsons's sense.

In addition to the works of Pope, Wallwork, and Turner, Steven Lukes's intellectual biography of Durkheim (1973) is another exception to the tendency to read current sociological concerns back into Durkheim's works. However, as Stjepan Meštrović (1985) pointed out, even Lukes has not been successful in understanding Durkheim as part of a philosophical tradition that was concerned with problems of mental representation. Meštrović (1985, 1988a,b, 1989a–c), Wallwork (1972), and Dominick La Capra (1972) perhaps have been more successful, but they have interpreted Durkheim as a philosopher interested primarily in moral theory and have not given sufficient attention to Durkheim's philosophical views on the nature of scientific knowledge and the methods for attaining it. Turner (1986) has provided the best account to date of Durkheim's philosophy of science and has shown its relation to the methods employed in *Suicide*. However, he has not worked out this relationship to Durkheim's other substantive sociological writings and thus has not shown the continuity of Durkheim's thinking through *The Elementary Forms of the Religious Life*.

Other interpreters have also failed to perceive the continuity of Durkheim's thinking. Even Pope at one time followed Parsons in perceiving a trend away from "materialism" in Durkheim's early works, where he supposedly emphasized social structure, to "idealism" in his later works, in which he emphasized collective representations (1973:410). More recently, however, Pope and Tendzin Takla (1985) have argued that Durkheim's social realism and conception of social forces provides some continuity to his thought. For Vogt (1979) this continuity is provided by a mentalistic conception of sociology. Nevertheless, most recent interpreters continue to follow Parsons in contrasting Durkheim's supposed emphasis on "material," "mechanistic," or "deterministic" factors in *The Division of Labor* with his later emphasis on cultural or ideational factors.[10] To defend the continuity of Durkheim's thinking, then, I will argue in chapter 3 that he identified sociology with the study of collective representations at even the earliest stages of his career. In chapter 5 I will show that, contrary to Parsons, Durkheim never held a positivist theory of meaning and reference that would have proscribed any appeal to unobservable entities such as collective representations. Also, in chapter 6 I will undermine the materialist interpretation of *The Division of Labor*. This misinterpretation is due to a serious misunderstanding of his causal claims, in which effects are confused with causes. Specifically, Durkheim's readers have mistaken the physical population density for the cause rather than the effect of what he calls the "moral" or "social density" of a society, which is a function of the number of social relations and is a property of a society's collective consciousness. Durkheim's concept of causality, in which he denied the possibility of a plurality of causes for any given effect and thus made causes and effects both necessary and sufficient conditions for each other, has been at least partly responsible for this confusion.

Among recent interpreters of Durkheim, perhaps Jeffrey Alexander (1982, 1986a, 1986b) comes the closest to Parsons, praising the latter's interpretation (1982: vol. 2, 403 n. 24) over what he regards as the "fashionable" interpretation that emphasizes continuity (1982: vol. 2, 82). Unlike Parsons, however, Alexander is aware of Durkheim's early concern with collective representations and does not maintain that Durkheim began his career as a positivist. In order to take into account Durkheim's early concerns, Alexander maintains that there were actually two turning points in Durkheim's intellectual development. He sees *The Division of Labor* as turning away from the "idealism" and "voluntarism" of Durkheim's earlier publications and toward a "determinism," "scientific positivism," and "instrumentalism." Leaving Durkheim's putative first turning point unexplained, Alexander accounts for the second in terms of Durkheim's professional and career interests. Durkheim turned away from materialism, Alexander alleges, in order to distinguish his sociology from the Marxist economic determinism and historical materialism of the socialists and thus to gain academic respectability for himself and for sociology and to advance his career:

> He broke with the *Division of Labor* for theoretical reasons and for ideological ones, for careerist motives and because of pedagogical ambition. Durkheim wanted to clear his good name, for his own conscience and for the good opinion of others. (1982: Vol. 2, 233)

Durkheim began to identify the subject matter of sociology with collective representations only later in his career, Alexander asserts, in order to make his thinking appear idealist and to distance it that much further from Marxist materialism (1982: vol. 2, 248–49). In the French Third Republic, reports Joseph Llobera, sociology was often confused with socialism and holding socialist opinions was an impediment to academic success (1980:386, 400). Victory Karady adds that Durkheim's collaborators on the journal *L'Année sociologique* adopted similar strategies, purposely avoiding the mixing of their social science with the political issues of the day, including socialism and the Dreyfus affair (1983:74).

Whether Durkheim's conception of the subject matter of sociology changed in the way that Alexander asserts will be taken up in chapter 3. Alexander's argument, however, reveals an assumption that appears to be typical of many interpreters of Durkheim. Durkheim's supposed theoretical changes cannot be explained in terms of method alone, Alexander argues, since his methods remained the same and because the changes did not result from new empirical observations (1982: vol. 2, 214, 471 n. 83). Alexander overlooks the fact that an important part of Durkheim's methodology is concerned with the formulation and comparative evaluation of explanatory theories and not just with the manipulation of empirical data, as I explain above. He both underestimates the importance of Durkheim's methodological continuity and fails to perceive that it is rooted in a continuity of explanatory goals. Alexander perceives a fundamental discontinuity in Durkheim's thinking just to the extent that he takes Durkheim's most basic assumptions to be about social order and action and not about method and explanatory goals (1982: vol. 1, 3; vol. 2, xix). Like Parsons, Alexander interprets Durkheim within a framework of contemporary sociological problems and concepts rather than within a historical, philosophical context.

In support of his interpretation, Alexander (1982: vol. 2, 232; 1986a:101–2) cites some early reviews of *The Division of Labor* as well as Mauss's introduction to Durkheim's *Socialism and Saint-Simon* (1928a). The reviewers in question, including Léon Brunschvicg and Élie Halévy (1894), Charles Andler (1896), Paul Barth (1897), and Georges Sorel (1895), did accuse Durkheim of things like "materialism" and "mechanism." However, as I argue in chapter 6, these concepts ought not to be confused with Marxist economic determinism. *The Division of Labor* was perceived by those who mattered as more Darwinian than Marxist.

The evidence Alexander draws from Mauss is just as equivocal. According to Alexander, "Mauss recounts Durkheim's belief that the attacks on him as a materialist had kept him from receiving the professorships in Paris which he so ardently sought and which he surely deserved" (1982: vol. 2, 232).

What Mauss actually said, however, is that Durkheim was reproached with "collectivism," not with materialism. The context provided by the rest of Mauss's preface makes clear that "collectivism" refers to Comtean sociology, which Mauss reported had fallen into ridicule among French academics at that time, and not to Marxism.[11] In 1877, Alfred Espinas had been forced by Paul Janet to suppress the introduction to his dissertation, *Les Sociétés animales,* because Espinas had discussed Comte in it (Lukes 1973:101 n. 8; Bouglé 1938:19). Espinas, incidentally, was the dean at Bordeaux when Durkheim was given his first university teaching position there in 1887 (Lukes 1973:100; Clark 1973:163). By 1902, however, Comte had found favor among Parisian academics and a statue of him was erected in the Place de la Sorbonne (Clark 1972:163). 1902 was also the year that Durkheim finally obtained a professorship in Paris. Far from having gradually moved away from a form of historical materialism in order to advance his career, it could be argued that Durkheim had stubbornly adhered to certain Comtean ideas initially at the expense of his career.

Although Alexander's interpretation is not supported by the evidence, I think he nevertheless raises the important issue of how *The Division of Labor* was understood. I would agree that Durkheim was dismayed by the way in which *The Division of Labor* work was interpreted, but not for the reasons that Alexander cites. I do not believe that Durkheim greatly feared being confused with the Marxists. The accusation of materialism disconcerted Durkheim because it indicated to him how unclear he had been. Durkheim after all was educated as a philosopher and there is nothing that philosophers—at least in the French and English tradition—pride themselves on more than their clarity. In *The Rules of Sociological Method* Durkheim tried to remove some of this ambiguity by clarifying his conception of causality. He also dropped the term *social density* and tried to clarify his position by substituting the concepts of integration and regulation in *Suicide* for the concepts of mechanical and organic solidarity in *The Division of Labor.*

In fact, some of the very same evidence that Alexander cites in support of his interpretation can also be understood as evidence that Durkheim was only striving to achieve greater clarity in expressing his sociological theories. Durkheim's increasing use of the term *collective representation* over the course of his career may be explained in terms of his desire to avoid the term *collective consciousness,* which had given rise to metaphysical misunderstandings. Since a collective consciousness was nothing more than a set of collective representations for Durkheim, he could make this substitution without changing his meaning. Alexander, however, is skeptical of Durkheim's claims in the prefaces to the second editions of *The Rules* and *The Division of Labor,* published in 1901 and 1902, respectively, that he had always believed that social life consists of collective representations.[12] Alexander takes this as evidence that Durkheim was attempting to "rewrite" his own intellectual history in order to distance his work from its historical materialist interpretations. The accusation of rewriting, however, appears wholly gratuitous. Durkheim may have written the new prefaces simply to make clear to everyone what was obvious only to professional philosophers. *The Division of Labor,* after all, was originally submitted as a philosophy dissertation in 1893 and *The Rules* was first published in a series of articles the following year in the *Revue philosophique.* At that time, as Durkheim pointed out, all academic sociologists in France were philosophers (1895e:89). By 1902 this was beginning to change, due at least in part to Durkheim's own efforts.

Goals, Methods, and Social Facts

Alexander's explanation of Durkheim's theoretical shifts in terms of politics, ideologies, and career ambitions is illustrative of the recent "externalist" fashion in the historiography of the natural sciences. *Externalism* takes its meaning largely through contrasting it with *interalism,* the competing approach. As is to be expected with academic disputes, there are differences of opinion even among various factions within each of the major camps as to how to define these schools of thought. In general, however, I think it is safe to say that the internalist approach is rooted in intellectual history and the philosophy of science while the externalist approach is rooted in social history and the sociology of knowledge. An internalist would seek to explain the development of a scientific discipline in terms of what are often called "cognitive" factors regarded as "internal" to science, such as its empirical findings and its assumptions about methods and explanatory goals. An externalist, on the other hand, would seek to explain the development of science in terms of noncognitive social factors that are "external" to science, such as the structure of the larger society and the social interests and religious affiliations of scientists.

Alexander is not the only Durkheim interpreter to offer an externalist account of some aspect of his career. Hanan Selvin, for instance as we shall see in chapter 7, explains Durkheim's not having adopted new developments in statistical methodology in terms of a lack of institutional support. Robert Alun Jones, as we shall see in chapter 8, appeals to Durkheim's status as a Jew in France in order to explain the fact that he differed from his contemporaries in anthropology in having believed that totemism is the oldest form of religion. These interpreters, to be sure, are representative of an older externalist historiography that turns to external factors to account only for what they perceive as error or for those aspects of someone's thought that they think internal factors fail to explain. Thus Alexander turns to politics when Durkheim's methods appear to fail to explain his theoretical shifts and Selvin turns to institutional factors and Jones to religion to explain what they take to be Durkheim's mistakes.

Sociologists of scientific knowledge have criticized this form of externalist historiography for being too fainthearted. David Bloor, one of the chief spokespersons for the self-styled "Strong Programme," argues that externalist accounts should be "impartial with respect to truth and falsity, rationality or irrationality, success or failure."[13] According to his principle of symmetry, the same kind of explanation should be provided for both sides of these dichotomies. The kind of explanation favored by the Strong Programme is typically one that invokes the social, political, or class interests of the people in question in order to explain their beliefs. Traditionally, the sociology of knowledge has explained people's political, religious, and other ideological beliefs in terms of such factors, but has exempted scientific beliefs from such accounts. Karl Mannheim's *Ideology and Utopia* (1936) provides a case in point. The Strong Programme, however, attempts to expand the explanatory scope of the sociology of knowledge to include the natural sciences as well. They consider mathematics and the physical sciences as providing the most difficult test cases for their approach. That is, they seem to regard the claim that scientists' wider social and political interests decide the fates of research programs as far more problematic for the natural than for the social sciences. They therefore devote much of their energy to persuading their readers of the viability of their approach for the historiography of the natural sciences, assuming that if they are successful their readers also will accept their approach for the historiography of the social sciences as a matter of course.

Although many sociologists of scientific knowledge reject the Strong Programme's interest model, they nevertheless subscribe to a form of sociological reductionism in which beliefs are to be explained in terms of an underlying social reality. Bruno Latour, for example, does not seem to realize that the criticisms he makes of the Strong Programme's explanatory concepts apply equally to his networks of social actors: they could be mere social constructs cobbled together by social scientists. Nevertheless, these social networks enjoy explanatory primacy for Latour. He proposes that we turn to what he calls cognitive factors "only if there is something unexplained once the networks have been studied." In fact, he goes so far as to call for a ten-year moratorium on cognitive explanations of science and technology.[14]

In this book I propose to turn the tables on these externalists. Where the Strong Programme takes mathematics and the natural sciences as the difficult test cases for the externalist approach, I am taking the social sciences as the difficult test case for an internalist approach to the historiography of science. Purposely violating Latour's moratorium, I hope to demonstrate the explanatory power of cognitive factors in precisely the domain where one would expect a thinker's wider social and political interests to be the most relevant: in the social sciences. I argue that the intellectual development of Durkheim—the very thinker whom David Bloor (1976, 1982) appropriates as the patron saint of the Strong Programme[15]—can be adequately explained in terms of his intellectual concerns and the core philosophical assumptions of his research program.

Explanations in terms of cognitive factors need be no less sociological than explanations in terms of noncognitive social factors. As I explained above, even Durkheim regarded assumptions about the goals and methods of scientific inquiry as social rules that govern scientific disciplines. To the extent that Durkheim held these beliefs, he makes an unsuitable candidate for the patron saint of an externalist research program in the sociology of scientific knowledge. Indeed, Durkheim's beliefs about the social character of rules of method in conjunction with his account of the mechanisms of academic specialization are of potentially greater significance than the tenets of the Strong Programme for understanding science. According to Bloor (1982), Durkheim's primitive classification hypothesis, which says that primitive classifications of things in nature reproduce classifications of people in society, can be generalized to include modern scientific classifica-tions as well. In chapters 8 and 9 I will present my reasons for thinking that this is not a fruitful hypothesis for the sociology of scientific knowledge. In chapter 9 I will also argue that Bloor and others are mistaken in their inter-pretation of Durkheim's sociological theory of meaning as a precursor to Wittgenstein's extreme nominalism and that Durkheim thus cannot be used to legitimize a relativist position in the sociology of knowledge.

To explain the direction taken by someone's research program in terms of its goals and methods, however, requires that one first provide a careful anal-ysis of these goals and methods. One must look to methods of persuasion and comparative theory evaluation as well as to methods of data manipula-tion. What may appear to be a mistake on a narrow interpretation of a thinker's methodology may turn out to be the correct decision under a broader interpretation of the thinker's goals and methods. In this light, Durkheim's creation of an intellectual niche in the sociology of knowledge can be seen as a rational move. To explain this move as rational, however, is not to deny that Durkheim faced unsolved conceptual problems. In fact, he was aware of these difficulties and changed the direction of his research be-cause of them. Of course, with the gift of hindsight we may interpret some-one's work as fraught with conceptual problems that were apparent neither to the author nor to his or her critics and that played no role in the reception of the work. Such an interpretation, however, would be critical without be-ing explanatory and would be unsociological as well as ahistorical. Avoiding this pitfall, I provide an account of Durkheim's creation of the sociology of knowledge that shows it to have been rational for him; yet it is a sociological account nonetheless. The appeal to cognitive factors does not violate the so-ciological principles of impartiality and symmetry as long as one interprets a thinker's decision as rational by his or her light and not by one's own.

2 Durkheim on the Division of Intellectual Labor

Toward the end of his now-classic *The Structure of Scientific Revolutions,* Thomas Kuhn suggests that science may increase in scope through the proliferation of progressively more narrow research specialties (1962:170). In this chapter I draw on a similar suggestion in Durkheim's *The Division of Labor in Society,* according to which the increasing specialization among scientists may be regarded as a Darwinian solution to the struggle for existence in the research environment. As I mentioned in the last chapter, Durkheim and his collaborators on the journal *L'Année sociologique* appear to have followed this strategy by dividing sociology into specialties and then having each assume responsibility for a different specialty. Durkheim's initially taking charge of the section reviewing works in the sociology of morality and the law and then creating a new section for the sociology of knowledge reflects the movement of his own research in this direction.

As I suggested above, Durkheim's ideas concerning academic specialization provide a hitherto unexplored set of potential explanatory concepts for the sociology of science. Durkheim considered science to be a cooperative endeavor governed by methodological rules that function as social norms. Science, however, was not a monolithic enterprise for Durkheim. It was divided into a number of overlapping and competing disciplines, each with its own subject matter, goals, and methodological norms. "We are far from the time when philosophy was the sole science," he said, "it has fragmented into a multitude of special disciplines" (1902b:2). What is true of science as a whole for Durkheim is also true of each scientific discipline. That is, specialization within each discipline allows for a greater number of practitioners to do research within that discipline. This belief is implicit in Durkheim's repeated criticisms of Comte for having failed to see that in order to establish a tradition of research in sociology, this discipline must continue to resolve itself into a variety of specific problems that would allow for a division of intellectual labor (1986a:214; 1888a:100; 1903c:471; 1915a:9). According to Durkheim, Comte's sociology reduces to the single problem of discovering the law according to which all societies develop. Once Comte solved this problem through proposing the three-state law, Durkheim said, there was nothing left for anyone else to do (1903c:470; 1904b:83; 1909e:145–47).

What is true for science and for each scientific discipline is also true for each research program within a discipline. In other words, scientists who share commitments to a research program may solve the struggle for existence among themselves by forging distinct intellectual niches within this program. As I defined this concept in chapter 1, an intellectual niche within a research program is individuated by an empirical subject matter. Because a niche is also characterized by the intellectual commitments of the research program to which it belongs, however, different niches can be created by different research programs around the same empirical subject matter. That is, different niches can use the same facts but for different purposes. As long as scientists in each of these niches are able to work toward their respective goals, they will all continue to thrive. When scientists in one of these niches are blocked from achieving their goals by empirical and conceptual problems, however, they may no longer be able to compete effectively in this empirical domain. Nevertheless, they can still contribute to the growth of their research program by temporarily setting aside this difficult niche and turning to ones in which success comes more easily. In fact, as I will demonstrate, this is what Durkheim did when he set aside the sociology of morality and the law and created a niche in the sociology of knowledge, specializing in the study of the formation of collective representations in primitive religions and societies. A research program can thus continue to grow by creating new niches for itself as long as it can achieve its goals in each of them, regardless what other research programs are doing.

According to Durkheim, the proliferation of research specialties is a response to the struggle for existence in science. He argued by drawing an analogy between social life and biological evolution that the division of intellectual labor will continue to increase. Organisms, he said, are in competition with each other only to the extent to which they depend upon the same resources. In the competition for resources, some organisms, of course, will either die or migrate to wherever resources are more plentiful. According to Durkheim's understanding of Darwin's account of the divergence of characteristics under natural selection, organisms may alternatively adapt themselves to new foods or different ways of living. The struggle for existence may in this way bring about the diversification of members of the same species. Darwin said that

the more diversified the descendants from any one species become in structure, constitution, and habits, by so much will they be better enabled to seize on many and widely diversified places in the polity of nature, and so be enabled to increase in numbers.[1]

The same principle, Durkheim believed, applies to people. Members of the same society who practice different professions are not in competition with one another. The more similar the markets for their work, however, the more they come into conflict. Durkheim asked the reader to imagine a manufacturing center supplying a local market with a certain product, which then has to compete for that market with another manufacturing center through the opening of some new means of communication or transportation. If one of these manufacturing centers is inferior to the other, it will be forced either to go out of business or to diversify and offer a different product (1902b:248–59). Although the example is taken from economic life, he said, "this explanation applies to all social functions without distinction. Scientific, artistic, and other work does not divide itself in any other manner nor for other reasons" (1902b:253). Durkheim was not implying here that all scientists are constantly trying to create new specialties in order to survive. Rather, he was saying that when scientists *do* diversify and create new specialties, that this is the explanation for it.

Dietrich Rueschemeyer objects to Durkheim's analogy, arguing that Darwin's principle of divergence concerns the competition for the *consumption* of resources and that Durkheim illegitimately used it to account for divergence in economic *production* (1982:582). In defense of Durkheim it must be said that economic production typically depends on the consumption of resources. In the case of science, the diversity of production can be explained as resulting from the exploitation of new kinds of resources. Scientists may develop new specialties in order to make use of hitherto untapped sources of empirical information for research. Professors may invent new disciplines and courses in order to compete more successfully for students or even to attract students who may not otherwise attend the university.

However, as I hope to show, Durkheim's attempt to create an intellectual niche does not reduce to an attempt to create a secure institutional basis. An intellectual niche is not merely an institutional one. To be sure, Karady points out that job prospects for philosophers were dismal in France around the turn of the century and argues that one could have improved one's prospects by switching to sociology (1983:78). Before Durkheim there were no academic sociology positions in France. Sociologists competed for chairs mainly in philosophy and history (Clark 1973:167). Although Durkheim and his associates were deeply involved in the creation of chairs in sociology in France, such institutional anxieties did not exhaust their professional concerns. Durkheim in particular considered his sociological research program to be in competition with research programs in Germany, Italy, Great Britain, and the United States, even though he was not in competition with foreign scholars for such things as university chairs or students.

Creating a Niche

Durkheim's beliefs about specialization in science, as I have suggested, are important for understanding his own career. His shift in empirical concerns over the course of his career was at least in part the result of intense negotiation among the community of scholars who founded the journal *L'Année sociologique,* which began publication in 1898. Although each issue contained some original research articles, the bulk of this annual was devoted to reviews of the current literature.[2] These reviews were divided by topic into seven sections, each of which was the responsibility of a different collaborator or group of collaborators who thereby laid claim to the respective specialty.[3] The structure of this journal reveals not only what Durkheim and his collaborators thought were the most important traditions of research in the social sciences but how they judged the relative merits of each other's work among themselves. Which collaborators were to be responsible for which section was determined by who among them was perceived to be in the strongest position to compete intellectually with the other traditions of social science research. Durkheim and his colleagues thus organized themselves in such a way that they would avoid competing with one another.

The seven divisions of *L'Année sociologique* and the persons responsible for each were as follows: (1) General sociology, which dealt with conceptual and methodological issues, was originally the sole responsibility of Celestin Bouglé. (2) The sociology of religion was a cooperative effort between Marcel Mauss and Henri Hubert and was concerned with historical, comparative, and ethnographic studies of religion. (3) Durkheim himself, along with Paul Lapie and others, took responsibility for what was called moral and juridical science, which investigated the causes and origins of moral and legal rules. (4) The section on criminal and moral statistics focused on the functions or effects of such rules and began under the direction of Gaston Richard, who was ultimately replaced by Maurice Halbwachs. (5) François Simiand took charge of the section on economic sociology. (6) Durkheim was also the first person to be responsible for the section on social morphology, which included the study of population distribution and its relation to social structure. (7) Various reviewers contributed to a final section of miscellaneous reviews of works on the sociology of art, technology, language, and so on. This description is, of course, highly simplified. Many others also contributed reviews, and because scholarly work does not always fit into neat categories, even those named above often crossed over and reviewed works in other categories.

What strikes one immediately about the organization of this journal is that Durkheim did not take charge of the section for criminal and moral statistics, in spite of the fact that he had just published *Suicide,* which falls squarely into this category, the year before the journal began. The reason, I believe, is that although *Suicide* may be regarded as a classic among sociologists today, at the time of its publication its statistical methods were already considered dated by Durkheim's collaborators. In a letter written that year (1897) to Bouglé, Lapie characterized Durkheim's methods as "sufficiently old."[4] To consider just one example, in investigating the effect of family size on suicide, Durkheim classified the eighty-two departments of France into six groups according to their suicide rates, and then compared the average family size for each group (1987a:209).[5] This method of grouping and averaging was popular among the Italian criminologists like Enrico Morselli (1882) from whom Durkheim also borrowed much of his data. It suffers from the obvious weakness that one can reclassify one's data to get the averages one wants. Of course, if Durkheim had simply compared suicide rates and average family size for each department taken singly, he probably would not have obtained a neat correlation. Then again, there were statistical techniques available that would have allowed him to determine which deviations, if any, were significant. Nowhere, however, do we find Durkheim using Quetelet's methods of error analysis, let alone Galton's methods of analyzing regression and calculating coefficients of correlation. Having followed the letters rather than the sciences track at the École Normale Supérieure, Durkheim lacked sufficient mathematical training to use these methods. Even those who had a background in mathematics had to teach themselves statistics, as it was not taught in the university (Clark 1973:125, 145).

To say that Durkheim used unsophisticated methods of statistical analysis, however, is not to suggest that other students of suicide rates always used more sophisticated methods than Durkheim did. On the contrary, one does not find bell-shaped curves or calculations of standard deviations either, for example, in Morselli's important work on suicide (1882). Methods similar to Durkheim's were used by members of the Italian school of criminologists and by German statisticians, as well as in France in government ministries such as the Bureau of Legal Statistics, headed by Durkheim's rival Gabriel Tarde.[6] What I am suggesting is that Durkheim was using methods that were little better than those of the average worker in a fairly crowded field.[7] I think Durkheim and his collaborators believed that in order to do more serious work in social statistics, one would have to rise significantly above this standard level of mathematical competence, something which Durkheim was unprepared to do. Furthermore, *Suicide* was published at about the same time that Pearson and Yule were introducing major changes in statistical methods, taking up and refining Galton's work on the coefficient of correlation and regression analysis (Selvin 1976). After *Suicide* Durkheim published only one paper in this area, a criticism of Bertillon's work linking divorce and suicide. This paper, published in 1906, drew all of its data from a single table in *Suicide* (1906d).[8]

Not only did Durkheim lack the preparation to follow new developments in statistical methodology, but, as I will discuss in chapter 7, he began to doubt that the pursuit of statistical studies was the best way to achieve his theoretical goals. In *Suicide* Durkheim used statistics for the polemic purpose of defending the reality of social causes and social forces. Ultimately, however, he wanted to get beyond polemics and to investigate the laws governing the collective representations that are responsible for social forces.

Durkheim's deficiency in mathematics may account also for the fact that he eventually turned over the section of *L'Année sociologique* devoted to social morphology to Halbwachs, Simiand, and others. Social morphology drew on demographics, history, ethnography, and geography, resembling the "political geography" of Freidrich Ratzel.[9] As this science relied at least in part on demographics, it required some facility with statistics. Durkheim's lack of mathematical training may explain, too, why he never contributed to the section on economic sociology. This field, as Durkheim characterized it, overlapped to some extent what was already being done in Germany by economists like Georg Friedrich List, the "socialists of the chair" Adolf Wagner and Gustav Schmoller, and Karl Bücher and Wilhelm Roscher (1903c:487–89). These economists had turned away from the study of individual economic behavior to the study of national economies and were considering the interaction of the economy with other sorts of social factors. Although some of these economists were making use of ethnographic studies in order to understand the economies of "inferior" societies, economic sociology included statistical studies of economic and social classes and other aspects of modern industrial society. One could not compete successfully in this field without an understanding of mathematical work in economics.

Durkheim faced potentially the most serious challenges, however, in the field of moral and juridical science. This specialty investigates the origins and development of moral and legal rules and is best represented among Durkheim's works by *The Division of Labor*. Drawing heavily on written codes of law for empirical support, this book competed in a field already crowded not only by law professors but by historians. Germany was home to a school of comparative law that was already making use of ethnographic as well as historical information. Durkheim cited the work of Rudolf von Jhering as well a that of Albert Hermann Post and his school of ethnological jurisprudence or juridical ethnology, which included Josef Kohler, France Bernhöft, and Sebald-Rudolf Steinmetz.[10] Such research was also beginning to be carried out in France among the law faculties (Vogt 1983:185–86). Historians in Germany, England, and France in those years had moved away from the mere narration of wars, treaties, the actions of statesmen, and so on, to the history of institutions. As an example, Durkheim cited the work of one of his teachers, Fustel de Coulanges, who drew on classical codes of law in his study of *The Ancient City.* Durkheim also named Georg von Maurer, Wilhelm Wilda, and Henry Sumner Maine as contributors to institutional history (1903c:486–87).

Of course, the fact that the field of comparative law was already crowded with historians and legal scholars would not alone have prevented Durkheim from competing in this field if he could have succeeded at doing something new with this data. However, Durkheim met with serious conceptual problems in trying to achieve the goals of *The Division of Labor.* Some of these problems were pointed out to him by philosophers. As I mentioned above, *The Division of Labor* was originally submitted as a philosophy dissertation. Durkheim wrote it with the aim of determining whether the division of labor in society is desirable from a moral point of view. During his oral defense and in a review by Léon Brunschvicg and Élie Halévy, Durkheim was criticized for committing what philosophers since G. E. Moore have called the naturalistic fallacy, that is, for attempting to derive a prescriptive "ought" statement from premises consisting only of descriptive "is" statements. At the dissertation defense, Henri Marion told Durkheim that his thesis did not "reach morality" but was only a "moral physics." In reply, Durkheim took the offensive, pointing to the weaknesses of more traditional approaches to moral philosophy. Paul Janet, however, continued Marion's line of objection, arguing that Durkheim failed to distinguish moral duties from social functions. In answer to Janet, Durkheim merely reaffirmed the duty of the modern individual to consent to fulfill his or her specialized role within society (Muhlfeld 1893:441; Lukes 1973:296–97.) Brunschvicg and Halévy (1894:581) raised similar criticisms, asking rhetorically, if the "juridical order" is part of the natural order, then why should one be obliged to obey?

In the original introduction to *The Division of Labor* Durkheim had tried to explain the role of empirical methods in moral philosophy. There he had declared the aim of what he called the "positive science of morality" to be that of arriving at general moral principles from the study of moral facts (1893b:20). In the second edition, however, he eliminated most of this introduction, about thirty pages, characterizing it as "useless" (1902b:i n. 1).

There were other conceptual problems with *The Division of Labor* in addition to Durkheim's failure to establish moral conclusions on a scientific basis, as I will demonstrate in chapter 6. Specifically, Durkheim did not always make clear such key concepts as his notions of a causal relationship and the social environment. Due to these ambiguities his readers understood him to be saying that specialization was somehow a *result* of factors in the *physical* environment rather than an *adaptation* to factors in the *social* environment (Andler 1896; Barth 1897; Brunschvicg and Halévy 1894; Sorel 1895). Also, Durkheim's reliance upon a dubious concept of abstract general ideas weakened his account of the collective representations shared among the members of advanced societies. Durkheim found these conceptual problems sufficiently serious, I believe, to warrant his setting aside the problem of the moral desirability of specialization and creating a new empirical focus for his sociological research program.

However, Durkheim never abandoned the project of trying to establish morality upon a scientific study of moral phenomenon and their causes. Rather, he set it aside until he had first solved what he regarded as a more general, fundamental set of problems.[11] Since moral phenomena were but a species of social phenomena for him, his science of morality depended upon a more general science of social phenomena. As I will argue in the following chapter, Durkheim thought that social phenomena exist in collective representations. His project for a science of morality thus had to wait until he had worked out a general theory of the formation of collective representations. As I mentioned in chapter 1, Durkheim believed that he had made some progress in working out a theory of collective representations in *The Elementary Forms of the Religious Life*. Thus toward the end of his life he returned to the problem of establishing morality on a scientific basis. In the introduction to his unfinished work *La Morale,* he affirmed that a knowledge of the causes of moral rules will provide the basis for a moral ideal from which we can deduce a practical morality. However, he never wrote the promised section of *La Morale* in which he was going to explain how to derive this moral ideal (1920a:93–94, 97). Although Durkheim maintained an interest in a science of morality throughout his life, it is nevertheless an exaggeration for Robert Hall to say that the idea of a sociology of morals provided Durkheim's "conception of the overall program of his research" (1987:13; cf. 211). For Hall, Durkheim's interest in collective representations and the sociology of knowledge was merely a parallel development to his main interest in the sociology of morals (1987:39–40, 141–43). Hall fails to realize that Durkheim's interest in the causes and origins of moral rules was simply a part of his interest in the causes and origins of collective representations generally.

Durkheim thought he could best achieve his goal of working out a theory of collective representations by creating a niche around the last remaining major empirical research specialty in sociology recognized by *L'Année sociologique:* the sociology of religion. He found that the comparative study of religions, originally founded by Max Müller, had in the works of James Frazer, William Robertson Smith, Frank Byron Jevons, and others, widened its field of comparison to include ethnographic as well as historical studies (1903c: 490–91). Durkheim then set out to use these studies in a new way, creating a niche in the ethnographic approach to the sociology of knowledge. Assuming that contemporary preliterate societies resemble our own primitive origins, Durkheim believed that we could analyze our contemporary collective representations by tracing their development back through the historical record and to their origins in primitive societies. A collective representation for Durkheim was something like a complex idea for Locke that over the millennia has been progressively compounded from other representations. The simple ideas from which collective representations are ultimately compounded, he believed, are most easily observed in primitive societies (1912:4–5). By studying the formation of collective representations in primitive societies, I believe, Durkheim hoped to find a way to account for the shared notions that hold a society together without relying on the problematic concept of abstract general ideas.

In an autobiographical essay, Durkheim's nephew and student Marcel Mauss reported on his uncle's encounter with some of these ethnographic studies of religion around 1894–95 (Mauss 1930), a period that we saw that Durkheim characterized as marking a turning point in his career (1907b). During this time, Durkheim was developing a course on the sociology of religion and guiding Mauss to choose a dissertation topic in this area. Mauss (1930) said that he and his uncle were generally satisfied with the comparative studies of religious rituals, as well as with comparative studies of institutions, the family, and law. Nevertheless, Durkheim and Mauss were critical of these comparative ethnographic studies on methodological grounds. Durkheim found them rather too descriptive and not sufficiently explanatory and found the scholars who made use of ethnographies in their comparative studies far too uncritical of their sources (1903c:489–90). Specifically, he found fault with the methods of Post and Steinmetz in their studies of comparative law, Edward Westermarck in his work on the family, and Frazer in his comparative study of religion. All of these writers were accused of attempting to demonstrate their conclusions merely through the accumulation of a large quantity of facts drawn from disparate societies instead of choosing their facts carefully in such a way as to present "crucial" or "well-made" experiments [*les expériences . . . bien faites*]."[12] That is, ethnographic facts were interesting for Durkheim only if they could be used to decide between competing sociological hypotheses.

He raised similar methodological criticisms against the studies of the development of social institutions by Charles Letourneau, whom Durkheim nevertheless considered the leading ethnologist in France at the time.[13] In a like manner, Mauss found Robertson Smith in particular and British anthropologists in general lacking in method and preoccupied with the mere accumulation of facts (Hubert and Mauss 1899:34).

Mauss (1930) went on in his autobiography to say that he and his uncle were least satisfied with Frazer and Robertson Smith on "oral ritual," that is, prayer, and what they called "religious ideation," or the formation of the collective representations of which Durkheim thought religions consist. Prayer became the topic of Mauss's dissertation, completed in 1909. The study of religious ideation gave rise to a new specialty: the ethnographic approach to the study of what Durkheim referred to as "the sociological conditions of knowledge," which concerns itself with the formation of collective representations generally. The development of this new sociological specialty was reflected in a change in the structure of *L'Année sociologique*. In the eleventh volume, published in 1910, a subsection concerning the sociology of knowledge was added to the section for general sociology. Although reviews did not appear under this new rubric until 1910, Durkheim said that this subject had always been an important one for them. He pointed out that reviews that had previously appeared under "Religious Representations of Beings and Natural Phenomena" in the section devoted to the sociology of religion could also be considered to belong to the sociology of knowledge. The study of the religious factors that have entered into our representations of the world belongs to the sociology of knowledge, he explained, since religion is a social thing that provides at least some of the social conditions of knowledge. (1910a(ii)(1):41.)

Indeed, Durkheim believed that our most fundamental concepts— including time, space, genus, number, cause, substance, and the self—owed their origins to religious representations (1912a:12–13). He defined the object of the sociology of knowledge as one of giving an alternative account of our basic categories of thought to that provided by "recent disciples of Kant," by whom I suppose he meant Charles Renouvier and Octave Hamelin. "It is first necessary to know" he said, "what they are, from what they are made, what elements enter into them, that which has determined the fusion of these elements into complex representations, what has been the role of these latter in the history of our mentality" (1909d:757–58). In other words, Durkheim's sociology of knowledge aims at an account of the genesis of the collective representations that serve as the categories.

Defining the sociology of knowledge in the way that he did, Durkheim was trying to avoid coming into conflict with epistemologists in the way in which *The Division of Labor* had brought him into conflict with moral philosophers. In his inquiry into the concept of cause in *The Elementary Forms,* he made it plain that his goals were quite different from those of philosophers concerned with the theory of knowledge:

> Undoubtedly, certain philosophers refuse any objective value to this concept; they see there nothing but an arbitrary construction of the imagination that should correspond to nothing in things. But we do not have to ask ourselves for an instant whether or not it is founded in reality: it suffices for us to show that it exists, that it constitutes and that it has always constituted an element of the common mentality; and this is recognized even by those who criticize it. Our immediate goal is to seek not what it may be worth logically, but how it can be explained. (1912a:519.)

Of course, the concept of causality continued to develop in literate societies, and the study of the history of science would be relevant to tracing this subsequent development. In a letter thanking Lucien Lévy-Bruhl for a copy of his recent book on Auguste Comte, Durkheim in fact recognized the importance of studying science as a historical and social phenomenon and even suggested that this is the best way to do methodology.[14] However, Durkheim's lack of a background in the natural sciences would have prevented him from taking up again the project of the first half of the *Cours de philosophie positive,* in which Comte presented an overview of the development of scientific method.[15]

According to Mauss's autobiography, then, he and Durkheim were consciously seeking to create what I have been calling a new niche. The niche they hoped to bring into being, however, was not a mere institutional one, as they could hardly be said to be in competition for academic positions with British scholars of comparative religion. It was, rather, an *intellectual* niche that they sought to establish, that is, a field of research where they felt it would be possible to make use of relatively untapped empirical resources in order to make positive contributions to knowledge with the methods at their disposal. By 1903, with their essay on the origin of our concept of a genus in primitive classifications, Durkheim and Mauss had forged this niche. Using ethnographic data to construct crucial experiments among theories in the sociology of knowledge, these scholars created a niche distinct from ethnography. In addition to their essay on primitive classifications and Durkheim's *The Elementary Forms of the Religious Life,* Durkheim cited Hubert's 1905 essay on time and Lévy-Bruhl's work on primitive mental functions as contributions to the sociology of knowledge (1915a:10).[16]

Mauss described what he referred to as the "French school of sociology" as continuing after the death of Durkheim to work through the categories using Aristotle's list as their point of departure. Durkheim, Mauss, and Hubert had inquired into the origins of the categories of class, space, time, and cause. After Durkheim's death, others investigated the concept of the soul, the world, and substance. The last contribution to this research program was Mauss's 1938 essay on the concept of the person or the self (Mauss 1938:1; 1930 [1979:218–19; 1983:149]).

Mauss's reference to the "French school of sociology" should not be taken as suggesting that sociologists in France were all in close agreement. Not even all those who collaborated on *L'Année sociologique* shared Durkheim's explanatory goals, methods, and theoretical beliefs. Intellectual conformity among this group was a matter of degree. Those like Mauss and Hubert who worked on problems in the sociology of knowledge and religion that were close to Durkheim's interests also tended to be more in harmony with his cognitive commitments. On the other hand, those like Halbwachs and Simiand who worked in areas that Durkheim avoided, such as criminal statistics, economic sociology, and demographics, also tended to be more independent in their thinking. Furthermore, as Pickering points out (1984:41), Durkheim was more tolerant of intellectual dissent with respect to particular theories than with respect to fundamental issues concerning method or the explanatory goals of sociology. Even here, however, some members of the group differed from Durkheim. Bouglé, for example, disagreed with Durkheim over the role of individual psychology in sociological explanation.[17] Unlike his uncle, Mauss at least claimed to have rejected the idea that so-called primitive or simple societies represent early stages of social evolution.[18] Membership in the *L'Année* group was as much a matter of personal and professional relationships as of intellectual consensus. Nevertheless, those who were most in intellectual disagreement, like Gaston Richard, often ended up leaving the group.

Through the Institut d'Ethnologie that Mauss founded in 1925 with Lévy-Bruhl and Paul Rivet, the Durkheimian tradition stimulated and influenced further work in social anthropology, including that of Claude Lévi-Strauss (Clark 1973:200, 214; Karady 1983:88). The shift of Durkheim and some of his closest collaborators to problems of ethnology or social anthropology is often characterized as a decline in this program of research, which is then attributed to various noncognitive social factors. Terry Clark, for example, lists three elements necessary for a successful program of research: good ideas, talented individuals, and adequate institutional support. Assuming that the Durkheimians continued to have good ideas to build upon, Clark then attributes what he considers the posthumous decline of the Durkheimian research program to their failure to attract sufficiently talented followers (Clark 1973:242). Victor Karady, on the other hand, seems to attribute this decline to a lack of institutional support: "Obviously," he says, "the causes of such an intellectual discontinuity must be traced to the institutional field proper" (Karady 1983:88.)

Clark and Karady, however, devote insufficient attention to the evolution of the ideas of the Durkheimians. It is dangerous simply to assume that Lévy-Bruhl, Rivet, Lévi-Strauss, and others associated with the Institut d'Ethnologie all shared Durkheim's and Mauss's conceptions of the goals and methods of sociology. Indeed, to some extent I think Clark and Karady are asking the wrong questions. They appear to assume that the Durkheimians' ideas remained fixed and then seek an explanation of the obvious fact that this fixed set of ideas has disappeared. A more fruitful approach, I believe, would be to investigate how Durkheim's original research program evolved into something quite different as certain key elements of the original program came to be recognized as untenable and were rejected. For example, in his 1938 essay we find Mauss tracing the development of the concept of a person from its simplest forms in Amerindian and Australian tribal societies through ancient Rome, modern India and China, the Christian Middle Ages, and finally to contemporary European society. Although French intellectuals at the turn of the century may have taken this developmental typology for granted, the assumption that, say, contemporary India is more on a level with ancient Rome than modern Europe was becoming increasingly untenable. What I hope to persuade the reader is that the tenability of such an assumption is a "social fact" that can serve just as well as a premise in a sociological explanation as facts about the structure of the French educational system.

Durkheim's turn to the study of the sociology of knowledge represented but a new niche in an already existing research program. His later works, which concerned the formation of the categories in so-called primitive Australian and Amerindian societies, reflected the same conception of sociology as did his early works, which concerned contemporary European societies. That is, Durkheim's conception of the subject matter of sociology, his explanatory goals, and his methods of investigation and theory evaluation remained the same. The next chapter will show that his conception of the subject matter of sociology did not change, while chapters 4 and 5 will be about his explanatory goals and methods, respectively.

II

*The Explanatory
Goals and Methods of
Inquiry of Durkheim's
Research Program*

3   Social Facts and Collective Representations

Throughout his career, Durkheim considered sociology to be the study of a type of unobservable, unconscious mental entity that he called collective representations. He conceived collective representations as real entities with causal powers that can be invoked to explain their effects, much as physicists explain natural phenomena through the postulation of theoretical entities. Unlike physical entities, however, collective representations, as a kind of mental entity, have both semantic and causal properties. For Durkheim, collective representations both contain the meanings and provide the constraining power of social facts.

Once the constraining power of social facts is understood to come from collective representations, much of the apparent ambiguity in his definition of social facts in *The Rules of Sociological Method* disappears. Such things as suicide statistics and legal codes, which some interpreters have taken to be the social facts that constitute the subject matter of Durkheim's sociology, are merely the external or visible effects by which collective representations are known. Durkheim consistently maintained a realist attitude toward unobservable entities such as collective representations and was never the positivist that some have taken him to be.

The Definition of Social Facts in *The Rules of Sociological Method*

Durkheim is usually understood as having suggested that what he called social facts form the subject matter of sociology. In the first chapter of *The Rules of Sociological Method* he asserted the existence of this class of facts, which he defined in terms of their ability to impose an external constraint upon the individual. After having provided a few examples of the way in which the individual is constrained by the laws and customs of society, he said, "there is, thus, an order of facts that present very special characteristics: they consist in manners of acting, thinking, and feeling, external to the individual, and that are endowed with a power of coercion in virtue of which they impose themselves on him" (1895a:8). He continued to use the same expression for social facts toward the end of his career, having referred to religious rites as "manners of acting" in *The Elementary Forms of the Religious Life* (1912a:13).

In *The Rules* Durkheim went on to say that social facts "consist of representations and actions" (1895a:8). His reference to actions however, should not be taken to suggest that Durkheim ever thought that one of the goals of sociology was to explain the actions of *individuals*. Durkheim was interested in individual actions only insofar as they were instances of different "manners of acting" or *types* of social actions. In *Suicide,* for example, where he referred to suicide and self-sacrifice as "manners of acting" (1897a:7), Durkheim was concerned only with explaining variations in the rates of suicide for different social groups and not with explaining individual suicides. In the preface to the first volume of *L'Année sociologique* he said that this journal was not interested in reviewing works in which historical individuals are the principal or exclusive object of study. Individual facts are of no significance for sociology, he asserted, "except when they are grouped into types and under laws" (1898a(i):iv–v). Science can neither explain nor even describe individual cases as such, he believed; only the poet and not the scientist would try to describe an individual thing without subsuming it under a type (1953a:35–36). Sociologists should even try to consider social facts in isolation from their individual manifestations, he said (1895a:57).

In *The Division of Labor* Durkheim had characterized moral facts in a way consistent with the way he characterized social facts in *The Rules*. That is, moral facts are not individual actions but *types* of social actions imposed on individuals: "The moral fact properly so-called does not consist in the act that conforms to the rule but in the rule itself" (1893b:30).[1] Durkheim's identification of moral facts with moral rules should not be understood as suggesting that a social fact is something that is written down in a code of conduct. On the contrary, as he explained in *Suicide,* these written rules are merely the expression or external sign of an underlying social reality:

These percepts themselves do nothing but express an entire subjacent reality of which they are a part; they result from it, but do not suppress it. At the base of all these maxims, there are actual and living sentiments that these formulae recapitulate, but of which these latter are only the superficial envelope. . . . If, then, we attribute a reality to them, we do not dream of making of them the whole of moral reality. That would be to take the sign for the thing signified. (1897a:356)

Without these "collective sentiments," as he called them in his lectures on moral education, a moral or legal rule would be "a purely verbal formula" (1925a:104). Of course, the English word *sentiment* is not an exact translation of its French cognate, which has the connotation of a sensation or perception held in one's mind, or, in other words, a mental representation. A collective sentiment, then, would be a collective representation. Elsewhere Durkheim said that moral rules "are given in representations" (1898a(ii):100 n. 2). Given his analogy between moral rules and rules of method in science, which I discussed in chapter 1, we can conclude that for Durkheim methodological rules also exist in collective representations.

In various places, Durkheim affirmed that social life is made up "entirely," "essentially," and "only" of representations.[2] If social life for Durkheim consisted entirely in representations, then it would seem that he believed that the manners of acting, thinking, and feeling that he identified with social facts also exist only in representations.[3] In order to avoid the rather obvious objection that actions exist not only in the mind, it must be emphasized that it is the *manner* or *type* of action and not the action itself that exists in collective representations for Durkheim. To classify actions as belonging to a certain type is of course to ascribe a certain meaning to them. In fact, what makes something an action in the first place and not just a motion is its meaning. Different motions can be classified as the same kind of action and the same motion can be classified as different kinds of actions. For example, two human beings moving rapidly on foot in a rectilinear direction, one behind the other, may also be described as the person in front fleeing from his enemy behind. Also, many different motions may be classified as fleeing from enemies. The kind of action that the motion constitutes and thus the meaning of the action exists for Durkheim in the human mind. When he said that manners of acting exist in collective representations he was trying to say that what makes the motion a kind of action is its shared meaning and not just the meaning that it may have for an individual.

One might also object that to identify the meanings of social actions with collective representations is to identify them with a type of mental entity of which we are not consciously aware. In Durkheim's defense, we could shift the burden of proof and argue that it is not at all clear that the semantic properties of our mental representations depend upon our consciousness. Furthermore, it could be argued that Durkheim's identification of meanings with unconscious mental entities should be judged in terms of the explanatory power it affords his sociology and not on purely philosophical grounds. Through his postulation of a kind of entity with both causal and semantic properties, Durkheim was trying to find a way that does not appear miraculous for meanings to have effects in the physical world. His solution to this problem was to have the meanings of social actions supervene on something that is causally relevant to their production. In order to be persuasive, however, this solution must include an account of the way in which collective representations interact with individual representations and produce action.

Durkheim's theory of the social forces that arise from collective representations, which I will discuss in the following section, was intended to provide that account.

According to Durkheim, society is a "collective consciousness," or a collection of beliefs, ideas, and sentiments shared by the members of a community (1925a:318; 1902b:99). In *The Division of Labor* he referred to the collective consciousness also as the "common" consciousness, which he distinguished from the "social" consciousness. The social consciousness consists of the mental life of everyone in a society, including those like scientists and lawyers who exercise specialized functions. The collective consciousness, on the other hand, consists of only those representations that are shared by everyone in a society (1902b:46).

To be sure, Durkheim also included such material objects as works of art and literature, tools, weapons, clothing, utensils, buildings, machines of various kinds, and so on among the elements of a society. These things are all the product and the legacy of generations of social activity, he said (1895a:138; 1897a:354; 1901a(ii)2; Gisbert 1959:355). Durkheim's inclusion of such things among the elements of society, however, does not contradict his position that social life is made up of representations. Just as in the case of social actions, what makes something a tool or a weapon and thus a social fact is the meaning it has for a society, which is held in the collective representations in the minds of the members of that society. In *The Elementary Forms of the Religious Life,* as I will show in chapter 8, Durkheim tried to explain how a physical object can acquire a social meaning through his account of how something can become a totemic object. An animal, plant, or other object becomes a totem, according to Durkheim, during periods of "collective effervescence," in which collective representations are formed from individual representations. As Durkheim had put it earlier in *The Rules,* although physical objects are "elements" of society, they alone do not give rise to "active social forces" (1895a:139). These social forces for Durkheim, as I will argue in the section below, arise from collective representations.

Collective ideas and sentiments for Durkheim can never be directly observed through introspection or other means and can be known only through their observable effects (1895e:98). Social solidarity, for example, according to Durkheim is "a social fact that one can know well only through the intermediary of its social effects" (1902b:31). Because it does not lend itself to observation and measurement, he said, we must substitute for this "internal fact" an "external" one that symbolizes it. "This visible symbol," he said in *The Division of Labor,* "is the law" (1902b:28).[4] In the introduction to the first edition of this work, Durkheim asked how we can know whether a rule is regarded as morally obligatory within even our own society. We cannot determine what is obligatory in our society simply through introspection, he said, due to differences in moral beliefs among individuals in contemporary society.[5] Nor can this problem be avoided through an appeal to the "normal" or average consciousness (1893b:24). Because it is impossible to see what is going on within the consciousness of even a single member of society, "in order to know in what manner the rules of conduct are represented there" it is necessary to be able to refer to some sort of external and visible sign that "reflects this internal state." The sanction can best play this role of external sign, since "every moral fact consists of a sanctioned rule of conduct." By having defined moral rules in terms of sanctions, he did not wish to suggest that the feeling of obligation is a result of the sanction. "On the contrary," he said, it is because the sanction derives from a collective sentiment of obligation that the sanction "can serve to symbolize" this feeling (1893b:24–25, 22).

Not all social facts are indicated by sanctions, Durkheim thought, as there are other sorts of social facts in addition to moral and legal rules. The presence of some sorts of social facts is indicated simply by their generality within a society. Rules of "economic technique" are social facts recognized in this way, he believed, as resistance to innovation in this sphere of life is not always enforced by easily visible sanctions (1895a:15–16; 1900c:29). Social facts for Durkheim also include all kinds of social currents, trends, fashions, or even waves of emotion in a crowd (1895a:9–10; 1897a:355–56). He explained that a society's manners of acting need not be "fixed" in any way (1895a:19). On the contrary, he said, the majority of them remain "diffuse" (1897a:355). When manners of acting and thinking take on fixed forms, according to Durkheim, they are called social institutions (1908a(3):241). In this sense, he said, even a language is a social institution (1917b(2):57). In the preface to the second edition of *The Rules* Durkheim suggested that sociology could be characterized alternatively as the science of social institutions (1901c:xxii–xxiii). He referred to an article by Fauconnet and Mauss, with whom he seems to have agreed, who wrote that social institutions exist only in the representations that society forms of them (Durkheim 1901c:xxii n.1).[6] Of course, these three thinkers were interested not in individual representations of institutions but only in the conceptions of them shared within a social group, which can be studied only through their "external signs" (1901c:xiv).

Various kinds of social statistics can also serve as indicators of social facts (1895a:13–14). For example, in *The Division of Labor,* as I will argue in chapter 6, the physical population density of a society for Durkheim is merely the sign of what he called its social or moral density, which itself is a property of the collective consciousness. In *Suicide* statistics are used to measure the strength of "suicidogenic currents" within the collective consciousness. The social rate of suicide, Durkheim said, is but the numerical expression of "the reality to which it corresponds" (1897a:333), which consists of various social forces that either cause or preserve the individual from suicide. Durkheim identified these forces, not the statistics themselves, with social facts; the statistics merely serve to demonstrate that social facts are "objective" (1897a:349).[7] In his earliest paper on suicide, published nine years earlier in 1888, Durkheim referred to suicide statistics themselves as "social facts." Even in this early paper, however, suicide statistics are characterized as but the numerical expression or "translation" of the state of social health (1888d:446–47). In a lecture published the same year, he spoke of statistics generally as "translating" social phenomena (1888c:271). The variation of the social suicide rate with the time of day or month of the year was said to "reproduce" the rhythm of social life (1897a:xi).

As I have been saying, Durkheim considered social forces as attributes of collective representations, which he regarded as real causal entities with coercive powers. Social forces are real for Durkheim in the sense that they have real effects on us:

Collective tendencies have an existence that is proper to them; they are forces as real as cosmic forces, although they may be of another nature; they also act upon the individual from without, although this may be by no other means. That which allows us to affirm that the reality of the former is not inferior to that of the second, is that they prove themselves in the same manner, to wit by the constancy of their effects. . . . They are so much indeed things *sui generis,* and not verbal entities, that one may measure them, compare their relative magnitudes, as one does for the intensity of electrical currents or luminous sources. (1897a:348–49; see also 1900b:130)

Durkheim believed that the different intensities of social forces in different societies can explain their different rates of suicides. He continued to regard social forces as real entities even in his later works (for example, 1911b:117).[8]

François Simiand, in his review of *Suicide* the year following its publication, argued that when Durkheim said that collective tendencies are forces just as real as "cosmic forces," he should have said "as little real." Science, he thought, uses terms like *attraction* or *current* as images or metaphors. It may be useful to talk this way, he said, but whether these things are "real" is of concern only to metaphysicians, not to scientists. So why, he asked, should it be necessary in sociology to consider "social forces" and "suicidogenic currents" as anything other than metaphors? To do so, he believed, would be to commit a "sociological metaphysics" or a "sociological realism" (Simiand 1898:649–50). A generation later Roger Lacombe tried to undermine Durkheim's analogy with the physical sciences by arguing that in these sciences, the methods of measurement are "the very definition" of the concepts involved (1926:69). More recently, Robert Alun Jones has gone so far as to dismiss Durkheim's use of terms like "force" and "current" as so much "obfuscatory language" (1986a:114).

For Durkheim, however, terms like *social forces* are not just metaphors or manners of speaking intended to conceal his meaning. Nor did he intend that they be understood merely through operational definitions, as Lacombe believed concepts in physics are understood.[9] Such terms as *collective representations, social forces,* and *suicidogenic currents* are theoretical terms for unobservable entities that have actual referents for Durkheim. For instance, he maintained a realist belief in suicidogenic currents:

It is not by way of metaphor that one says of every human society that it has a more or less pronounced aptitude for suicide: the expression is founded in the nature of things. Every social group really has a collective penchant for this act that is proper to it and from which the individual penchants derive, rather than the former proceeding from the latter. That which constitutes it are the currents of egoism, altruism, or anomie that work at the society under consideration, with the tendencies toward languorous melancholy or active renunciation or exasperated lassitude that are the consequences of it. They are tendencies of the collectivity that, by penetrating individuals, determine them to kill themselves. (1897a:336; cf. 345)

Durkheim even suggested that it is the use of theory and not of observation that makes sociology scientific. In several places, he drew an analogy between sociology and the natural sciences to make his point. What gives the latter their scientific status, he seems to have believed, is their use of hypotheses that appeal to unobservable things such as energy or electrical currents in order to explain our sensations of heat, light, and so on (1912a:597). The sociologist, the chemist, and the physicist alike must regard the phenomena that they study as the product of "unknown forces" (1900b:130; cf. 1920a:88 n. 2). "If sociology exists, it can only be the study of a still unknown world, different from those explored by the other sciences . . . If we had really only to open our eyes and take a good look to perceive at once the laws of the social world, sociology would be useless or, at least very simple" (1897a:349, 351).

Durkheim in fact did not merely allow the postulation of unobservable entities, but even found them necessary for providing scientific explanations of natural phenomena. In the following passage he defended the hypothesis of a luminiferous ether in this way:

We do not know what an imponderable material medium is and we cannot form an idea of it; however, the hypothesis of it is necessary for giving an account of the transmission of light waves. . . . Thus, even though a phenomenon is not clearly representable to the mind, one is not however within rights to deny it, if it manifests itself by definite effects which, themselves, are representable and which serve it as signs. . . . Indeed, there is no science that may not be obliged to take this detour for attaining the things of which it treats. (1898b:33)

Durkheim's attitude toward the postulation of unobservable entities in scientific hypotheses represented a significant departure from the positivist philosophy of Auguste Comte, the inventor of sociology. Comte had rejected the hypothesis of a luminiferous ether as being "radically inaccessible to observation and to reason" (1830–42: vol. 1, 457–59). However, Comte did not entirely rule out the use of hypotheses postulating theoretical entities.[10] In fact, he considered the molecular theory of matter to be a "model hypothesis, provided that one never attributes a vicious reality to it" (vol. 2, 736). The molecular hypothesis conformed to Comte's requirements that a hypothesis be only a "simple anticipation of that which experience and reason would have immediately been able to unveil, if the circumstances of the problem had been more favorable" (vol. 2, 457). In other words, hypotheses about things that are unobservable simply because they are too small to see were allowable for Comte. Theories that postulate the existence of forces, on the other hand, Comte would not have allowed, as these cannot in principle be observed. Such theories he regarded as mere holdovers from metaphysics. *Gravity,* for example, was not the name of a force for Comte, but only a general name for class of gravitational phenomena or, as he called it, a "general fact" (vol. 1, 26).[11] Comte would thus have rejected Durkheim's hypothesis of social forces for failing to conform to the requirements of a positive hypothesis.

Durkheim's more liberal attitude toward hypotheses in science can be explained at least in part, I believe, in terms of the history of science in the nineteenth century. At the beginning of the century, due to the legacy of Isaac Newton, scientists were still being subjected to criticism for using hypotheses that postulated unobservable entities and processes to explain the observable phenomena. The criticisms of Dalton's atomic theory provide a case in point. By the end of the century, however, it appeared that the use of such hypotheses had been a successful strategy in sciences ranging from physics and chemistry to geology and biology.[12] Objections on purely epistemological grounds to hypotheses postulating unobservables were becoming increasingly untenable. Although Durkheim was trained in neither the sciences nor the history of science, he could have become aware of these developments in science through his training in philosophy, on which they had a profound effect.

The Coercive Power of Collective Representations

In *The Rules of Sociological Method*, we recall, Durkheim said that social facts can be characterized by their property of having the power to impose an external constraint on the individual. However, Durkheim never equated social facts with these external constraints. External constraints for Durkheim are only the effects that social facts have on individuals. His conception of social life as consisting in collective representations explains this constraining power of social facts. Once the power is understood as arising from collective representations, the apparent ambiguities that other interpreters have found in his notion of social constraint disappear.

For Durkheim, social constraints are external to each individual in the sense that they are the product of society and thus come from outside the individual to be represented within him or her. As Durkheim explained, "the majority of our ideas and tendencies are not elaborated by us, but come to us from without" (1895a:9). He further clarified his concept of an external social constraint in *Suicide*, where he explained that he means external to each individual taken singly and not to all individuals in a society taken collectively (1897a:356). The fact that physical as well as social facts constrain us, Durkheim said in the preface to the second edition of *The Rules*, shows merely that both sorts of facts are equally real (1901c:xxii).

According to Durkheim, social facts constrain individuals in virtue of the social forces to which collective representations give rise. He said that normally a moral authority "invests the products of social activity," endowing them with a "prestige" that "inspires in us sentiments of deference and respect" (1900c:26). A thing inspires respect in us when "the representation that expresses it in consciousnesses is endowed with such a force that, automatically, it gives rise to or inhibits actions" (1912a:296). These mental or moral forces "draw all their power of action from representations, from states of consciousness" (1910b:60). If the constraint imposed by social facts is not always felt, he argued, this only attests to the power over us of education in the broadest sense of this term, which he said "consists in a continuous effort to impose upon the child ways of seeing, feeling, and acting" (1895a:11).

When Durkheim said that social facts constrain in virtue of "the prestige with which certain representations are invested" (1901c:xxi) and when he said that collective representations give rise to social forces, he meant these things not only in a figurative sense. To argue, as Keat and Urry do, that Durkheim "misleadingly" characterized representations as coercive (1975:83) is to impose our present concepts on this thought. That is, it is to assume what we mean by the term *representations* and to overlook the meaning this term assumes within the context of Durkheim's explanatory theories. In a passage in *The Division of Labor* that translators and commentators have failed to understand, Durkheim said that a mental representation is not merely an image or a shadow of reality but "a force that raises up around itself an entire vortex [*tourbillon*] of organic and psychical phenomena" (1902b:64). The term "tourbillon" was carefully chosen here. It alludes to

Descartes's mechanical philosophy, in which this term is used to name the vortices of invisible matter that carry the planets in their orbits around the sun (Descartes 1647: bk. 3 prop. 65ff.). It was still in use in the physical sciences in France at the time Durkheim was writing as a name for vortical motions generally.[13] To translate this term as "turbulence" or "whirlwind" is to overlook Durkheim's Cartesian allusion and thus to obscure part of his meaning.[14] He wanted to convey to the reader the sense that mental representations are not mere ghostly pictures but real things with real effects.

Collective representations, Durkheim believed, are endowed with a "psychological energy" superior to that of individual representations that allows collective representations to impose themselves on us and constrain our thoughts and actions (1955a:173). That is, due to their superior energy, collective representations are able to overwhelm individual representations and give rise to beliefs and actions in place of those that would result from individual representations alone. The superior energy of collective representations, Durkheim explained, is due to the fact that a collective representation has the combined energy of all the individual representations from which it was formed (1912a:297). This energy arises through a process that he described in *The Division of Labor* as the "fusion" of similar mental representations in the minds of individuals:

> If someone expresses before us an idea that was already our own, the representation that we have of it comes to be added to our own idea, superimposes itself on it, is confounded with it, communicates its vitality to it; from this fusion arises a new idea that absorbs the preceding ones, and that, consequently, is more lively than each of them taken separately. (1902b:67)

The process that he called "fusion" is characteristic not only of Durkheim's early work. He made use of the concept of the "fusion" of mental images as late as in *The Elementary Forms of the Religious Life* and in his subsequent defenses of this work.[15]

When Durkheim said that social phenomena are characterized by the property of imposing an external constraint on the individual, then, he did not mean that individuals are physically compelled to conform to certain manners of acting. The "force" that makes the individual conform to society he said is "not necessarily material." Nor did he believe that "the collective practices or beliefs must necessarily be inculcated in men by violence or coercion" (1900c:25).[16] To be sure, in his discussion of the forced division of labor he used the concept of constraint in the sense having to do with physical force. However, he made it clear that he thought that the forced division of labor occurs only in an abnormal form of society (1902b: bk. 3, chap. 2). In the normal state of affairs, society compels us by means of a set of mental representations that have been inculcated in us through our education.

An apparent ambiguity in Durkheim's concept of a social constraint has arisen chiefly from the examples that he provided. These include performing certain duties defined by law or custom as a brother or a husband, having to subscribe to certain religious beliefs and to follow certain religious practices, having to follow certain professional practices, being compelled to speak French and to use French money in France, suffering penalties or public disapprobation and ridicule for violating rules of law and morality, having to use the latest methods of industrial production on pain of financial ruin, being overtaken by the emotions of a crowd of people, and even being limited in one's actions and choices by the structure of society, the distribution of population, the means of communication and transportation, and styles of architecture and dress (1895a:6–18). What Durkheim meant to say, I think, is that all of these seemingly different kinds of constraint are but various outward expressions of the constraining effect of the superior force or energy possessed by collective representations. For example, we are not constrained to dress in a certain style by the articles of clothing themselves but rather by social forces that induce conformity to that style of dress.

Parsons, however, saw Durkheim as having made use of three different concepts of constraint here: being constrained by certain socially constructed facts of the environment, being constrained by society imposing sanctions in order to enforce certain norms, and being constrained only through one's voluntary consent to the binding authority of moral norms that one has "internalized" (Parsons 1937:378–90; 1975:107).[17] Steven Lukes distinguishes five different concepts of constraint in Durkheim's list of examples: the constraint exercised through the authority of legal and moral rules and social conventions that is manifested through sanctions when these rules are violated, the need to follow certain rules to accomplish certain tasks, the causal influence of the structure of society on social life, the psychological pressure of the crowd on the individual, and the effect of socialization and acculturation on the individual. Lukes considers only the first sense to be a legitimate use of *constraint,* which for Lukes carries the connotation of being compelled to do something against one's wishes, a connotation absent from the last three senses. The second sense Lukes argues is not a form of constraint at all but represents merely a type of reasoning about means and ends (1973:12–13).

It was never Durkheim's intention, however, to say that society compels the individual, through the threat of sanctions, to do otherwise than he or she wishes. As we saw above, a sanction is merely the sign of the presence of a social fact. Furthermore, at least some of the individual's wants and desires are the product of his or her society in the first place. Also, as Stephen Turner argues, as I mentioned in chapter 1, Parsons's distinction between "internalized" constraints and those that are merely imposed on the individual from "without" through his or her fear of sanctions begs the question (Turner 1986:126–28). Parsons assumes that internalized norms constrain through the individual being able to introspect or reflect upon their meaning. For Durkheim, on the other hand as we have seen, collective representations produce their effect through a process of which the individual is not necessarily aware.

In general, Parsons and Lukes have mistaken the superficial manifestations of the coercive power of collective representations for social constraint. Even things in the socially constructed environment such as styles of architecture or dress do not constrain the individual directly but rather are manifestations of a social constraint that operates on the level of collective representations. For Durkheim, all social facts constrain individuals in the same way, namely, through the social forces to which collective representations give rise.

Individual Representations and Collective Representations

Durkheim, I have argued, maintained a realist attitude toward collective representations and the social forces they engender. His realism is manifest also in his belief that collective representations form a natural kind distinct from individual representations. According to Durkheim, "states of the collective consciousness are of a different nature than states of the individual consciousness; they are representations of a different kind" (1901c:xv–xvi).[18] In the later part of his career, Durkheim said that the duality between collective and individual representations has its outward expression in the sacred and the profane (1899a(ii):157, 162–63; 1912a:304).

For Durkheim there were actually at least three basic kinds of mental representations: sensations, images, and concepts (1901c:xviii; 1955a:200). Sometimes he also included perceptions as a fourth type of representation, but these were characterized as the combination of sensations with images. Images themselves are nothing but residual sensations, he believed (1914a:316 n. 1). According to Durkheim, sensations and images are private; concepts alone are universalizable and can be communicated from one person to another through language (1955a:204; 1914a:317; 1912a:619). Concepts, in other words, are the bearers of meaning. Only concepts are collective representations; the other kinds of representations, he said, have their source within the individual (1914a:331; 1912a:618–20). Concepts, of course, for Durkheim include moral and legal rules. Society, he believed, is the source of anything having to do with morality, religion, or even conceptual thought. "If one removes from a man everything that comes from society, nothing remains but a being reduced to sensation" (1906b:74). Durkheim did not wish to suggest that individuals are wholly incapable of forming concepts. Rather, he said that individuals are constrained to construct their concepts in such a way that they can be universalized or communicated (1914a:318 n. 1).

Collective representations are distinct from individual representations for Durkheim in the same way that individual mental representations are distinct from neural states. According to Durkheim, many psychologists would refuse to identify mental representations with states of the nerve cells (1898b:16, 37–38). He professed to be asking merely that we grant sociologists the same courtesy that we grant psychologists and other scientists and allow sociologists to take as their subject matter a group of phenomena that is formed through the synthesis of elements of a "lower" order (1897a:362). Specifically, he asked that sociologists be allowed to study the collective representations that are the result of a "chemical synthesis" of individual representations (1900b:128), much as the latter are formed through reactions among neural states (1898b:39). There is the same "break in continuity" between psychology and sociology, he affirmed, that there is between biology on the one hand and chemistry and physics on the other (1895a:128). If one were to deny the independent existence of collective representations, he reasoned, to be consistent one would also have to believe that nerve cells can think:

Those, then, who accuse us of leaving social life in the air because we object to reabsorbing it into the individual consciousness, have not, undoubtedly, perceived all the consequences of their objection. If it were well-founded, it would apply just as well to the relations between the mind and the brain; consequently, it would be necessary, in order to be logical, to reabsorb thought into the cell and withdraw from mental life all of its specificity. (1898b:43)

Among the earliest critics of this group of analogies, Charles Elmer Gehlke argued that Durkheim never explained the sense in which a representation is like a chemical element or a cell (1915:95).[19] Durkheim seems to have thought that atoms combine to form molecules, molecules combine to form nerve cells, the states of nerve cells combine to form individual representations, and individual representations combine to form collective representations, with new properties emerging at each level as the result of these syntheses. To accept Durkheim's argument we would have to regard individual mental representations as both the product of a reaction among neurons as well as elements that can enter into further combinations. As Gehlke argued, "to insist so vigorously on a distinction between the individual-psychic and the social-psychic and then calmly to derive the psychic in its elementary forms from matter, seems, to say the least, to be straining at a gnat and swallowing a camel" (1915:96). In Durkheim's defense, one can say that he never intended these analogies to *demonstrate* the existence of collective representations. Instead, Durkheim drew these comparisons only to argue that it was plausible to consider the possibility of such representations and that the investigation of this possibility would be a worthwhile endeavor.

Durkheim's Social Realism

Durkheim's realist attitude toward unobservable entities is apparent not only in his beliefs about collective representations and social forces. It is also and above all manifest in his belief that a society is a real entity. For Durkheim, this type of entity consists in a set of collective representations shared among the members of a particular society: "The eminent role of collective representations is to 'make' this superior reality that is society itself," he said (1955a:174). Another name he gave to this superior reality is the "collective consciousness," which for Durkheim is a "a psychic individuality of a new kind" that arises through the "fusion" of individual consciousnesses (1895a:127).

Durkheim's belief that societies are real things irreducible and "external" to the individuals who compose them has been called his social realism by his critics, beginning with Simon Deploige around the beginning of this century (Durkheim 1907b:402). In answer to Deploige's charges that Durkheim's social realism represents the importation of German metaphysical ideas into French thought, Durkheim responded that his concept of society belonged to an Anglo-French tradition that included the thought of Comte, Spencer, Renouvier, and Espinas (1907b:402; 1913a(ii)15:326). A misunderstanding of Durkheim's social realism opened him up to the charge of "hypostatizing" his collective consciousness as some sort of group mind.[20] Gabriel Tarde attacked Durkheim's concept of society as "metaphysical" and accused him of "scholastic realism" (Tarde 1898:73; Durkheim 1904b:86–87). Durkheim's "collective souls" were ridiculed by Charles Andler as "large shadows . . . hovering above every society of men" (1896:245). In his early writings on Schaeffle's concept of the social consciousness, however, Durkheim tried to remove this objection. Although for Schaeffle society is some sort of "being" with its own life and its own "consciousness," Durkheim said, "this being has nothing metaphysical about it. It is not a more or less transcendent substance; it is a whole composed of parts" (1885c:632; cf. 1885a:84, 92; 1888a:97). Similarly, Durkheim made it clear that although he uses expressions like "the collective soul" in his own work, "we do not at all intend to hypostatize the collective consciousness. We no more admit a substantial soul in society than in the individual" (1897a:14 n. 1).[21]

Unlike many of his critics, Durkheim understood that to regard something as real is not necessarily to hypostatize it.[22] To hypostatize something entails regarding it not only as real but as a separately existing *substance*. Thus, for example, when Durkheim criticized Albert Regnard for "hypostatizing" the concept of the general interest, he accused him of treating it like a "metaphysical entity" or "substance" (1886a:199–200). Similarly, he described the process by which the idea of a god is formed in primitive societies as hypostatization (1912a:295). By *substance* Durkheim meant some sort of I-know-not-what that supports properties. Thus he said that sociology has no more need of an "invisible substratum" than psychology does of a soul: "All of these hypotheses are objects of faith, not of science" (1885a:92). Sociology, he believed, requires no other substratum than "that which forms through the uniting and combination of all the individual consciousnesses. This substratum has nothing substantial or ontological about it, since it is nothing other than a whole composed of parts" (1897a:361). As he wrote in a letter to Bouglé, "a phenomenon of individual psychology has for its substratum an individual consciousness, a phenomenon of collective psychology a group of individual consciousnesses."[23] Durkheim did not believe that societies can exist without their individual members. "Society," he said, "needs individuals in order to exist" (1913b:74; see also 1897a:x). For Durkheim, society "can exist only in and through individual consciousnesses" (1912a:299).

The existence of individuals, however, is only a necessary and not a sufficient condition for that of society. Individuals must also be associated in a certain way (1895a:127). Society, for Durkheim, may best be described as an emergent reality. That is, society forms a whole with properties that cannot be reduced to the properties of its parts:

> It is indeed true that society does not contain any other active forces than those of individuals; but individuals, in uniting, form a psychic being of a new species . . . The social fact does not emerge except when they have been transformed by association, because it is only at this moment that it appears. Association itself is also an active factor that produces specific effects. (1897a:350)[24]

To deny that society is a distinct kind of entity and to say that a society is merely the sum of its individual members, Durkheim contended, would be to assume that the whole is equal to the sum of its parts, which he denied (1895a:126; 1887c:37–38). He repeatedly argued by way of analogy with what he believed to be true in chemistry, biology, and psychology that new properties can arise through the combination of elements. The chemical properties of compounds can be different from those of the elements from which they were synthesized, he said.[25] As an example he cited the fact that soft metals like copper and tin can combine to form bronze, which is hard (1925a:293). For the biologist, he added, a living cell is more than a mere collection of atoms and molecules; it is also the way in which they are associated. Biological phenomena, he maintained, cannot be "analytically explicated" by inorganic phenomena (1895a:126). In other words, one cannot *define* the terms of biology in terms of physics and chemistry.

One of the emergent properties of a society is the social relations that exist among individuals. For Durkheim, these social relationships exist neither independently of individuals nor in any single individual.[26] As a philosophical realist, Durkheim believed in the real existence of relations among particulars without maintaining that these relations can exist independently of the particulars that participate in them. Where, then, and how do these social relationships exist for Durkheim? They exist, I believe he intended to say, as collective representations in the collective consciousness.[27]

The collective consciousness is nothing more than Durkheim's term for the group of collective representations held in the minds of the members of a society. In fact, I suspect that in his later works he preferred to use the term *collective representation,* shunning *collective consciousness* to avoid just the kinds of misunderstandings we have been discussing. The existence of collective representations implies the existence of a group mind no more than the existence of individual representations implies that of a soul. Although Durkheim spoke of two "beings" in each of us, an individual and a social being, he said, "I do not wish to say that there are in us two substances, in the metaphysical sense of the term, but simply two circles of internal life, *two systems of states of consciousness*" (1913b:73). When Durkheim spoke of the "soul" of a society he was speaking only of "the whole of its collective ideals" (1911b:116). Even though he wrote of society becoming conscious of itself, he appears to have meant by this no more than that the individual members of society acquire the "common sentiment" of their "moral unity" (1912a:329–31).

Durkheim presented an interesting argument here. What he was saying, in brief, is that just as psychology has been able to dispense with the concept of the soul at least since Locke, sociology should be able to dispense with the need for a metaphysical substratum. Such a notion cannot be known by science anyway and the possibility of a science that studies the relevant class of phenomena does not depend on it. Durkheim could have just as easily argued that the possibility of physics does not depend upon the existence of material substance. In fact, Berkeley extended Locke's argument against the existence of mental substance in just this way to argue against the existence of matter. Durkheim drew the analogy between the concepts of physical and mental substance in an early account of the concept of self:

In order to be able to represent to ourselves the connection among material phenomena, we are obliged to form a concept that permits us to link them one to another, and this concept can contain nothing phenomenal, since its content must by definition by inaccessible to the senses. This is the concept of substance. Once this notion is formed, the mind is led to apply it equally to internal phenomena, for which it is not however made and for which it is not suitable. (1887c:128)

In other words, according to Durkheim our concept of the soul or of spiritual substance is something that we have invented in order to represent the notion of the self or personal identity to ourselves. For Durkheim, this notion is nothing but one of those common-sense preconceptions that science should try to avoid, as I will show in chapter 5 in my discussion of Durkheim's methodology. Thus to argue that Durkheim's treatment of the individual as an abstraction is a violation of common sense, as Pascual Gisbert does (1959:365), is to beg the question. From Durkheim's point of view, the self [*moi*] could even be considered a society: that is, both individual and collective life are made up of representations (1895a:136; 1898b:14).

What philosophers have referred to as the "soul" is for Durkheim nothing but the collective consciousness "incarnated" within each individual (1913b:74). Durkheim was fond of trying to explain his concept of society by saying that there exists in each of us two beings or consciousnesses that, although they cannot be separated except by abstraction, are nevertheless distinct. One is comprised of those individual representations that arise from physiological states, including sensations and psychological tendencies or appetites. The other is made up of those collective representations that are shared among the members of a society and represent that society within each individual.[28] To say that there are two consciousnesses in each of us is something of a simplification, he added, for there are actually as many consciousnesses in each individual as there are social groups to which he or she belongs (1902b:74 n. 1).

The Development of Durkheim's Conception of Sociology

I have been arguing that throughout his entire career Durkheim identified the subject matter of sociology with collective representations. In support of my position I have shown that this conception of the subject matter of sociology is consistent with his definition of social facts in *The Rules of Sociological Method*. One may object to my interpretation by arguing that I have interpreted *The Rules* largely in light of his later writings. At least one influential interpretation of Durkheim asserts that he altered his conception of social facts after *The Rules*. I would thereby beg the question against this interpretation by trying to understand Durkheim's definition of social facts in the context of his later works.

Gabriel Tarde was the first to argue that Durkheim had moved closer to a psychological conception of social facts since the founding of *L'Année soci-ologique* (1901:460 n. 1).[29] In fact, Tarde claimed that Durkheim made this change in order to accommodate the views of the philosophers who were to become his collaborators on this journal. Now Durkheim did in fact defend a psychological conception of sociology in his correspondence with Céles-tin Bouglé in regard to their journal. In these letters Durkheim said that because he conceived of social facts as a kind of mental representation, he regarded sociology as a kind of psychology—specifically, a social, collective, or group psychology as distinguished from individual psychology.[30]

However, Durkheim did not adopt a psychological or mentalistic con-ception of sociology only to suit Bouglé or the others associated with *L'An-née sociologique*. In his published response to Tarde, Durkheim said that this has always been his view (1901d). Indeed, Durkheim expressed much the same conception of the subject matter of sociology in some of his earliest publications. In the first year that he published anything, Durkheim de-clared that social psychology is "the very substance" of sociology (1885c:633). Durkheim characterized social psychology in his opening lec-ture for his very first course in sociology at Bordeaux, published in 1888, much as he did thirteen years later in his preface to the second edition of *The Rules*. He explained that social psychology is based on a comparative study of myths, legends, religious traditions, political beliefs, languages, and so on, and has as its goal the discovery of the laws governing the for-mation of the collective ideas that hold society together (1888a:101–2; 1901c:xviii.)[31] He saw the laws of social psychology as analogous to the laws of the association of ideas: they will describe how collective representa-tions "attract and exclude each other, fuse with or distinguish themselves from one another" (1901c:xviii). Durkheim saw philosophy as splitting into two separate disciplines: psychology, which studies individual con-sciousness, and sociology, which studies the consciousness of society (1888a:106).

Durkheim seems to have used "social psychology," "sociology," and "general sociology" as equivalent expressions. In an early essay he defined one of the explanatory goals of what he calls "general sociology" as "the for-mation of the collective consciousness," much as he characterized the goal of social psychology (1886a:214). In later writings he described "general soci-ology" as being "like the philosophical part" of sociology, with no specific subject matter distinct from that of the other social sciences (1899a(i):i; 1902a(ii):168; 1909e:152). Its goal is to "reconstitute the unity of the whole thus decomposed by the analysis" of each of the specialized social sci-ences (1902a(ii):168). This synthesizing discipline, which he also referred to as the "philosophy of the social sciences," draws general conclusions from the various social scientific disciplines, which in turn use these conclusions to guide their subsequent research (1898a(i):iv; 1905c:257; 1909e:152). He characterized "sociology" without qualification in similar terms, as taking for its subject matter "the totality of social facts without exception," embracing all of the individual social sciences, which are but branches of it (1903c:485; 1900c:33).

Given Durkheim's apparent terminological indecision, it is ironic that he complained of a lack of precision in the way in which others have used the term *general sociology* (1902a(ii):167). But I think that this apparent confusion in Durkheim's terminology can be resolved if we understand that sociology for Durkheim had as its goal to explain the formation of the collective consciousness. He sometimes called it "social psychology" in order to allude to the analogous task of individual psychology of accounting for the individual consciousness.[32] When he referred to it as "general sociology" or "philosophy of the social sciences" he was emphasizing that sociology draws on all of the social sciences for what they can tell us about the formation of the collective consciousness. For example, the sociology of law and morality will tell us how the legal and moral concepts that make up an important part of the collective consciousness are formed. His use of the term *philosophy* here was to make a historical connection with the discipline that had traditionally been concerned with problems of consciousness and mental representation.

Terms like *collective consciousness* and *collective representation* actually predate Durkheim's first use of the term *social fact*. The first term appeared as early as 1885. In a review of Albert Schaeffle's *Bau und Leben des sozialen Körpers,* Durkheim compared him favorably to Alfred Espinas, whom he cited as the source of his view of society as an "organism of ideas" (1885a:84–86).[33] Espinas had used the term *collective consciousness* in his 1877 philosophy dissertation, *Les Sociétés animales* (1878:514f.). In another review, Durkheim characterized Schmoller and Wagner as working with a similar concept of a social being and said that terms like the *social consciousness* and the *collective spirit* were part of the "current language" (1887c:37–8). Also, as I showed above, in his opening lecture for his first course in sociology, Durkheim defined the explanatory goals of this science in terms of the collective consciousness (1888a:101–2). The term *collective representation* made its first appearance in 1887, in a review in which religion was characterized as a "system of representations" (1887b:308). It was not until the following year, however, in an essay on suicide, that the term *social fact* appeared (1888d:446), and it is not at all clear that this term was being used here in the technical sense in which it was used in *The Rules*.

Durkheim's early papers reflect the philosophical tradition in which he was educated. The concept of mental representation has played an extremely significant role in philosophy since the days of ancient Greece. More recently, Kant had used the equivalent German word "Vorstellung" as a generic term for mental states of all kinds (1787:A320/B376), much as Durkheim did. Kant of course informed much of nineteenth-century French philosophy, most notably Renouvier, who began his monumental *Essais de critique générale* with an analysis of the concept of representation (1854–64: part 1; see Allcock 1983:xxxi). Durkheim's close friend Octave Hamelin continued Renouvier's project, working twenty-five years to complete *Les éléments principaux de la représentation* (1907; see Durkheim 1907e). To borrow an apt expression from Stjepan Meštrović, the concept of representation was "pedestrian" in Durkheim's milieu (1985:203).[34]

Some recent commentators, echoing Tarde, have insisted that Durkheim's identification of social life with collective representations is but a later addition to his thought and a departure from his original conception of social facts. Lukes is skeptical of Durkheim's reply to Tarde and his letters to Bouglé, and affirms that "the focus of Durkheim's attention shifted" to collective representations only from the publication of the *Rules* in 1895. He cites what he takes as evidence for this interpretation the letter to the editor published in the *Revue néo-scolastique* in 1907 that I quoted in chapter 1. In this letter Durkheim described his reading of Robertson Smith and others in preparation for a course on the sociology of religion given in 1895 as marking a "line of demarcation in the development of my thought" (1907b:613; cited by Lukes 1973:6).

Not only does Lukes appear to overlook Durkheim's early concerns with the collective consciousness but he also seems to conflate Durkheim's conception of social facts as collective representations with his interest in religion. Benoît-Smullyan (1948:510) avoided this confusion but nevertheless saw this conception of social facts as dating only from Durkheim's essay "Individual Representations and Collective Representations" (1898b). To be sure, it is in this essay that Durkheim first described how collective representations, once they are formed, begin to take on an independent life, forming new syntheses or combinations in accordance with their own laws (1898b:46; cf. 1912a:605). However, as I shall argue below, Durkheim never discovered these laws of collective ideation.

Like Lukes, Jeffrey Alexander believes that "representation became important to Durkheim only after his encounter with religion" around 1895 and that he subsequently "discovered" its importance for secular life (1982: vol. 2, 248, 488 n. 176). Alexander's externalist account of Durkheim's intellectual career, as we saw in chapter 1, takes this presumed shift in Durkheim's conception of sociology as the fact to be explained and then invokes noncognitive social factors to explain it. Unfortunately for Alexander, however, this supposed change in Durkheim's conception of sociology never took place.

Throughout his career Durkheim conceived sociology as being concerned with collective representations, a type of unobservable mental entity with both causal and semantic properties. In the next chapter we shall see that the dual nature of collective representations is reflected in his conception of a sociological explanation, which combines causal explanation with the explication of concepts.

4   *Durkheim's Concept of Sociological Explanation*

Durkheim conceived the goal of sociology to be to explain social phenomena in terms of their hidden essences, which he identified with collective representations. He believed that collective representations not only are real entities but can be grouped into distinct natural kinds. Different species of social phenomena are to be explained as resulting from different kinds of collective representations. His reasons for denying that a kind of phenomenon can have a plurality of causes reveal that for Durkheim, to provide a causal explanation of a kind of phenomenon and to state the real essence of that kind of phenomenon are one and the same thing. Natural kinds ground the existence of causal, explanatory laws for Durkheim. That is, he believed that each kind of phenomenon is governed by laws that derive from the essence of that kind.

Natural kinds for Durkheim exist at different levels of generality, that is, at both the genus and the species levels. His belief that individual and collective representations formed two distinct genera of mental entities provided the basis for his restrictions against psychological explanations of social facts. In addition to uncovering the collective representations that explain particular kinds of social phenomena, another of Durkheim's goals for sociology was to discover the sociological laws that govern collective representations in general. According to Durkheim, these "laws of collective ideation" would be analogous to the psychological laws of the association of ideas yet distinct from them (1898b:47 n. 1; 1901c:xvii–xviii).

Durkheim's concept of sociological explanation is modeled on the traditional concept of explanation through essences. According to this model, to give the essence of a type of phenomenon is both to define it in terms of its genus and species and to give its function and its cause. He insisted that social causes be clearly distinguished from social functions as well as from psychological causes and that social functions be clearly distinguished from individual goals and purposes. For these reasons he rejected sociological explanations premised upon the social, economic, or political interests of social actors. The fact that something, if it were to come to pass, would do somebody some good could never be the explanation of how it came to be in the first place.

Rightly or wrongly, Durkheim believed the essentialist model of scientific explanation to be that which is followed by the natural sciences and he patterned his concept of a sociological explanation after what he thought an explanation in the natural sciences was like. He held that the very idea of a scientific explanation requires that there be real essences in nature. Whether or not he was right about the natural sciences, however, his adherence to the essentialist model of explanation in the social sciences may have been advantageous for him. His essentialism provided a role for the meanings of social phenomena in causal explanations of them. For Durkheim, to show that the essence of a kind of social phenomenon is a collective representation was to give both the meaning and the cause of this kind of social phenomenon.

Far from imposing an antirealist interpretation on the natural sciences, as contemporary sociologists of scientific knowledge tend to do, Durkheim, in subscribing to an essentialist model of scientific explanation, risked going too far in the realist direction. Nor did he ever abandon his scientific realism, even after he took up his research in the sociology of knowledge. In fact, the very next year after he published his masterpiece in the sociology of knowledge, *The Elementary Forms of the Religious Life* (1912a), he gave a series of lectures in which he excoriated William James's pragmatist conception of truth and defended the idea that truth is a copy of reality (1955a). In his sociology of knowledge, Durkheim never failed to maintain a distinction between the epistemological thesis that even our most basic explanatory concepts are social in origin and the metaphysical thesis that there are natural kinds. The social construction of scientific categories for Durkheim does not preclude their gradually coming to carve up reality in a way that reflects how things are actually divided in nature. He believed that the very fact that science is a social process that involves an ever-expanding international community of scientists tends to correct any influence that social interests may have originally had on scientific thinking.

Durkheim's Essentialism

Durkheim discussed his concept of sociological explanation in a chapter of *The Rules of Sociological Method* devoted to the rules for the *explication* of social facts. This French term has two different senses that today we would distinguish clearly but that Durkheim did not keep separate. In one of these senses, to explain a *fact* means to subsume it under a more general expression, often one that provides the cause of the fact in question. To explain the meaning of a *concept*, on the other hand, means to analyze it or to provide a definition.[1] Durkheim seems to have had both senses of explanation in mind in the following passage in which he said: "to explain an institution is to give an account of the various elements that serve to form it, it is to show their causes and their reasons for being" (1909e:154). As we saw in the last chapter, Durkheim believed that social institutions exist in collective representations. Once these collective representations are resolved into their elements, a sociological explanation for Durkheim would then show how these elements combine in accordance with what he called the laws of collective ideation. Hence, Durkheim's conception of the sociological explanation of social facts would seem to involve the analysis of ideas and thereby resemble a traditional conception of definition.

Durkheim's ambiguity concerning the nature of explanation carried over to his concepts of definitions and laws. He wrote about definitions in terms that are appropriate for discussions of empirical hypotheses and characterized laws in ways that make them sound like definitions. For example, he thought of definitions as standing in need of "confirmation" (1902b:73) or "verification" (52) and compared definitions with regard to their ability to account for the facts.[2] The reason he provided for preferring Gaston Richard's definition of crime to Raffaele Garofalo's, for instance, is that Richard's has greater explanatory scope. In Durkheim's words, Richard's "definition is larger . . . and gives a better account of the facts" (1893a:293).

Laws for Durkheim express the nature or essence of a kind. He clearly recognized the knowledge of explanatory laws to be the goal of science (1900b:113) and even praised Comte for having extended the notion of natural law to the social realm (1903c:485). However, Durkheim criticized Comte for having failed to recognize that there are different species or types of societies, each governed by its own set of laws (1895a:95; 1903c:469). In addition, Durkheim conceived these laws as entailed by the essence of each type. Sociology, he believed, could not have become a science "before one had acquired the sentiment that societies, like the rest of the world, are subject to laws that derive necessarily from their nature and that express it" (1909e:138). To explain the facts in any science, according to Durkheim, "is to establish among them relations that make them appear to us as functions of one another, as vibrating sympathetically according to an internal law, founded in their nature" (1912a:339). For Durkheim it is not necessarily the number or frequency of cases in which two phenomena are found to be related that makes their relationship a law, but rather "that it be of a nature to repeat itself" (1908a(3):232–33).

Durkheim found the earliest expression of the idea of a sociological law in Montesquieu, who defined laws as "necessary relations that derive from the nature of things" (1909e:139). In his Latin dissertation, Durkheim explained that for Montesquieu the laws of society are no less natural than other laws of nature, differing only in that the former derive from the nature of the social organism. According to Montesquieu, Durkheim said, there are laws that can be found in all societies and that can be deduced from the very definition of society: "they are implied in the very notion of society and can be explained by it" (1960b:22; see also 19–21).

Laws for Durkheim, again in contradistinction to Comte, can express causal relationships. "Every relation of causality is a law," Durkheim said (1908a(3):232). Causal laws for Durkheim are distinct from empirical generalizations. His well-known generalization that Protestant suicide rates are greater than Catholic suicide rates, for example, should not be mistaken for a law stating a causal relationship between religion and suicide. This "law" is merely an empirical regularity that serves as corroborating evidence for a more fundamental causal theory of suicide. Nor should Durkheim's use of social statistics in *Suicide* be misunderstood as indicating a belief that causal laws are statistical in form. As Stephen Turner argues, Durkheim conceived statistics only as measures of the strength of suicidogenic currents and social forces (1986:134–35, 154).

Durkheim described causes as well as laws in terms relating to natures and essences. For example, in *The Division of Labor* he described crime as a cause that brings about punishment as its effect. Not only did he expect the definition of crime "to give an account of all the characteristics of punishment" (1902b:52), but he believed that to inquire into the causes of punishment is to investigate "that in which crime essentially consists" (1902b:35). In *The Elementary Forms* he analyzed our ordinary, common-sense notion of causality in the following terms: "The cause is the force before it has manifested the possibility [*pouvoir*] that is in it; the effect is the same possibility, only actualized" (1912a:519). "Possibility" may be a better translation of *pouvoir* than "power" in this context, as "power" does not provide the needed contrast with "actuality." That is, according to the common-sense notion of the causal relationship, the effect a thing has is conceived as the realization of its potential and the cause is conceived as the essence or nature that defines that potential. In spite of his warnings against prenotions in *The Rules* (1895a:20ff.),[3] Durkheim seems to have accepted something like this common-sense notion for his own account when he said that "the manner in which a phenomenon develops expresses its nature" (160).

Durkheim's essentialism has been criticized by a variety of thinkers. Georges Sorel was among the first to argue that Durkheim's social science should be concerned with only the relations among phenomena and not their essences (1895:3). More recently, Russell Keat and John Urry have faulted Durkheim's identification of causes with essences for implying that only ultimate explanations, or those for which there are no further explanations, are truly explanatory (1975:81, 85, 163). Whether or not Durkheim's essentialist model of explanation can be justified, however, it is important to understand this model and its limitations in order to understand the course of his intellectual career.

Durkheim's Realism

Durkheim never pretended that the essences of social facts were easy to know. He distinguished the definitions that express real essences from those preliminary or provisional ones that are necessary to guide research. Provisional definitions merely express certain superficial characteristics that allow one to recognize the presence of a certain kind of fact; they "do not directly touch the foundation of things," he said (1899a(ii):140).[4] Only definitions that provide the "essential elements" and "profound characteristics" of things, he believed, are "truly explanatory" (1912a:6, 31–32).

It is tempting to compare Durkheim's distinction between provisional definitions and definitions that are truly explanatory to Locke's distinction between the nominal essences by which we recognize and classify things into kinds and the real essences that make things what they are (Locke 1690: bk. 3, chap. 3, sec. 15). There are, however, important differences between Locke and Durkheim. Durkheim believed that kinds enjoy real existence and are not merely nominal, as they are for Locke. Durkheim's provisional definitions are nominal only in the sense that they provide the criteria for the application of a name. Far from a knowledge of real essences being unattainable in principle as it is for Locke, such knowledge is the goal of science for Durkheim. "Scientific theories have for their unique goal to express the real" (1925a:2), he said.

In some ways, Durkheim's philosophical position is actually closer to the rationalist tradition of Descartes than to the empiricist tradition of Locke. Durkheim accepted what he called the "rationalist postulate" that "there is nothing in reality that one may be justified in considering as radically refractory to human reason" (1925a:4, 291). By endorsing the rationalist postulate, Durkheim did not mean to suggest that science is complete or even approaching completion in the sense of exhausting reality (5, 304). He intended merely to deny that any category of facts is "invincibly irreducible to scientific thought" or "irrational in its essence" (5). In Durkheim's mind, rationalism denies that there is some limit to the progress of science such that one can say, "you will go this far and no further" and that beyond this point all is mystery and irrational (304).

Durkheim, in fact, rejected empiricism on the grounds that it would make scientific knowledge impossible. In spite of the fact that the empiricists are generally regarded "as the protagonists of positive science," he accused, "under the color of positivism, they introduce mystery everywhere" (1897f:288). Durkheim considered empiricism and mysticism to be "two closely related ways of thinking" (1906b:88) and reproached empiricists with "obscurantism" (1925a:303). Empiricism, he charged, results in irrationalism (1912a:20), clarifies nothing (1897f:288), and makes the world "unintelligible" (1925a:303).[5] Unlike rationalism, according to Durkheim, empiricism does not assume that there are any "internal connections" or "logical relations" among things that would "allow one to think of some with the help of and as a function of the others" (1897f:287). For the empiricists, he affirmed, concepts like causation "are only more or less elaborated sensations" (1914a:321), "purely artificial constructions" (1912a:26 n. 2), illusory appearances "that correspond to nothing in things" (1912a:19–20). On the assumption of empiricism, he argued, each fact would be like a "stranger" to those that precede or follow it. Science would be reduced to supplying mnemonic devices for remembering the usual order in which one thing follows another; in other words, he said, it would be "blind" (1897f:287–88).

Having repudiated empiricism, Durkheim adopted the realist belief that there are real relations among things that exist in virtue of the fact that there are natural kinds. He argued that the history of science has "progressively demonstrated" that the phenomena of nature can be linked to one another in accordance with "rational relationships" through the discovery of these very relationships (1925a:5). That natural phenomena can be linked in this way, he believed, has been shown to be the case first in chemistry and physics, then in biology, and was being demonstrated in psychology. Therefore, he thought, there is no reason to believe that moral or social facts cannot be linked in this way as well (6).

In his argument that the history of science has progressively demonstrated the existence of relations among phenomena, Durkheim was attempting to provide empirical evidence for what is in fact a metaphysical position, much as contemporary scientific realists do. There are, however, some important differences between Durkheim's "rationalism" and contemporary realism. The latter employs a strategy of inference to the best explanation, arguing that the success of science can best be explained only on the realist assumption that the terms of our scientific theories refer to real things in nature.[6] Whatever the merits of this form of argument—and they are not clear—it is not Durkheim's. He was attempting instead to generalize inductively from the "lower" sciences a conclusion that he carefully said there is no reason not to apply to the "higher." The weakness of his argument stems from its assumption of some sort of Comtean hierarchy of the sciences, according to which the phenomena of nature form distinct kinds corresponding to each of the sciences in the hierarchy.[7] That is, Durkheim's argument presupposes a type of realism and thus would seem to beg the question.[8]

Durkheim's commitment to rationalism was not total, however. He questioned the a priorism that one often finds associated with the rationalist belief that "the world has a logical aspect that reason eminently expresses" (1912a:20). Because the a priori philosophy holds that "the logic of things is identical to that of the mind," he argued, it assumes that in order to do science it is sufficient to retreat into oneself and to know one's own nature. Durkheim, however, said that he knew "that things are not so clear, so transparent, so easy to penetrate" (1897f:287, 289). He criticized what he called Cartesianism for overlooking the role of observation and experiment and attempting to reduce all of science to a universal mathematics (1925a:300–1). In addition, he found fault with philosophers who believed that general laws governing all of human history "are inscribed in the constitution of man" and can be deduced from our concept of the nature of man (1895a:94–95).[9]

Durkheim also accused rationalism of what he called "simplism" (1897f:290) or the tendency to believe that that alone is real that is perfectly simple (1925a:286). This attitude, in his opinion, is best represented by the philosophy of Descartes. According to Descartes, Durkheim believed, nothing is real save what can be the object of a clear and distinct idea and nothing can be the object of a clear and distinct idea that does not possess mathematical simplicity. Complex things and even secondary qualities such as color and taste are represented to the mind only in a confused manner and thus have no basis in reality for Descartes, at least as Durkheim interpreted him (1925a:287, 289). For the rationalist, Durkheim thought, neither complex things, particulars, concrete individuals, nor sensations are real, but are only deceptive appearances or illusions. The rationalist, he said, identifies reality with "a system of logically linked concepts" and believes that between this reality and the world of appearances there is a "break in continuity" (1897f:287; 1925a:289).

Durkheim, on the other hand, did not drive a wedge between reality and appearance and identify the real exclusively with universals or simples. Nor did he believe that universals exist independently of the things that they characterize. Although he considered himself a realist, he criticized the Platonic version of realism, according to which, for example, "the idea of whiteness in general constitutes a sort of distinct entity" (1898b:25). For Durkheim, "The general exists only in the particular" (1912a:617). Furthermore, he believed that reality is complex and not simple. Whether the cause of complexity is inside or outside the human mind, complexity is nonetheless real, and there is no reason to believe that complexity exists only in virtue of the human mind. However, even if it were true that the complex were only an appearance, an appearance is nevertheless "a real phenomenon," much as a reflection in a mirror is a real thing. Colors, odors, and sounds, Durkheim said, are the "realities in which I live" (1925a:292–94). What he found true of nature generally he also believed to be true of society: "Social reality is essentially complex, not unintelligible, but merely refractory to simple forms" (1904b:84).

The mitigated rationalism Durkheim adopted may be compared with the "rationalist empiricism" of Hippolyte Taine. Taine, like Durkheim, accepted that there are logical relations among things (1897f:287). Unlike the Cartesian rationalists, however, Durkheim said, Taine denied that the sensible world is but an appearance: "It is reality, all of reality" (1897f:288). Elsewhere Durkheim wrote that "the world exists for us only to the extent to which it is represented" (1909d:756).[10] Durkheim claimed to have accepted this philosophy for his own account but added that it needed to be "refounded and rethought anew by a more vigorous mind" (1897f:290). Although Durkheim was not explicit about where he disagreed with Taine, I think he would have rejected Taine's monistic identification of reality with the sensible. Durkheim adopted instead a kind of dualism that he distinguished from what he called "empirical monism" (1914a:321). His "dualism" included within reality a kind of mental representation that is not met with in our sensory experience. In other words, his philosophy went beyond the rationalist empiricism of Taine in that Durkheim allowed for the real existence of collective as well as individual representations.

Tom Gieryn (1982) argues that Durkheim abandoned the "realist" views characteristic of his early works for a "relativist" position in his later writings in the sociology of knowledge. Gieryn has in mind particularly Durkheim and Mauss's "Primitive Classification" paper of 1903 and Durkheim's *The Elementary Forms of the Religious Life* of 1912. Both works defend the position that social groups serve as our model for the concept of a genus. Gieryn seems to think that Durkheim's early belief in the independent existence of natural kinds was undercut by his later belief in the social construction of our basic categories of thought, especially our classificatory concepts.

One can in fact find passages that would seem to support a nominalist interpretation of Durkheim's thought even in works as early as his 1898 essay "Individual Representations and Collective Representations." Here Durkheim had maintained that our conscious experience consists of a continuous flux and that distinctions among our representations "are our work; far from finding them there, it is we who introduce them into the psychic continuum" (1898b:25). In their essay on primitive classification, Durkheim and Mauss said that during the early stages of individual development, before one has learned the concepts and classifications of one's culture, that one is not even capable of making such distinctions (1903a(i):399–400).

Nothing in our experience, neither of the external world nor of our internal mental life, Durkheim believed, could suggest the concept of a kind to us. Our classificatory concepts, he said, are "instruments of thought that have been constructed by men" or "by human groups," taking their own social groups as archetypes:

Now, one cannot see where we would have found this indispensable model, if not in the spectacle of collective life. A genus, in fact, is an ideal but clearly defined grouping of things among which there exist internal connections analogous to the bonds of kinship. But the only groupings of this sort that experience makes known to us are those that men form through association. (1912a:209–10; cf. 27)

In anticipation of the objection that the notion of a class was suggested by nature, Durkheim argued that a heap of sand or a pile of rocks lacks the necessary organization and definition. It is significant that he cited examples only of aggregates here and failed to consider examples such as a herd of animals or a flock of birds, although elsewhere he did concede the existence of animal societies (1912a:523 n. 1). To be fair, he did argue that totemism does not see animals as forming distinct kinds but includes them in the same totemic groups into which human beings are divided (1912a:210; 1903a(i):396–97). If indeed we could accept his conjecture that totemism constituted humanity's earliest system of beliefs, the fact that totemic groups include animals as well as humans would refute the hypothesis that biological species and not human social groups were the source of our concept of a class.

Perhaps a better reason Durkheim provided for rejecting nature as the source of our classificatory concepts is that a classification includes a concept of a hierarchy, which must be of a social origin. "Neither the spectacle of physical nature," he said, "nor the mechanism of mental associations can furnish us the idea of it. The hierarchy is exclusively a social thing" (1912a:211). That is, as we shall see in chapter 8, Durkheim thought that the notion of subsuming a species under a genus could not have been suggested by nature but must have arose from the idea of subsuming a clan under a tribe. The fact that Durkheim believed that *all* societies subsume clans under tribes and species under genera, however, would seem to call into question Bloor's (1982) relativist interpretation of Durkheim's primitive classification hypothesis.

Whatever the merits of Durkheim's arguments, to think that his sociology of knowledge conflicts with his realism would be a mistake. To deny that biological species are the source of our concept of a kind, as Durkheim did, is not to deny that living species are real things. Durkheim recognized the danger that one could confuse the epistemological question concerning our knowledge of kinds with the metaphysical question concerning their real existence and then mistakenly give his theory of the categories a nominalist interpretation:

> But if the categories originally translate only social states, doesn't it follow that they cannot apply to the rest of nature except by right of metaphors? If they are made solely for expressing social things, they cannot, it seems, be extended to the other realms except by way of convention. . . . One thus returns, by another road, to nominalism and empiricism. (1912a:25)

He offered two replies. First of all, he asserted that societies are real things and a part of nature. Durkheim found it unlikely that kinds would exist within one part of nature, that is, in societies, but not in the rest of nature. He then argued that if classificatory concepts contain any artifice, "it is an artifice that follows nature closely and that is forever forced to come nearer to it" (1912a:25–26). Socially constructed concepts are repeatedly submitted to the control of experience, he affirmed, and would not "acquire an extended and prolonged empire over the minds" of people "if they were inconsistent with the nature of things" (1912a:625). Categories may, in some sense, be works of art, but he said it is "an art that imitates nature with a perfection susceptible of increasing without limit" (26 n. 2). Science, according to Durkheim, is responsible for bringing our system of representations into an ever-increasing harmony with nature through a social process of verification. "Verification," he said, "takes place mutually: the experiences of all individuals are criticized by each other" (1955a:205). The verification process in science for Durkheim involves a method of hypothesis and test based upon Mill's methods of eliminative induction, as we will see in the following chapter.

Opting for a relativist position himself, Bloor finds inadequate Durkheim's replies to the charge of reintroducing nominalism. However, Bloor also misreads these replies. According to Bloor's interpretation of a passage from *The Elementary Forms*, "scientific concepts are said progressively to escape the grip of society as they become responsive to nature alone" (Bloor 1982:292 and n. 70; citing Durkheim 1915d:493). When we turn to the passage Bloor cites, however, we find that Durkheim was saying something quite different:

> If logical thought tends to rid itself more and more of the subjective and personal elements that it still carries from its origin, it is not because extra-social factors have intervened; it is much rather because a social life of a new kind has expanded more and more. It is a matter of this international life that has already had the effect of universalizing religious beliefs. . . . Consequently, things can no longer be held within

the social framework in which they were primitively classified; they demand to be organized according to principles that are proper to them and, thus, logical organization differentiates itself from the social organization and becomes autonomous. . . . Truly and properly human thought is not a primitive given; it is a product of history; it is an ideal limit toward which we are always approaching closer but which in all likelihood we will never succeed in attaining. (1912a:634–35; cf. 1915d:493)

Bloor's oversight is significant, as Durkheim's argument that scientific classifications change as social life becomes more international raises a serious problem for the theory that putative knowledge claims reflect social or political interests. Interest accounts in the sociology of scientific knowledge look much less plausible when we consider how scientific ideas are appropriated by a growing international community. A knowledge claim may come to be accepted as knowledge by a group with very different social interests than the interests that this claim was originally proposed to further. What allows a scientific claim to serve so many different interests is the fact that it has passed through a social process of testing and criticism. As we shall see in the next chapter, Durkheim believed that the aim of this social process is to bring the extensions of our concepts into closer agreement with the extensions of natural kinds.

In sum, Durkheim believed that real kinds exist and that we are gradually approaching a knowledge of them. Durkheim recognized that the metaphysical question about the existence of real kinds must be distinguished from the epistemological question over whether our systems of categories always adequately represent these kinds. Our common-sense categories may not, while our scientific ones at least aim at providing an adequate representation of real kinds. "Two representations can be alike," Durkheim said, "as the things that they express, without our knowing it. The principal discoveries of science consist precisely in perceiving overlooked analogies among ideas known to everyone" (1898b:29). This distinction between metaphysical and epistemological questions, of course, is important not only for the purpose of interpreting Durkheim. Durkheim's lesson for the contemporary sociology of scientific knowledge is that the fact that science is a social construction has no antirealist, metaphysical consequences.

Durkheim's Conception of Causality

For Durkheim, the relationship between cause and effect is necessary rather than contingent. This necessary relationship, he believed, makes it possible to explain an effect by deducing it logically from its cause, much as the properties of a thing are deduced logically from its definition or essence. His essentialism also appears to underlie his belief that there can be only one cause for any given effect and only one effect for any given cause. An analysis of his arguments concerning the plurality of causes shows that for Durkheim, X is the cause of Y if and only if X is necessary and sufficient for Y in the normal state of affairs, and X is temporally antecedent to or concomitant with Y.

The Plurality of Causes

In the chapter on explanation in *The Rules*, Durkheim never really told us what causes *are*. He told us only what they are *not:* they are not functions. To understand what a cause is for Durkheim, one must turn elsewhere. Durkheim revealed his concept of causality in the chapter of *The Rules* concerning "proof" in which he argued that Mill's belief in the plurality of causes is "a negation of the principle of causality" (1895a:156). Durkheim's denial of the plurality of causes entails that the cause must be a necessary and sufficient condition for the effect.

Durkheim's first argument against the postulate of the plurality of causes is that it is

in contradiction with all the results of science. This conception of the causal bond . . . would render it nearly inaccessible to scientific analysis. . . . If an effect could derive from different causes, in order to know that which determines it within a group of given circumstances, it would be necessary that the experiment take place under conditions of isolation that were practically unrealizable, above all in sociology. (1895a:155)

In other words, Durkheim believed that it would be almost impossible to achieve knowledge about causes on the assumption of the plurality of causes. As Durkheim said, on this assumption we could never be sure that we had correctly picked out the cause of some event from its group of antecedent circumstances. His argument, I think, overestimates the degree of certainty scientists hope to achieve through their experiments. Nor are the practical problems that Durkheim raised quite as serious as he makes them out to be. Typically, scientists do not pretend to be able to achieve the total experimental isolation of a single condition. They would be satisfied merely with the isolation of a putative causal condition from what they take to be other relevant conditions, relying on accepted scientific theories to tell them what is relevant and what is not. A biologist studying the metabolism of plants, for example, need not worry about isolating his or her experiment from changes in the phases of Venus.

Another argument he raised against Mill's belief in the plurality of causes is premised upon Durkheim's essentialism. Mill's position, Durkheim said, "is linked to the fundamental principles of his logic," which by implication Durkheim would also seem to want to reject (1895a:155). The term "logic," I believe, refers to Mill's nominalist conception of reasoning, according to which all inferences are based ultimately upon a knowledge of particulars.[11] Although Mill accepted the real existence of natural kinds as well as of particulars, what defines a natural kind for him is not an essence but rather that all of its members share an indefinite number of empirically discoverable properties. If all these common properties followed logically from some small number of such properties, he believed, then the kind would not be a real or natural kind (1843: bk. 1, chap. 7, sec. 4; cf. Hacking 1991). For Durkheim, on the other hand, a natural kind seems to be individuated by an essence that is causally responsible for the properties of that kind. In other words, Durkheim appears to have grounded the existence of the causal relationship in an essentialist notion of natural kinds. These reasons for rejecting Mill's nominalism are consistent with Durkheim's rejection of empiricism. Empiricism, he thought, undercuts the possibility of real causal relations in nature. If the cause and the effect were "absolutely heterogeneous," if there were no "logical relationship" between them but if this relationship were "purely chronological," then there would be nothing contradictory in Mill's position, Durkheim reasoned (1895a:156). On the other hand, he asserted, if the link between cause and the effect "results from the nature of things, the same effect can sustain this relation with only a single cause, because it can express only a single nature" (1895a:156).

Durkheim also argued that what he called the "intelligibility" of the causal relationship is assumed by the method of science: how else, he asked, can one explain "the role of deduction in experimental reasoning" and the principle of the proportionality of cause and effect (1895a:156)? However, the principle of the proportionality of cause and effect, which is assumed by the method of concomitant variation, does not seem to require that the causal relationship be "intelligible" in Durkheim's sense. Nor does the use of deduction in experimental reasoning require the intelligibility of the causal relationship. Durkheim seems to have thought, however mistakenly, that experimental reasoning must proceed through the deduction of effects from causes alone. Even if experimental reasoning proceeded in this way, it would not rule out the possibility of a plurality of causes, as all that would be required for deducing the effect would be that the cause be a sufficient condition for it. But we do not even need to think about experimental reasoning in this way. We may conceive of it instead as beginning with a causal hypothesis along with premises that state the presence of conditions that satisfy the antecedent of this hypothesis and concluding that an instance of the effect is to be expected. We do not necessarily need to solve the philosophical problems concerning the nature of the causal relationship in order to explain how we derive experimental predictions from a hypothesis in conjunction with statements of initial conditions.

Durkheim, however, seems not to have conceived of predictions or explanations as collections of laws and conditions that provide sufficient conditions for expecting a given effect to occur. His concept of explanation appears to have been quite different from that made familiar in this century by the logical empiricists, according to whom a phenomenon is to be explained by deriving it from laws and statements of initial conditions.[12] Unlike the logical empiricist concept of explanation, Durkheim's essentialist concept of explanation, in which a single cause is a necessary and sufficient condition for a single effect, does not provide a way to take into account the contributions of a variety of circumstances to producing an effect. In an early review of *The Division of Labor*, Léon Brunshvicg and Élie Halévy criticized Durkheim for having overlooked the contributions of particular causes and conditions specific to time and place, including perturbations and accidents that have the individual as their source. Citing the single footstep that causes the avalanche, they argued that nothing should be considered insignificant in a scientific explanation (Brunschvicg and Halévy 1894:567–68).

In a final argument against the plurality of causes, Durkheim claimed that if there seems to be more than one cause for any given effect, either these apparently different causes are in reality one and the same or there must be more than one type of effect. When scientists are confronted with an apparent plurality of causes, he affirmed, they either try to reduce all of these seemingly different causes to one or try to show that what seems to be one type of effect is actually different in the case of each of the known causes (1895a:156–57).[13] As an example of the former situation, he said that scientists have attempted to reduce the production of heat through friction, percussion, and chemical action to a single cause (1895a:156). Similarly in sociology, he believed, although punishment appears to have more than one cause, these seemingly different causes may all have an element in common that is responsible for the effect (1895a:158). He also provided examples of the situation where the unity of the effect is merely apparent. For instance, if fever seems to have a plurality of causes, he said, this is only because there are a multitude of types of fever. Analogously, he thought that if suicide and crime seem to depend on more than one cause, this is because there are different kinds of crimes and suicides (1895a:157; 1897a:141).

More recently, philosopher Rolf Gruner has also argued that a given effect may only seem to have a plurality of causes, unless or until we further analyze the effect. For instance, we may be able to analyze "death" into "death by arsenic," "death by falling from a height," and so on. A problem arises, he pointed out, as to how one can know when one has analyzed the effect sufficiently so that one can stop and conclude that for the effect described in a certain way, there can be only one possible cause (Gruner 1967:369). There is, of course, a trivial sense of the denial of the plurality of causes that amounts to saying that there is only one particular cause for each particular event. For example, we could say that the cause of *this* person's death was *this* bottle of arsenic. But I do not think that this is what either Gruner or Durkheim intended. Given Durkheim's belief in natural kinds, he seems to have been suggesting not merely that there is only one particular cause for each individual fact, but rather that there is only one *kind* of cause of each *kind* of effect. But Gruner's problem returns again in a new form: how does one know when one has arrived at a natural kind?

One could, perhaps, reply that Durkheim did not really need to specify criteria for recognizing which kinds are natural. In his defense one could argue that he was merely proposing a rule of method that says *whenever* one can draw a distinction with regard to the cause of some effect, one should attempt to draw a distinction with regard to the effect. Durkheim could then have said that one never knows for sure when one has arrived at a natural kind but merely accepts what the best available explanatory laws or theories tell one are natural kinds. In fact, he did say that we should allow our classifications of types of society to grow up together with our explanatory theories in sociology (1895a:99). However, as I will show below, he did not always follow his own advice of letting his classifications develop along with his theories. In particular, when evaluating putative sociological explanations, he often simply assumed a classification of causes into sociological and psychological categories.

Durkheim and Natural Necessity

Durkheim's position that there is a logical connection between the cause and the effect that allows one to derive one from the other seems to assume a principle of natural necessity or the belief that there is a kind of logical necessity in nature. Durkheim, however, seems to have dismissed the question as to whether the laws of sociology are supposed to be contingent or necessary truths as an irrelevant metaphysical issue. In the preface to the first volume of *L'Année sociologique*, he said,

we do not wish to say that it is necessary to deny all contingency in order to be a sociologist; sociology, like the other positive sciences, does not have to pose this metaphysical problem. It supposes only that social phenomena are linked in accordance with relations that are intelligible and accessible to scientific investigation. (1898a(i):v)

To suggest that sociology need not assume that the relations linking social phenomena are necessary connections, however, is only to raise the question as to what licenses the inference of effects from causes for Durkheim. Durkheim may have meant to say that sociology need assume that these necessary relations exist only in our descriptions of social phenomena and not in the phenomena themselves. One could argue, in fact, that he was attempting to draw something like this distinction in "The Determination of Moral Facts" (1906), in which he tried to distinguish moral rules from rules of hygiene.

When one violates a rule of hygiene, he said, unfortunate consequences

result mechanically from the act of violation. If I violate a rule of hygiene that orders me to preserve myself against suspect contacts, the consequences of this act, namely sickness, are produced automatically. The completed act engenders by itself the consequence that results from it and, by analyzing the act, one may in advance know the consequence that is analytically implied by it. (1906b:59–60)

When he said that the results follow "mechanically" or "automatically" it seems that he was trying to say that the relation between cause and effect here is independent of our descriptions of these events. He only confused the issue, however, when he added that the consequence of violating a rule of hygiene is "analytically implied" by the act of violation. The term *analytic* normally applies only to the connections among concepts or propositions and not to those among actions or events. In Kant's distinction between analytic and synthetic statements, for example, analytic statements are logical truths, depending solely on the principle of noncontradiction. One may try to defend Durkheim by arguing that once we have described an act as a "suspect contact," an analysis of the meaning of this term may lead to the prediction of illness. Of course, it does not follow with logical necessity, but only with a statistical regularity, that one will become ill as a result of contact with an infected person.

Durkheim seems to have believed that the way in which unfortunate consequences follow upon the violation of a moral rule is quite different from the way in which sickness supposedly results from the violation of a rule of hygiene:

But when I violate the rule that orders me not to kill, I analyze my act in vain, I will never find either blame or chastisement there; there is between the act and its consequence a complete heterogeneity; it is impossible to draw out *analytically* from the notion of murder or homicide the least notion of blame or stigma. The bond that unites the act and its consequence is, here, a *synthetic* bond. (1906b:60)

If anything is clear, however, it is that the *notion* of blame is in fact implied in the *notion* of murder. The act of killing another human being does not imply anything; but the *description* of this act as murder entails blame or censure. The notion of blame, in fact, distinguishes murder as a special class of felonious, as opposed to justifiable, homicide. Once again, Durkheim's use of the term *analytic* is obscure.

The reader's confusion is only compounded by Durkheim's use of *synthetic* in this quoted passage. He went on to explain that sanctions are joined by a "synthetic" link to the act that violates the moral rule: "It is not the intrinsic nature of my act that leads to the sanction. Thus homicide, stigmatized in ordinary times, is not stigmatized in times of war because there is then no precept that forbids it" (1906b:60). What Durkheim appears to have meant is that since the notion of *homicide* alone does not imply the notion of blame or censure, the notion of blame is *added* to that of homicide to create the *moral* notion of *murder*. To interpret the passage in this way at least makes his distinction between moral rules and rules of hygiene clear. Durkheim's point then appears to be that the consequences that follow upon the violation of only moral rules and not rules of hygiene depend upon society.

In his replies to the objections to this paper on moral facts, however, Durkheim hedged somewhat on this distinction between two kinds of rules. Here he asserted that moral rules become more like rules of hygiene just to the extent to which the former take on religious overtones:

Very often, the sanction that is attached to the violation of ritual precepts is completely analogous to that which is attached to the violation of rules of hygiene. The imprudent person who has exposed himself to a suspect contact contracts an illness that results analytically from this contact. In the same way the profane person who has improperly touched a sacred thing has let loose upon himself a redoubtable force that leads to illness and death. (1906b:92)

But by qualifying his remarks in this way Durkheim unfortunately succeeded only in muddying the waters. Here we have the acts that violate religious rules as well as those that violate rules of hygiene leading "analytically" to their consequences. Properly speaking, only rules and not the relations among events may be characterized as analytic. Even rules may be characterized as analytic only when they are phrased in such a way that the predicate is implied in the subject. For example, the statement "Murder is wrong" is analytic; the moral imperative "Thou shalt not kill" is not. Durkheim's use of the term *analytic,* however, is consistent with a belief that there are logical relations among events themselves and not just among our descriptions of them.

Earlier in his career Durkheim had at least attempted to distinguish logical from causal relationships. In a review of a work by Gaston Richard, for example, he said that "that which is important to sociology, is to know what are the relationships that actually exist among things and not those according to which concepts must be logically arranged" (1893a:239). Thus when Durkheim said in *The Rules of Sociological Method* that the relationship between cause and effect is a "logical" one, he must mean by this term something other than logic in the ordinary sense captured by the rules of deductive inference. Indeed, given the fact that Durkheim denied not only the plurality of causes but the plurality of effects, as I will discuss below, it seems that he must have had in mind an alternative to our contemporary concepts of extensional logic, in particular the concept of material implication. In material implication, the same antecedent can have any number of different consequents and the same consequent can have any number of different antecedents. If Durkheim had conceived the causal relationship in terms of extensional logic, then, the same cause could have a plurality of effects and the same effect could have a plurality of causes.

After drawing a distinction between "logical" and "real" necessity in his dissertation on Montesquieu (1892a [1960b:47]), Durkheim did in fact go on to hint at an alternative logic:

We are not saying that social phenomena as such are illogical. But though they may have a certain fundamental logic, it is not the logic to which our deductive reasoning conforms. It has not the same simplicity. Perhaps it observes other laws. If we are to learn this logic, we must consult the things themselves. (1960b:54)

Durkheim, however, never articulated his concept of logic and alternatives to extensional logic were, of course, developed only after his death.

The Plurality of Effects

As I mentioned above, Durkheim denied not only the possibility of a plurality of causes for the same effect but the possibility of a plurality of *effects* from the same cause. He thought that his denial of the plurality of effects, like his denial of the plurality of causes, was a necessary consequence of his realism: "There can be no types of things unless there are causes that, though operating in different places and at different times, always and everywhere produce the same effects" (1960b:12).

In *Suicide,* Durkheim even seems to have argued that his denial of the plurality of causes is entailed by his denial of the plurality of effects:

A single antecedent or a single group of antecedents cannot produce sometimes one consequence and sometimes another, because, then, the difference that distinguishes the second from the first would itself be without cause; that would be a negation of the principle of causality. Every established specific distinction among the causes thus implies a similar distinction among the effects. (1897a:141)

The second sentence is simply a non sequitur. From the proposition that the same cause cannot have more than one effect it does not follow that the same effect cannot have more than one cause. To put it another way, the claim that there can be no uncaused differences in effects says that there must be a sufficient reason for any difference in effects. The denial of the plurality of causes, however, goes beyond the denial of the plurality of effects and says that a given cause is *necessary* as well as sufficient to produce a given effect.

In the first sentence of the passage quoted above, Durkheim was arguing that from the proposition that nothing happens without cause, it follows that the same cause cannot have more than one effect. Of course, there is a sense in which this conclusion is true. The truth of the proposition that the same cause cannot have more than one effect depends on what one means by the phrase "same cause." Durkheim, however, seems to have thought that two causes can be of the same type regardless of differences in local circumstances. In his dissertation on Montesquieu, for example, Durkheim asserted that "if the same citizens under a different ruler could produce a different state, it would mean that the same cause, operating under the same circumstances, had the power to produce different effects; there would be no rational tie between social phenomena" (1960b:12). What is troubling about this passage is that he seems to have assumed that the same circumstances would obtain in spite of the fact that there would be a different ruler. For Durkheim, the situation would be the same because the same social structures with the same social roles would be in place. Which individuals fill which roles seems to have been irrelevant for him. It is not hard to imagine another social theorist disagreeing with this position and thus disagreeing with regard to what counts as the "same" circumstances.

Elsewhere in his dissertation, however, Durkheim said that a society "may affect another form than that which rationally results from its nature" (1953a:92),[14] which would seem to suggest that the same cause or "nature" may give rise to more than one result. I suspect, however, that if we had confronted Durkheim with the charge of contradiction here, he would have replied that the same cause always yields the same effect but only in the *normal* state of affairs. In *pathological* cases, on the other hand, he would have allowed that we may get different results from the same cause or even the same results from different causes, as I will attempt to explain below.

Normal and Pathological

Durkheim's desire to get clear about how there can be exceptions to his rule that the same cause always produces the same effect seems to have been his original motivation for drawing the distinction between "normal" and "pathological" cases.[15] In the introduction to the first edition of *The Division of Labor,* he said that "the moral conscience of societies is subject to self deception. It may attach the external sign of morality to rules of conduct that are not by themselves moral, and, on the contrary, leave without sanctions rules that ought to be sanctioned" (1893b:33). In order to guard against the dual errors of accepting as moral facts that are not so and of overlooking facts that are indeed moral facts, he went on to suggest, sociologists should follow the example of biologists who distinguish normal from pathological physiology (1893b:33).[16] Although these passages were removed from the second edition, he nevertheless made a similar point in another passage that remained in this edition. In explaining the need to talk about "abnormal forms" of the division of labor, Durkheim reasoned that it is important to investigate these pathological forms in order that the division of labor not be suspected "of logically implying them." Furthermore, this investigation will help us to understand better the conditions upon which the normal form of the division of labor depends (1902b:343).

To say that a society may "deceive itself" and take on an abnormal form not implied in its essence, however, is not to say that there is anything unnatural or miraculous about pathological cases, he explained in *The Rules of Sociological Method*. These cases are "equally founded in the nature of things," although not their "normal nature." Nevertheless, Durkheim's primary concern was with the normal state of affairs: "it is the immediate subject matter of science" (1895a:72, 70). That is, he seems to have been suggesting that one must first study the causal relations implied in the essences of things and only subsequently investigate the conditions under which these causal generalizations will not hold.

Durkheim identified this "normal nature" with the "average type." The way in which he characterized the construction of the concept of the average type brings to mind the methods of Quetelet: one "assembles . . . into a sort of abstract individual, the most frequent characteristics in the species" (1895a:70).[17] This method is obviously not sufficient, for if one were to use it in biology, one would be forced to conclude that a human being with blond hair and blue eyes would not only be abnormal but, given Durkheim's usage, sick. Durkheim tried to clarify his position by saying that the generality of some phenomenon is merely a "sign" of its normality. He said that one can verify that some general characteristic is indeed normal by showing that it is causally connected with the "nature" of the species in question: "The normal character of the phenomenon will be, in effect, more incontestable, if one demonstrates that the external sign that had first revealed it is not purely apparent, but is founded in the nature of things; if, in a word, one can raise this normality of fact to a normality of law" (1895a:74).[18] This demonstration, he continued, may proceed in either of two ways: by showing that this trait is useful, in the sense that it allows the species to adapt to its conditions of existence, or that, regardless of its utility, "it is necessarily implied in the nature of the being" or that it is "a mechanically necessary effect" of the conditions of existence (1895a:74–75). Similarly, he maintained in *The Division of Labor* that the presence of a sanction is not always a sufficient indicator of the presence of a moral fact. He said that one also needs to compare this fact to other known moral facts in order to determine whether "it plays the same role, that is to say whether it serves the same ends," and whether "it results from causes from which other moral facts likewise result" (1893b:36; see also 1902b:8).

If these external signs or indicators may be present without their usual causes, however, then, contrary to what he said elsewhere, a plurality of causes for the same effect is in fact possible. That is, in pathological cases there may be causes of the same social facts other than those found in normal cases. Causes for Durkheim would then seem to be necessary and sufficient conditions for their effects only in the normal state of affairs. Similarly, effects would be necessary and sufficient conditions of their causes in this normal state. However, the cause and the effect are each necessary for the other only in the normal state. For Durkheim, the cause and effect each in conjunction with a special set of circumstances forms an unnecessary but sufficient condition for the other. This is to make the cause and the effect insufficient but necessary parts of unnecessary but sufficient conditions of each other.[19]

The logical symmetry of the relationship between causes and their effects in Durkheim has led to a common misunderstanding concerning Durkheim's concept of the relationship between social facts and their signs. A social fact and its sign are also necessary and sufficient conditions for one another, at least in the normal state of affairs, and there has been a great deal of confusion as to whether these signs are supposed to be the effects or the causes of social facts. This confusion has led to the misinterpretation of *The Division of Labor* as offering explanations of such social facts as the number of interpersonal relationships into which each individual enters in terms of such physical factors as population density. Many of Durkheim's readers have failed to see that he believed that the physical population density is only a *sign* and not a cause of what he called the "moral density" of human relationships, as I will discuss in chapter 6.

Given the logical symmetry of the causal relationship for Durkheim, it seems that the only thing that could distinguish the cause from its effect would be the cause's temporal antecedence. However, temporal antecedence alone for Durkheim is not sufficient for causality. This would seem to be implied by his argument that Mill would have been correct about the plurality of causes if the causal relationship were "purely chronological." Yet for Durkheim temporal antecedence does not seem to be necessary for causality either, for a cause may also be temporally concomitant with its effect. When the cause and the effect are temporally concomitant, their relationship would appear to be completely symmetrical. The only temporal constraint he imposed on the causal relationship is that the effect cannot be temporally antecedent to the cause. This requirement is implied in the way Durkheim distinguished causes from functions.

Causes and Functions

In *The Rules of Sociological Method* Durkheim distinguished causal explanations from explanations in terms of functions: "when one undertakes to explain a social phenomenon the efficient cause that produces it and the function it fulfills must be investigated separately" (1895a:117).

As stated, the rule does not proscribe functional explanations from sociology but merely distinguishes them from causal explanations. He added that it is appropriate to investigate the cause of a social fact before we investigate its effects, for indeed sometimes a phenomenon will have the effect of preserving its cause. In such cases, he reasoned, knowing the cause will help us to discover the function more easily. The concepts of crime and punishment from *The Division of Labor* serve as his example. An act is criminal, he contended, when it offends strong and well-defined states of the collective consciousness. Punishment, he asserted, consists of a passionate reaction on the part of society against those who have violated its rules of conduct. This reaction in turn serves to reinforce the collective consciousness (1895a:118; cf. 1902b:47, 64f.). However, even though the function of some social phenomenon is investigated only secondly, it is still necessary to include it in its explanation, he maintained. Although its usefulness to society may not be what brought an institution into existence, he affirmed, it is generally what allows it to continue to survive. If, on the other hand, it is not useful, he believed, it will be a drain on society (1895a:119).

Durkheim explained that he used the term *fonction* instead of *fin* or *but* in order to avoid any connotation of intentionality. The term *function* for Durkheim seems to connote merely the "correspondence" between the social fact and "the general needs of the social organism" (1895a:117). Durkheim's reasons for keeping functional explanations distinct from intentional explanations have to do with his desire to keep sociological explanations distinct from psychological explanations.

Intentional explanations, Durkheim believed, typically seek the purpose of a social fact among its useful effects. However, he contended, the usefulness of some fact cannot explain what originally brought it about. Even the fact that we desire or need something because it is useful, he thought, cannot bring it into existence. Human tendencies, needs, and desires cannot create something out of nothing, he argued (1895a:111–13; 1901a(i):260). For example, Durkheim considered the notion that a legislator can create a new institution through an act of sheer will to be little more than a holdover from a belief in miracles (1912a:37).[20]

In the essay "Two Laws of Penal Evolution" he provided another argument for ruling out individual needs or desires from sociological explanations. This argument is premised on the fact that what one desires may not be possible: "In order to account for an institution, it does not suffice to establish that at the moment in which it appeared it responded to some useful end; because from the fact that it was desirable, it does not follow that it was possible" (1901a(i):260). Although the premise of this argument may be true, the argument appears to be entirely beside the point. Presumably, sociology aims at the explanation of actually existing institutions. If the institution exists, it must have been possible. The fact that people also desire things that are not possible would seem to bear little relevance to the question whether desires play a role in bringing about things that are possible.

In *The Division of Labor* Durkheim argued for reasons having to do with the temporal order of cause and effect that the desire to achieve some useful effect cannot be the cause of that effect. He gave the following reason for denying that the desire to increase human happiness is a cause of specialization: "an anticipated representation of the effects that the division of labor produces" could not be the cause of the division of labor because these effects could not become perceivable until after the division of labor was already rather far along (1902b:211). I grant that an effect must not be temporally antecedent to its cause. From this assumption, however, it does not follow that one's anticipations of the results of one's actions cannot be among the causes of one's actions. The act of anticipating results may be regarded as a mental event that is temporally antecedent to the action itself (Davidson 1963).

Durkheim's real concern in ruling out anticipations as causes had less to do with insuring that effects are not temporally antecedent to their causes than with distinguishing sociological from individual, psychological explanations. I will have more to say about the latter distinction below. Nevertheless, by having ruled out explanations of social facts expressed in terms of anticipated results, Durkheim actually strengthened his position having thereby avoided the problem of the unanticipated or unintended results of human actions. That is, the best argument against explaining social facts in terms of anticipations of their beneficial effects is that many if not most social facts occur as the unintended consequences of human actions.

The problem of unintended consequences is contained in the following argument: Durkheim said that "a fact can exist without serving any purpose" either because it never had one or because it did once but no longer does and continues only through custom. Alternatively, he pointed out, the same institution may come to serve a new function. He provided the example of how the rule that a legal marriage with the mother establishes the father's rights over the children now serves to protect the children's rather than the father's interests. Another example he gave concerns the fact that oaths began as a kind of judicial proof but now serve only to solemnize the occasion (1895a:112–13). The cause that gives rise to the existence of some social fact, he concluded, is independent of the purpose it serves. This conclusion appears to conflict with his concept of causality. Presumably, for some social fact to continue to exist through serving some useful purpose would be for it to have causal consequences that would in turn be causally responsible for maintaining it. But if a social fact can serve a new purpose, the cause that maintains it could differ from that which originally brought it about and it would have a plurality of causes. To resolve this contradiction, I suggest that for Durkheim to maintain a fact in existence is a different effect than to bring it about in the first place.

Explanations of social facts premised upon social interests face the same problem of unintended consequences as other kinds of intentional explanations. Durkheim's argument thus distinguishes his sociology from the kind of interest accounts in the sociology of knowledge that have been defended by his champion David Bloor. Like other social facts, the fact that a putative knowledge claim comes to be accepted within a social group may be unintended. In particular it may be unintended by the person who originally proposed it. The growing internationalization of science to which Durkheim called our attention increases the likelihood that a knowledge claim will be accepted by a social group with completely different social interests than those of the scientist who first put it forth. Hence, from the fact that a scientific theory may serve some political or social interest, one cannot conclude that it was originally proposed to further that interest. In short, interest theories fail to heed Durkheim's warning that one not confuse the purpose served by a social fact with the cause that brought it about. Bloor in particular ignores this warning in his attempt to use interest theory to argue for a relativist interpretation of the Durkheim-Mauss primitive classification thesis, as I will show in chapter 9.

Sociological versus Psychological Explanation

Durkheim's prohibition of intentional explanations in sociology, including interest accounts, is also a consequence of his more general restrictions against psychological explanations in sociology. According to one of Durkheim's well-known rules, every time a social phenomenon is "directly explained" in terms of a psychological phenomenon, the explanation is false (1895a:128).[21] "The determining cause of a social fact," he said, "must be sought among antecedent social facts and not among the states of the individual consciousness" (1895a:135).[22] As we saw in the last chapter, social facts for Durkheim exist in collective representations, which are formed through the psychological mechanism of the "fusion" of individual representations. Individual consciousnesses, he said, are the "theater" of these reactions (1902b:72). Once these collective representations are formed, however, they take on "a life of their own" (1898b:46). Although individual representations may be necessary conditions for collective representations, he seems to have believed, they do not provide sufficient explanations of social phenomena. Social phenomena constitute an emergent reality that cannot be reduced to psychological phenomena.

Durkheim's rule against explaining sociological phenomena purely in terms of psychology is grounded in his realist philosophy of science, in which individual and collective representations form two distinct natural kinds. As he explained in a letter to Bouglé,

I have defined social facts as actions and representations, but of their own kind [*sui generis*]; I have said that the social being was a psychic individuality, but of a new kind (p. 127). Supposing only that, I conclude from it that one does not have the right to treat collective psychology as a prolongation, an enlargement, a new illustration of individual psychology.[23]

Sociology, he said in *The Rules of Sociological Method*, is not a "corollary" of psychology but has its own laws (1895a:125; cf. 1897a:352; 1901d:52). Another reason he gave for the independence of the laws of sociology is that the constraint that characterizes social facts cannot arise from individual consciousnesses (1895a:124–25). Also, he found the tendency to give psychological explanations of social facts to be bound up with the tendency to explain them in terms of final causes (1895a:120). Finally, he argued, when psychological explanations appeal to instincts to account for such social facts as marriage, incest taboos, and the bond between parent and child, they are offering mere verbal explanations that explain nothing. Durkheim compared such accounts to the explanation of the soporific powers of opium in terms of its dormitive virtue (1895d:609).

In a published letter to the editor of the *Revue néo-scolastique,* written in defense against Deploige's charges of social realism, Durkheim explained that he learned from Émile Boutroux the Aristotelian conception of science whereby each explains by its own principles. He claimed that he was "confirmed" in this opinion by his reading of Comte, for whom each science is irreducible to the ones below it in his hierarchy (1907b:613). This remark appears to be somewhat disingenuous, as twelve years earlier in *The Rules of Sociological Method* Durkheim had criticized Comte's concept of the relationship between sociology and psychology, attributing to Comte the view that this relationship is "analytic." Durkheim quoted Comte to the effect that historical generalizations cannot be accepted into sociology unless they can be "rationally linked" to a theory of human nature. Psychology for Comte, Durkheim asserted, always has "the last word." However, Durkheim also recognized that for Comte "the theories of psychology cannot suffice as premises for sociological reasoning" but serve merely as a "touchstone" from which to test the validity of inductive generalizations in sociology (1895a:122).

Durkheim's account of Comte's philosophy of sociology is not quite accurate here, as Comte conceived the theory of human nature as a part of biology, not psychology. For Comte, in fact, psychology is not a distinct science, as it has no subject matter distinct from that of biology on the one hand and sociology on the other. To say that a sociological hypothesis must be tested against a theory of human nature means little more in Comte's philosophy than to require that sociological hypotheses must be consistent with biology. Durkheim's use of biology as a source of additional confirmations for sociological hypotheses, in fact, differs little from Comte's. For instance, where Comte found his theory of the mind or brain confirmed in the phrenology of Gall and Spurzheim, Durkheim asserted that the skull measurements carried out by Le Bon and Broca confirm his hypothesis that primitives resemble each other more than contemporary human beings do (1902b:103–4).

Durkheim's prohibition against grounding sociology in psychology, however, does not rule out any appeal to psychology in sociological explanations. As a matter of fact, in *Suicide* Durkheim suggested that his sociological explanations differ from Tarde's theories of social imitation precisely because Durkheim allows a mediating role for psychological processes:

> between the representation of the act and the execution is intercalated an intellectual operation that consists of an apprehension, clear or confused, rapid or slow, of the determining characteristic, whatever it may be. The manner in which we conform to the morals or to the fashions of our country thus have nothing in common with the mechanical ape-like imitation that makes us reproduce the motions of which we are the witness. (1897a:113)

In other words, in order for society to produce its effects on individuals, individuals must be aware of the collective representations responsible for these effects. Of course, as we have seen, collective representations are supposed to be unobservable entities. Perhaps what Durkheim meant is that the collective representation gives rise to an individual representation of the action to be performed.

According to Alexander Rosenberg's interpretation of *Suicide,* Durkheim implicitly relied on some such psychological mechanism to mediate the effects of society on the individual. Durkheim could not possibly have meant that the sheer *number* of children that one has determines the degree to which one is protected from suicide, Rosenberg argues. He thinks that what Durkheim must have had in mind is that it is one's *belief* that one has a certain number of children that preserves one. A man's unknown illegitimate children could have no such effect. (Rosenberg 1980:33–34.) Nor does it even seem likely that one would be protected from suicide by the mere *knowledge* that one has had some number of illegitimate children, I would add, if one had given all of them up for adoption and subsequently lost all contact with them. That is, Durkheim must assume not only psychological causes but psychological causes of a particular type in order to account for the effect of the number of offspring on suicide rates.

Some of Durkheim's interpreters have criticized him for having relied on psychology in his sociological theorizing in spite of his injunctions against psychological explanations in sociology. For instance, Roger Lacombe charged that Durkheim was constantly obliged to have recourse not merely to psychology but to unscientific, common-sense "folk" psychology (1925:383). To be sure, especially in his early writings, Durkheim made use of the psychological concept of habit formation in order to explain the origins and causes of customs, moral rules, religious beliefs, laws, social functions and their relationships, the values society assigns to things, and even the structure of the family and society.[24] However, to say that Durkheim relied on psychological explanations is not to suggest that Durkheim either contradicted himself or violated his own rules governing the explanation of social facts. Durkheim's *The Rules of Sociological Method* sets out an entire program of research, including explanatory goals that could be attained, if at all, only after much intellectual labor on the part of a whole community of sociologists. Much of Durkheim's intellectual development, in fact, can be interpreted as a struggle to find genuinely sociological laws that would allow him to explain social phenomena without relying on psychological mechanisms. His exclusion of the concept of habit from his explanatory repertoire in his later works, for example, may have been motivated by a desire to avoid psychological explanations of social facts (Camic 1986:1050–57). Nevertheless, as we shall see in chapter 8, Durkheim was frustrated in his search for the laws of collective ideation. Before we turn to a discussion of the development of Durkheim's sociological theories in his major works, however, we need to understand his concept of sociological method.

5 Durkheim on Method

For Durkheim, sociology is the study of a type of unobservable entity that he called collective representations. This chapter will show that Durkheim held a concept of scientific method appropriate for the study of such entities. According to Durkheim, one can acquire knowledge of collective representations and the relations among them only by formulating hypotheses about them and submitting these hypotheses to empirical tests. Of course, to attribute a method of hypothesis and test to Durkheim is not to suggest that he embraced hypothetico-deductivism in its contemporary sense, which involves drawing experimental predictions from hypotheses in conjunction with statements of initial conditions. Just as in his concept of explanation, in his concept of hypothesis testing Durkheim appears to have believed that effects can be logically derived from their causes alone. Nevertheless, Durkheim's method of hypothesis, unlike naive methods of induction by generalizing from observations, allows the postulation of unobservable causes in order to explain observable events.

Because collective representations are a type of mental entity, they cannot be observed even through introspection, Durkheim believed. By drawing an analogy with what he thought to be the method of psychology, he argued that sociology must study these entities through what he believed to be an "experimental method." Psychologists, he claimed, long ago abandoned the method of introspection as a means for investigating the motives of human actions. The reason they abandoned introspection, he argued, is that this method can at best reveal only facts, not their causes: "Every causal relation is unconscious, it is necessary to divine it after the event" (1908a(3):230). Furthermore, he maintained, even the facts that are revealed by introspection are "too rare, too fleeting, too malleable" to serve as controls or counterweights to the corresponding ideas that habit has fixed in us. Philosophers such as Locke and Condillac who relied on introspection ended up studying not sensation but only their ideas of sensation, he said (1895a:38). For these reasons, he thought, psychologists must use the "complicated and roundabout procedures" of the natural sciences instead of the introspective method (1897a:351; 1898b:36).[1] If psychological facts cannot be studied through introspection, he reasoned, it must be even more true that social facts, "of which the causes even more evidently escape the consciousness of the individual" (1908a(3):230),[2] cannot be investigated through introspection either. As he explained in a letter to Célestin Bouglé, "society is not in any individual, but in all of the individuals associated under a determinate form. It is thus not through the analysis of the individual consciousness that one will be able to do sociology" (6 July 1897 [1975a: vol. 2, 400]).

The experimental method of the natural sciences, according to Durkheim, consists of formulating theoretical causal hypotheses and submitting them to empirical tests. As we saw in the previous chapter, Durkheim believed that there exist real relations in nature. Because these relations cannot be perceived, he thought, they must be hypothesized: "Now these relations and internal connections, sensation, which can only see things from without, cannot discover to us; the mind alone can create the notion of them" (1912a:339–40). In a discussion of Wilhelm Wundt, Durkheim appears to have accepted his position that science "creates hypothetical concepts" or "postulates that are not immediate givens of experience, but . . . are added by the mind in order to render it intelligible" (1887c:114–15).

In addition to allowing sociologists to form conjectures concerning unobservable entities and causal processes, Durkheim's method of hypothesis has another advantage over inductivism. His conception of scientific method reflects his understanding of the social character of conceptual thought and scientific research. The use of the method of hypothesis and test on the part of an expanding community of investigators allows the extensions of our explanatory concepts to achieve progressively closer approximations to real kinds, he believed. Durkheim's appreciation of the fact that these concepts have been socially constructed over the course of hundreds of generations led him to take precautions with these concepts. His very preoccupation with avoiding "prenotions" in science was due to his concern that the extensions of these concepts, because of the way in which they have been formed, are very unlikely to correspond to natural kinds.

Durkheim did not pretend to any original insight with regard to the problem of prenotions or preconceptions but recognized that such canonical philosophers as Bacon and Descartes were aware of it as well. The problem, as Durkheim saw it, is one of how to achieve a sociological analysis of the formation of our concepts without being misled by these concepts in the process. His solution to this problem started with the recognition that, as the meanings of these concepts have changed over time, so too have their extensions. He then consciously avoided granting a privileged status to the extensions of our concepts under their current interpretations. This line of thinking, I will argue, underlies his sometimes surprising definitions of such concepts as religion and suicide. If we are to understand the social causes that lead people to kill themselves, Durkheim reasoned, we cannot let ourselves be deceived by the fact that only certain acts of self-destruction have been socially constructed as suicides, while others have been socially constructed as noble acts of self-sacrifice. Similarly, he thought, it would be a mistake to allow our present, Western notion of religion to exclude non-theistic systems of belief from the sociological analysis of religion. In order to explicate the concepts that constitute our social reality, Durkheim prescribed a method of historical or comparative analysis. This method, through a liberal use of hypotheses, investigates the way in which concepts have been socially constructed over the course of generations and attempts to reveal what has been preserved in these concepts through all of these changes.

Durkheim's combination of hypotheticalism and realism with regard to both the natural and the social sciences constitutes a coherent, defensible position. What permits Durkheim's realism to be compatible with his belief in the social construction of scientific concepts is not only that his realism is a metaphysical thesis while his constructivism is an epistemological one, as I argued in the previous chapter. It is also his belief that the social process of science involves the criticism and testing of hypotheses with the aim of bringing the extensions of our socially constructed concepts into closer agreement with natural kinds.

Durkheim's Method of Hypothesis

Although there is nothing in the secondary literature like a consensus about Durkheim's methods of inquiry, according to at least one fairly common interpretation his professed methodology would rule out the use of hypotheses entirely. *The Rules of Sociological Method* is often read as advocating the idea that scientists construct theories through a process of inductive generalization on the basis of prior observations that were made in an objective, unbiased manner.[3] Critics like Paul Hirst, for example, accuse Durkheim of having been too much of a naive inductivist and of having overlooked the role of theories in the observation of social facts (1975:97).[4] Steven Taylor believes that Durkheim professed a "crude empiricist" and "inductive epistemology" that is "clearly at odds with the realist ontology of his major studies" (1982:203 n. 26, 204 n. 27). George Catlin went so far as to deplore Durkheim's postulation of an unobservable collective consciousness as a deviation from his methods (1938:xiv). Similarly, Jack Douglas says that although Durkheim pretended to use an inductive method, in *Suicide* he in fact used a method of hypothesis, which, Douglas asserts, is "anything but scientific" (1967:153).

Far from believing that a scientist must begin with unbiased observation, however, Durkheim thought that one must begin with some concepts that allow one to distinguish which are the relevant facts. As he explained in *The Division of Labor,*

one deceives oneself with a vain hope when one believes that the best way to prepare the coming of a new science is first to accumulate patiently all the materials that it will use, because one cannot know which ones it will need unless the science already has some sense of itself and of its needs (1902b:xliii)

Durkheim in fact seems to have thought that something like a hypothesis is necessary even for making observations in the first place.[5] To describe a fact, Durkheim believed, one must first have some sort of preconceived ideas. "One cannot describe a fact well which is singular or of which one possesses but scanty examples because one cannot see it well" (1898a(i):32). Unless one has something with which to compare an observation, he thought, one is not able to perceive it clearly: "Supposing that one discovered an entirely new being, without analogy in the rest of the world, it would be impossible for the mind to consider it; one could not even represent it to oneself except in terms of some other thing which one already knew" (1888a:92). Similarly, Durkheim's colleagues Paul Fauconnet and Marcel Mauss believed that there are no brute facts in sociology. Sociology does not merely give a narrative description of the facts, they argued, it "constitutes" them (Fauconnet and Mauss 1901:173).

To be sure, Durkheim did on occasion use *hypothesis* or *conjecture* as a term of abuse. Typically, he did this in two contexts. In one he would attack an opponent's speculative theory by calling it, for instance, an "arbitrary and unverifiable conjecture" (1912a:267). For Durkheim, as we shall see below, a hypothesis must contain only terms defined in such a way that the hypothesis will be testable. Unobservable entities, for example, must be defined in terms of their observable signs.

A second context in which he used *hypothesis* as a term of abuse occurred when Durkheim would attack other sociologists, such as Herbert Spencer, for what he considered their misuse of Darwin's theory of evolution. He stated that "to make sociology rest on Darwinism, is to establish it on an hypothesis, which is contrary to any good method" (1895d:608). The evolutionary hypothesis may be a "suggestive and fecund idea from the back of the mind," he said, "but it is neither a method nor an axiom" (1887c:279). In *The Division of Labor* Durkheim found evolutionary theory to be a helpful model for avoiding teleology in functional explanations in sociology. However, he believed that sociological hypotheses suggested by biology stand in need of sociological verification. Durkheim allowed sociology to use biological analogies for what he called "secondary verifications" but he stipulated that one should never attempt to "induce" sociological laws from biological ones (1898b:13).

In order to support an interpretation according to which Durkheim was opposed to *any* use of hypotheses, some of his commentators have quoted out of context passages where Durkheim used *hypothesis* as a term of abuse. Jeffrey Alexander, for example, quotes the following statement from Durkheim's criticism of evolutionary theories in social science: "A science has for its point of departure facts and not hypotheses" (Durkheim 1887c:279).[6] Immediately following this sentence, however, Durkheim added, "without doubt, when it is just born, subjective and conjectural views occupy nearly the entire domain, and it is good that such may be the case; but to the extent that it is raised up and consolidated, hypotheses recede from the base to the summit" (1887c:279). Note that hypotheses never disappear for Durkheim: they are needed where science is still building. Once these hypotheses are verified, Durkheim seems to have assumed, they can be accepted as "laws" or "axioms." Thus the hypotheses "recede from the base," which consists of laws accepted as "facts," to the "summit." What Durkheim seems to have meant in this passage is that only laws verified as facts and not hypotheses borrowed from another science can serve as the most basic explanatory principles of sociology.

The Role of Experience in Durkheim's Methodology

Durkheim was enough of a logician to know that one cannot prove a thesis simply by marshaling facts that agree with it. "To illustrate an idea is not to demonstrate it," he said (1895a:165).[7] Specifically, Durkheim singled out Herbert Spencer and Montesquieu for having cited facts that merely illustrated their conclusions but established nothing (1902b:181, 198; 1888a:94; 1960b:52–3). He attacked economists on similar grounds (1895a:33).[8] Likewise, in *The Elementary Forms of the Religious Life* he criticized philosophers for only illustrating their definitions of religion (1912a:6). In his criticism of Belot's definition of socialism, Durkheim pointed out that the fact that Belot could cite examples that agreed with his definition did not at all show that all socialist ideas without exception have conformed to it (1893c:229). He rebuked previous sociologists for merely having cited an imposing number of favorable facts and having ignored contradictory ones (1902b:xliii) and criticized Westermarck, Post, and ethnographers in general for their concern with the mere accumulation of facts (1895d:608; 1907a(10):586).

Durkheim recognized that induction does not establish its conclusions with certainty and that this weakness cannot be overcome simply through increasing the number of confirmatory instances. The quantity of observations that one accumulates, Durkheim believed, is of only "secondary" concern (1895d:608). Of greater importance than the mere number of facts is their "value," he thought (1912a:134). Truly significant facts are those that are learned through what he called "well-made experiments." Through such experiments "a single fact can bring a law to light, whereas a multitude of imprecise and vague observations can produce only confusion" (1912a:135).[9] In response to the anticipated objection that in *The Elementary Forms of the Religious Life* he has drawn an inductive generalization from a single religion, he argued that "when a law has been proven by a well-made experiment, this proof is universally valid" (1912a:593). "It is not true," he argued, "that science can institute laws only after having passed in review all the facts that they express" (1895a:97).

What Durkheim seems to have meant by "well-made experiment" is a "crucial experiment" or an experiment designed to decide between two competing explanatory hypotheses, confirming one and disconfirming the other. The true scientific method is concerned only with what Bacon called "decisive" or "crucial facts," he said (1895a:98).[10] For Bacon, a fact is crucial only relative to a pair of competing hypotheses. Bacon provided the example of whether high and low tides take place at the same time on the opposite sides of the isthmus of Panama or whether they are reversed. For Bacon, this fact was crucial for deciding whether high tides were due to a general rising of the seas or to a sloshing back and forth of the seas in their respective basins (1620: bk. 2, sec. 36). A crucial fact, like a crucial experiment, decides between competing explanatory hypotheses.[11]

Durkheim's emphasis on crucial facts suggests that for him the scientific method begins not with facts alone but with causal hypotheses that offer putative explanations of the facts. Indeed, one of the reasons Durkheim provided for his journal's policy of not accepting submissions of biographical articles was that facts of individual biography will become of interest to sociology only when it is ready to offer an explanation of them:

Without doubt, nobody can say that these diverse particularities will always be refractory to science; but the time when it will be perhaps possible to attempt an explanation, even a partial one, is so distant that it would be wasting one's time to interest oneself in them. After all, that which one calls a scientific fact is quite simply a fact ripe for science. (1898a(i):35)

For the most part Durkheim placed more confidence in disconfirmation than in confirmation, arguing that while several facts may not establish an hypothesis a single fact may suffice to refute it (1895a:163). Durkheim's beliefs about scientific method actually resemble Karl Popper's method of bold conjectures and severe tests more than the inductivist views of Popper's logical positivist opponents. In the preface to the first edition of *The Division of Labor* Durkheim said that "there is, in fact, only one way to do a science, and that is to dare it, but with method" (1902b:xliii). This passage suggests that Durkheim recommends an even greater boldness than Popper does, as Durkheim endorsed the audacious venturing of entire sciences and not just of mere hypotheses.

Roger Lacombe has criticized Durkheim for having believed that hypotheses can be eliminated with crucial facts. Lacombe pointed out that a fact can have more than one interpretation, that one can never eliminate all competing hypotheses because new ones can always be dreamt up, and that even if a hypothesis is contradicted by some fact, it can still be saved by the additional assumption of some "secondary" hypothesis (1925:382; 1926:89). However, Durkheim was no more a naive falsificationist than a naive inductivist. Elsewhere he said somewhat more carefully that if a fact seems to contradict a hypothesis supported by many facts, this fact should be put aside but not forgotten (1902a(i):321). Such a fact, he explained, may have been misinterpreted. Facts may be unreliable, Durkheim argued, if they were recorded under adverse conditions or colored by subjective biases (1888c:265–66; 1895d:608). Because facts are not always reliable, he said in his lectures on the family, it is sometimes necessary even to "discard" some facts: "It is better to neglect some facts than to use doubtful ones" (1888c:265, 271).[12] For one fact to prevail against a proposition supported by many other facts, he argued, we must be certain that the putative disconfirming fact "allows of only a single interpretation and that it is not possible to account for it without abandoning the proposition that it seems to disconfirm" (1912a:515).[13] Nor did Durkheim think that a crucial fact could establish the truth of some hypothesis once and for all. The results achieved by science are always provisional and could be overthrown tomorrow, he believed. In fact, in his lectures on moral education he recommended that this important truth be taught to school children. He proposed that the student be shown how over the course of the history of science a succession of hypotheses have replaced one another as the result of observations and experiments (1925a:300–1). Elsewhere he claimed that all any science can hope to achieve is to establish probabilities (1908a(3):236).

Like his contemporary and countryman Pierre Duhem, then, Durkheim recognized the possibility of an unlimited number of different interpretations of the same experiment or observation (Duhem 1906:152). More recently, of course, Quine has made a similar point in a slightly different way, arguing that any scientific claim can be defended from any attempt at empirical refutation if we are willing to make the appropriate changes elsewhere in our web of belief (1951:43). Durkheim, however, did not draw from the so-called Duhem-Quine thesis the conclusion that crucial experiments in science are impossible. Unlike contemporary relativists who draw on the Duhem-Quine thesis for support, Durkheim does not seem to have confused the logical underdetermination of theories by the evidence with the underdetermination of theory choice by rules of method.[14] In other words, Durkheim seems to have realized, a so-called crucial experiment may provide us with good reasons for preferring one theory over another even if it cannot definitively prove one and falsify the other.

The relativist line of argument rests on a number of confusions. First, it confuses the problem of the *proof* of a single putative scientific claim with the problem of *rational choice* among two or more competing claims. Second, and similarly, it confuses *proving* a claim with having a good reason to *accept* one. Certainly, if claims in science could be proven true, their proofs would constitute good reasons for accepting them. But in the absence of proof, there may still be good reasons for choosing one claim over another. Finally, but also important, the relativist argument confuses questions of what is logically possible with questions of what is practically possible. That is, although there may be an infinite number of claims that are compatible with any given finite piece of evidence, there may nevertheless be but a very small number of claims compatible with the evidence in any real epistemic situation.

In any real situation scientists confront, one hypothesis may stand out as unequivocally the best given the evidence at hand in conjunction with a set of cognitive goals and rules governing the use of evidence in adjudicating among putative scientific theories. As Duhem himself pointed out long ago, it is not logic and experiment alone but "good sense" that tells the scientists which hypotheses to accept and which to reject. Duhem, however, left the notion of "good sense" unanalyzed and gave scientists themselves the task of making it "more lucid" (1906:216–18). In *The Rules of Sociological Method* and elsewhere, Durkheim tried to explicate what constitutes "good sense" in his own discipline of sociology.

Ideas and Things

In the chapter of *The Rules of Sociological Method* devoted to the observation of social facts, Durkheim proposed the rule that we consider social facts as things (1895a:20). This rule is sometimes misconstrued as prohibiting the use of hypotheses. For example, D. A. Nye and C. E. Ashworth, in rejecting previous hypothetico-deductivist interpretations of Durkheim, argue that he clearly stated in *The Rules of Sociological Method* that science proceeds from things to ideas and not vice versa (Nye and Ashworth 1971:135–36). In fact, if one were to give a quick reading to the passage in question in the English translation, it does appear as if Durkheim had been attacking the method of hypothesis:

One may appeal to the facts in order to confirm one's hypotheses or the final conclusions to which they lead. But in this case, facts intervene only secondarily as examples or confirmatory proofs; they are not the central subjects of science. Such a science therefore proceeds from ideas to things, not from things to ideas. (1938b:14–15)

In the French, however, where Sarah Solovay and John Mueller have seen fit to introduce the term *hypotheses,* we find instead the term *notion* (1895a:21). Ironically, the immediate context from which this passage is taken is one in which Durkheim was warning us to be extremely careful with our terms and the concepts they denote. As I mentioned above, Durkheim was concerned about the fact that our ordinary notions of things may not correspond to natural kinds. *Notion* refers to Francis Bacon's "notiones vulgares" or "praenotiones," which Durkheim regarded as referring to crudely formed concepts, mingled with religious ideas, that should be replaced with scientifically defined concepts (1895a:23). "Preconceptions" would have been a better translation than "hypotheses"; nowhere else in this chapter of *The Rules* do Solovay and Mueller translate the French "notion" as "hypothesis."[15]

To interpret Durkheim's rule that the sociologist treat facts as things as prohibiting the use of hypotheses in sociology is also to ignore the larger context provided by his metaphysics. This rule actually contains a statement of his realism, stipulating that social facts or collective representations are real things. In the preface to the second edition of *The Rules of Sociological Method* Durkheim tried to clarify his concepts of ideas and things in order to remove the kinds of misunderstandings that nevertheless seem to have persisted until the present day. To say that social facts are things, he explained, is not necessarily to say that they can be reduced to material things but rather that they are just as real as material things. A thing, he said, is "any object of knowledge which is not naturally penetrable by the understanding," of which "we cannot form an adequate notion through a simple process of mental analysis." In other words, a "thing" cannot be known through introspection alone. The mind must go "outside of itself" by means of observation and experiment (1901c:x–xi).

To treat social facts as things rather than ideas, then, means to study social facts with the empirical methods of the natural sciences rather than the analytical methods of philosophy. Durkheim believed that the sciences in the beginning pass through an "ideological phase" in which people analyze and speculate upon their ideas of things, rather than observe, describe and compare the things themselves (1895a:20–21; 1895e:94). Science in this phase, Durkheim said in the passage cited by Nye and Ashworth, goes from "ideas to things, not from things to ideas," and appeals to the facts only to confirm its ideas or the conclusions it draws from them.

In the early stages of the development of every science, Durkheim believed, the phenomena that it takes as the object of its study are already represented in the mind not only by "sensible images" but by "coarsely formed concepts." This has been true for the physical sciences: before physics and chemistry became sciences, Durkheim said, people already had "notions that went beyond pure perception" of the relevant phenomena and which were often mixed with religious ideas (1895a:20). Our sensations provide only confused, passing, subjective impressions of heat, light, sound, and electricity, he argued, and not clear and distinct notions or "explanatory concepts" (1901c:xii). These confused ideas, he believed, were formed haphazardly and without method through the course of our everyday activities, "under the control of a thousand circumstances without relation to the intrinsic reality of the object to which they correspond" (1893c:228–29):

These ordinary concepts are necessarily coarse, because they are formed day after day, in the course of quotidian experience, without method and without criticism. They express things exactly as the sensibility shows them to us, as it is useful to represent them to ourselves in order to adapt ourselves without suffering, and not as they are. (1895e:95)

Because our preconceptions of things are more familiar to us than the things themselves, Durkheim believed, we mistake these ideas for the real things. Thus these prenotions form a "veil" between things and ourselves, he said. Although these notions may work for all practical purposes, Durkheim recognized, a representation can perform this useful function and still be false. He cited the obvious example of geocentric astronomy as a case in point (1895a:20–22).

The problem of substituting preconceptions for things has been even more serious in philosophy and the social sciences than in the physical sciences, Durkheim thought. To the first social scientists, he asserted, social reality had seemed to be nothing but the realization of people's ideas about society, the state, justice, religion, the family, and so on (1895a:24; 1895e:95). For example, Durkheim criticized Comte and Spencer for beginning with analyses of our ordinary notions of progress and society, respectively (1895a:25–29; 1895e:95). Durkheim also criticized Tönnies, Belot, and philosophers generally for their use of a "dialectical" method of analyzing concepts (1889b:422; 1893c:222; 1912a:6). For example, Durkheim rejected Belot's definition of socialism for being nothing more than the result of his analysis of his *idea* of socialism, rather than of socialism itself. This idea of socialism, Durkheim argued, is the result of mere historical knowledge of actual socialist doctrines rather than a scientific investigation of the causes and functions of socialist ideas (1893c:226).

Because of the danger of being led astray by our preconceptions, Durkheim proposed as a "corollary" to the rule that one should treat social facts as things that "it is necessary to systematically discard all prenotions" (1895a:40).[16] The sociologist, Durkheim said, must avoid concepts that have been formed outside of science (1895a:40–41). Durkheim warned against substituting the philosophical analysis of our ideas or prejudices for the study of things themselves. For example, in studying social deviance in another culture, one cannot readily transport one's values to that society and simply assume that, say, polygamy is a crime. He provided the example of being misled by one's common-sense concepts into objecting to the idea that crime is a normal social phenomenon (1895a:v–vi; see also 81f.). Durkheim informed the reader that there is nothing exclusively Baconian about this rule against preconceptions but that it is common to empiricists and rationalists alike. Descartes's method of systematic doubt, he thought, is also but an application of this rule (1895a:40). Descartes's procedure involves not merely wholesale skepticism but rather the questioning of our reasons for our beliefs in order to assure ourselves that the foundations of our science are secure. Durkheim's rule against preconceptions was similarly motivated by foundational concerns. Methodical doubt, he explained, is to be applied not to "moral reality" itself but only to our *explanations* of it (1902b:xlii).

Durkheim also found fault with political economists for their use of what he called a method of "ideological analysis" (1895a:30–32; 1895e:95; 1890a:219). Not even the famous law of supply and demand, he said, has ever been studied inductively in order to determine whether *in fact* this is how the economy operates. Instead, he argued, economists have merely demonstrated "dialectically" how individuals *should* act if they were being rational and understood what was in their own self-interest. Economists, he believed, fail to distinguish questions of logical necessity from questions of natural law (1895a:33–34).

What Durkheim was rejecting was not the use of hypotheses in the social sciences but the use of a method modeled on the construction of axiomatic deductive systems in mathematics. As Durkheim saw it, the "ideological" method of philosophers typically begins with the introspective analysis of their ideas or concepts and then uses the results of this analysis as a basis from which to elaborate a deductive, a priori system of thought, which they generally fail to submit to the control of facts. Mathematics provided Durkheim with his clearest example of a body of knowledge that is concerned with "ideas" rather than "things." In mathematics, he argued, because we construct the concepts ourselves and know what we put into them, the analysis of our concepts is sufficient to construct the science. However, he pointed out, the kind of analysis we use in mathematics is not sufficient in the case of facts or things that are not made by us and are hence "unknowns" for us (1901c:xi–xiii).

Durkheim considered the objection that because social facts, just like mathematical ideas, have been created by us, we should know what we put into them. He replied by arguing that this objection is based upon a false supposition. Most social institutions, he believed, have not been created by us but have been handed down for generations. Even when we are present for and participate in the formation of a new institution, he added, we may be confused about our real reasons for what we are doing. Since we could be mistaken even with regard to our personal motives for our actions, he asked, how can we expect to understand truly the causes that led a group of people to do something (1901c:xii–xiii)?

In his criticism of the methods of political economy Durkheim prescribed an alternative to the method of the analysis of ideas. This method involves carefully defining one's concepts in terms of empirical characteristics, classifying one's subject matter into kinds, finding the causes of each kind through eliminative induction, and comparing the results of these inductions:

> If "value" had been studied by it as a reality ought to be, one would see the economist first indicate by what means one may recognize the things called by this name, then classify its species, try to find by methodical inductions the causes as a function of which they vary, and finally compare these diverse results in order to extricate a general formula. (1895a:32–33)

Granted, this passage would seem to support an interpretation according to which Durkheim endorsed a naive inductive method of generalizing from observations. One must be careful, however, not to interpret this passage in an ahistorical, anachronistic way. By "induction" Durkheim did not mean what philosophers today call enumerative induction or the counting up of empirical facts and the drawing out of generalizations from them. Rather, he meant Mill's methods of induction, which are methods for the elimination of putative causal hypotheses through comparing them with the facts. In sum, *methodical inductions* for Durkheim refers to procedures for testing hypotheses.[17] In fact, as we shall see, not only induction, but every aspect of the method Durkheim endorsed in the passage just quoted, including also definition, classification, and comparison, makes extensive use of hypotheses.

It is also not clear that Durkheim intended that these four procedures of definition, classification, induction, and comparison are necessarily distinct and to be used in succession. To define something, for Durkheim, is to explain how it is to be distinguished from other kinds of things and thus seems to be intimately bound up with classification. Also, comparison appears to be used not only in defining and classifying things but, as we shall see below, in carrying out methodical inductions. It also seems that the whole procedure, for Durkheim, is supposed to be an iterative one, with science getting progressively closer to the truth with each iteration.

Definition

In order for a hypothesis to be testable, Durkheim believed, we need to know in advance which states of affairs it is supposed to rule out and will thus count as genuine counterinstances to it if they are met with in experience. We should also be able to determine which facts will support the hypothesis. To make this possible, Durkheim believed, one must carefully define the terms used in a hypothesis so that their extensions or denotations are clear.[18] It is only by carefully defining their terms in this way that scientists can agree whether a hypothesis applies to any given case:

> Every scientific investigation bears upon a determinate group of phenomena that answers to the same definition. The first step of the sociologist must then be to define the things of which he treats, in order that one knows and knows clearly what is the subject matter. This is the first and the most indispensable condition of every proof and of every verification; a theory, in fact, can be verified only if one knows how to recognize the facts of which it must give an account. (1895a:43–44)

Sociologists, however, Durkheim complained, have the bad habit of using terms without defining them,

> neither determining nor methodically circumscribing the order of things of which they intend to speak. . . . Under these conditions, the idea ends up by developing an ambiguity that defies discussion. Because, having no defined contours, it is able to be transformed nearly at will according to the needs of the case and without it being possible for its criticism to predict in advance all of the various aspects which it is capable of taking on. (1897a:108)

Thus, for example, in his discussion of Tarde's theory of imitation, Durkheim said, "before examining the facts, it is advisable to fix the sense of the word" (1897a:108). Similarly, he felt the need to analyze the concept of a race before examining the theory that the social suicide rate can be explained in terms of the racial characteristics of the societies in question (1897a:54–58).

As we have seen, Durkheim believed that the introspective analysis of our concepts of social facts will not suffice for arriving at definitions of them. In order to move a social science out of the ideological phase, he said, one must begin by making a "tabula rasa" of one's ordinary notions of things (1909e:158). Furthermore, he asserted, scientific concepts should be defined "without our worrying whether the class thus formed does not contain all the cases that one usually names in this way or, on the contrary, contains those that one usually names otherwise" (1897a:2). "Our goal," he explained, "is not simply to state precisely the usual sense of the word, but to give ourselves an object of research that can be treated by the ordinary procedures of science" (1899a(ii):141). For Durkheim it is only "natural" that an everyday concept will not "coincide exactly" with the scientific one (1895a:47). Ordinary and scientific concepts generally do not coincide, he thought, because the terms of common discourse do not always correspond to real kinds but often include either too much or too little. In the case of ordinary concepts, he said, "as the classification of which they are the product does not proceed by a methodical analysis, but only translates the confused impressions of the masses, it unceasingly happens that categories of very different facts are indistinctly combined under the same rubric, or that realities of the same nature are called by different names" (1897a:1).[19] Failure to recognize this problem has led Spencer, for instance, to confuse monogamy in fact with monogamy in law in his attempts to explain the evolution of the marriage relationship (1895a:48–49).

An Example: Durkheim's Definition of "Suicide"

Durkheim's concern with defining his terms in such a way that their extensions correspond with natural kinds can be illustrated through his controversial definition of suicide. Durkheim characterized suicide not as the intentional taking of one's life, but as "every case of death that results directly or indirectly from a positive or a negative act, performed by the victim herself and that she knows ought to produce this result." Thus Durkheim's definition includes even those cases of altruistic self-sacrifice in which the victim does not seek death itself but regards death merely as the "regrettable, but inevitable" consequence of the pursuit of some other end (1897a:5, 4).

A very common criticism of this definition is that self-sacrifice is not normally regarded as suicide. Because his definition does not refer to the intentions of the agent, it is argued, his definition disagrees with the common-sense notion of suicide assumed by the public officials who collected the suicide statistics Durkheim used. These official statistics, it is claimed, do not include the "altruistic" self-sacrifices subsumed under his definition of suicide. Because he relied upon these statistics, Durkheim is then accused of making an inductive generalization about suicide from facts with which it disagrees.[20]

This standard line of objection can be attributed to a misunderstanding of Durkheim's methods and explanatory goals. First of all, his critics overlook the fact that Durkheim was not trying to provide a precise expression of our ordinary sense of suicide. Rather, he was trying to provide a definition that corresponds in its extension to a natural kind:

That which is important is not to express with a little precision the notion that the average intelligence has made of suicide, but to constitute a category of objects, which on the whole can be labelled under this rubric without inconvenience, that may be objectively founded, that is to say corresponds to a determinate nature of things. (1897a:2)

Durkheim had further methodological reasons for defining suicide without reference to intention. The victim's intentions, he said, are not accessible to observation (1897a:4). Earlier theorists, including Morselli and Wagner, had already pointed out that suicide statistics were useless for getting at the intentions of the victims (Morselli 1882:7; Durkheim 1897a:144). One could object that what the victim *knew* is no more open to observation than what he or she *intended*. Keat and Urry (1975:83–84), for example, argue that Durkheim's definition of suicide fails to eliminate reference to mental states. In reply, however, one could say that we can *impute* certain mental states to suicide victims on the basis of observations. But if this is the case, as Maurice Halbwachs maintained, one can impute intentions as well as knowledge (1930:452). In answer to Halbwachs I would say that we cannot impute intention as easily as we can impute knowledge. For instance, how do we know whether the soldier who saved his companions by throwing himself on a live grenade intended only to save others from death or whether he seized this as an opportunity to end what he felt to be an unbearable existence? However, we could impute knowledge to the soldier in this case and say that he "should have known" that he would die.[21]

Clearly the knowledge that death will result is necessary for intending it by one's actions. Hence the extension of the traditional definition of suicide would be included as a subset within that of Durkheim's. Durkheim, in other words, regarded suicide and self-sacrifice as but two varieties of the same species and simply named the species after the first variety (1897a:4–5, 32).[22] In this way, Durkheim made the explanatory scope of his theory of the causes of suicide greater than that of its competitors.[23] That is, Durkheim was offering a hypothesis about the existence of a type of social action to which suicide, as it is commonly understood, belongs. Furthermore, he believed that to distinguish suicide from altruistic self-sacrifice in terms of societal approval of intention would be to draw an arbitrary line between what are in fact similar phenomena.[24] "When does a motive," he asked, "cease to be sufficiently praiseworthy so that the act it inspires may be considered a suicide?" (1897a:262.)

Provisional Definitions

Durkheim never pretended that we can provide perfectly adequate definitions at the outset of our research; these are rather what we hope to attain at the end. We must begin our scientific investigations, he believed, with provisional definitions expressed in terms of observable characteristics that tell us how to recognize those things to which the terms in question apply. He thus proposed as a second "corollary" to his rule stipulating that we treat social facts as things that we should "never take for an object of research but a group of phenomena previously defined by certain external characteristics that are common to them and include in the same investigation all of those that respond to this definition" (1895a:45).[25]

Such provisional definitions are not intended "to express the essence of reality," he claimed, and for this reason they are of little explanatory value. Their purpose, he explained, is merely to help us to arrive ultimately at the real essences of things by allowing us to "make contact" with them (1895a:53).[26] These provisional definitions may be thought of as hypotheses about the existence, as well as the extension, of real kinds. In "On the Definition of Religious Phenomena" he said one cannot hope to find the "exact frontiers" of natural kinds right at the start.[27] That which should be included under the definition of religious life, for example, "can be determined only slowly and progressively" as science advances (1899a(ii):140, 154).

According to *The Rules,* provisional definitions or hypotheses should be expressed in terms of "external," "objective," or "visible characteristics" or "signs" that are "borrowed" from experience (1895a:44, 54).[28] The observable characteristics of a social fact, he believed, are not merely accidental but must be caused by its nature. Hence, once the real essences of social facts are known they will be able to explain their signs, he thought. One formulates provisional definitions, Durkheim proposed, by comparing the observable characteristics of social phenomena and classifying them according to their resemblances and differences. If one is careful to group together under the same definition all and only those phenomena that share certain external characteristics in common, he assumed, one can be sure that they will also share the same nature or essence. According to Durkheim, this inference is justified if one assumes the denial of the principle of the plurality of causes or, as he stated it, if one assumes that "the principle of causality is not a vain word" (1895a:53–54).[29] In justifying his method of defining religion in terms of a set of visible characteristics shared by all religions, for instance, he argued that "these external resemblances suppose others that are profound" (1912a:6). Durkheim thus attempted to define even socialism in terms of its "external characteristics" (1928a:49).

The observable signs in terms of which the sociologist expresses provisional definitions, he said, must be "sufficiently objective in order to be able to be recognized by every observer of good faith" (1897a:2). Durkheim believed that defining one's terms in this way will make our hypotheses intersubjectively testable and allow science to exist as a collective enterprise. The way in which the facts are arranged, he said, will not depend on "the particular turn of mind" of any individual sociologist and "the affirmations of an observer can be controlled by others" (1895a:46). Providing definitions in terms of observable characteristics, he thought, will make sociological concepts "capable of transmission," permitting them to be "taken up and pursued." In this way, he argued, "a certain continuation is rendered possible in scientific work and this continuity is the condition of progress" (1897a:vii). At Durkheim's dissertation defense, Boutroux had criticized Durkheim's methods for condemning one to reason about only the signs of reality and not reality itself. Durkheim replied by pointing to the fact that progress in science depends upon its social character. He argued that it is as a consequence of the social organization of science that we can hope to obtain definitions that express better and better approximations to reality. The social nature of science allows for specialized research, he maintained, which makes it possible for scientists to penetrate to the causes of a relatively small number of these objective signs, rather than adumbrate global theories (Muhlfeld 1893:442).

Harry Alpert (1939:118) and Roger Lacombe (1926:103, 108–9) have criticized Durkheim for never revising his provisional definitions and thereby disregarding his own methodological prescriptions. I find this objection to be less than fair. Durkheim never pretended to full knowledge of the real essences of social facts or even of the true extensions of their kinds. This is a goal to be approached gradually, over the course of generations of research, through refining our concepts and our theories.

The Operationalist Misinterpretation

Durkheim's requirement that concepts in sociology be given provisional definitions in terms of observable characteristics is sometimes interpreted as a demand for operational definitions. There does appear to be some textual basis for this interpretation. He explained his concept of a provisional definition through an analogy with what he took to be the practice in physics or chemistry. In these sciences, he said, one may measure heat with a thermometer, electrical currents with an electrometer, and physical forces in terms of the motions that they cause (1895a:55; 1902b:30; 1900b:130; 1897a:349). Sociologists, he stipulated, ought to adopt a similar procedure:

In order to submit an order of facts to science, it does not suffice to observe them with care, to describe them, to classify them; but, that which is much more difficult, it is still necessary, according to the word of Descartes, to find *the bias by which they are scientific*, that is to say, to discover in them some objective element that allows for an exact determination, and, if it is possible, measurement. (1902b:xlii)[30]

Edward Tiryakian even goes so far as to regard Durkheim's definition of social facts in terms of external constraint as an "operational" definition (1978:202; see also Keat and Urry 1975:86). But it is hard to understand the sense in which social constraint can be observed, let alone measured, even when such constraint makes use of physical force. After all, it is never one's eyes alone that inform one that a certain person is a police officer. This is a judgment that presupposes concepts learned within the context of a particular society. Other commentators have criticized Durkheim for having failed to live up to his rules of method by *not* providing operational definitions of key theoretical terms. For example, there is a considerable literature that complains of the lack of operational definitions of his concepts of social integration and regulation, to which he appealed in his accounts of the social rate of suicide.[31]

However, to interpret Durkheim as having endorsed operationalism would be make his theory of meaning resemble those of the nominalist, empiricist, and pragmatist philosophers he attacked. After all, a realist like Durkheim could insist upon the importance of observation and measurement in science without requiring that every scientific concept be *identified* with or be considered equivalent to a set of observations or measurement procedures. As François-André Isambert (1982:177) argues, what Durkheim meant by his observability condition upon provisional definitions is only that they should allow for the deduction of observable consequences from any hypothesis containing the term in question. Although this requirement on definitions is fairly rigorous, it is not necessary to restrict oneself to operational definitions of one's terms in order to satisfy it. To be logically precise about this, an operational *definition* (as opposed to a reduction sentence) would involve a two-way entailment relationship between a concept on the one hand and an operation or set of operations on the other. All that is necessary to satisfy Durkheim's observability condition, however, is a one-way entailment from concepts to observable states of affairs.

Nevertheless, as I mentioned earlier, at least in Durkheim's early works there was some confusion over the relationship between social facts and their signs. If these observable signs were not only the effects of social facts but supposed to serve as *sufficient* conditions for the application of the relevant concepts, there would indeed be the sort of two-way entailment relationships that operationalism requires. However, as we have seen, Durkheim recognized that in certain "pathological" cases similar external signs may arise from different causes. If one identifies sociological concepts with such observable signs, then, one runs the risk of confusing pathological with normal cases and getting the extension of one's terms wrong. For this reason, he believed, one must be careful to distinguish normal from pathological cases. Furthermore, he cautioned, what is normal for a society depends upon the type or species of society to which it belongs (1893b:34; 1895a:70–71, 94).[32] In order to determine what is normal, then, we need minimally to proceed to the next stage of classifying societies and social facts generally. According to Durkheim, as we saw in chapter 4, in the final analysis, whether some phenomenon is to be considered normal depends upon the causes from which it arises and the functions that it serves. To achieve adequate definitions of social facts thus requires that we are in possession not only of some system of classifying them but of some sociological explanations of them. The ultimate explanation of a social fact, for Durkheim, is to be expressed in terms of his theories about collective representations. His sociological concepts take their meaning from these theories and not from operations concerning observable states of affairs.

Classification

The method Durkheim described for classifying societies into types also makes liberal use of hypotheses. At first, he said, it would seem that we should proceed by studying each society, describing each one in exhaustive detail in a series of monographs, and then classifying these societies on the basis of a comparison of these monographs. This method, he pointed out, would seem to be what is called for by a science of observation that proceeds from the particular to the general. However, he asserted, this procedure only appears to be scientific. For Durkheim, as we have seen, the true scientific method is concerned only with crucial facts or facts that decide among competing hypotheses (1895a:97–98). Science, in other words, is interested only in those facts that are relevant to the testing of hypotheses and it is these hypotheses that should guide our classification of the facts.

Thus, for example, in *The Division of Labor* Durkheim classified societies as organic or mechanical in accordance with whether the bulk of their laws are restitutive or repressive (1902b:32–34). As we shall see in the next chapter, Durkheim believed that this classification of the facts will allow him to test his hypothesis that the function of specialization is to create social solidarity. For similar reasons, in *Suicide* Durkheim classified what he believed to be the different types of suicide in accordance with his theory of its causes (1897a:139–43).[33] When Steven Lukes criticizes this procedure as question begging (1973:201), he seems to reveal a profound misunderstanding of the methods and goals of Durkheim's investigation of suicide. The only classifications that interest him, Durkheim said, are those that are relevant to explaining the social suicide rate and its variations. What these are, he argued, will never be learned from even complete descriptions of individual cases (1897a:142–43). We do not classify social facts just for the sake of classifying, he said, but to reveal explanations (1888c:263). The mere classification of a group of phenomena, however, is not alone sufficient to discover their explanation. Classification, he stated, is only a means for "facilitating" the "interpretation" of social facts (1895a:110).

Durkheim argued that in classifying things it is not even possible to begin with an exhaustive description of each individual because its characteristics are infinite in number. Furthermore, he maintained, even if a classification based on a complete description were possible, such a description could be arrived at only at the *end* of our researches, which does not solve the problem of the need for a classification for getting started. For this purpose, he said, we need to select some small number of "essential" characteristics. But this only raises the problem of establishing some criterion for determining which properties are essential. According to *The Rules* it is our "explanation of the facts" that tells us which ones are essential. Thus, he concluded, the classification of societies into types will grow up hand in hand with the development of explanatory theories in social science (1895a:98–99).[34]

Because Durkheim believed that social facts are to be explained in terms of the structure of society, he proposed that societies be classified into types or species according to their structure. The defining characteristic of a social species, for Durkheim, appears to be the degree of complexity of social structure. His classification begins with the concept of a "horde," a perfectly simple, undifferentiated society. A combination of hordes constitutes a "clan," a combination of clans forms a more complex society, and so on (1895a:102–4; 1902b:149–53).

Charles Elmer Gehlke objected to the concept of a horde as being "an exception to Durkheim's principle of method. The unit of classification should be one that may be objectively studied. The 'horde' is not such a unit. They why use it?" (1915:137). To raise this objection, however, is to overlook what Durkheim said about the role of hypotheses in science. It made no difference to him whether one regarded the horde "as an historical reality or as a postulate of science" (1895a:103). Durkheim found it a useful hypothesis for explaining the structure of clans, which are not purely hypothetical but typify Australian and Amerindian societies. He even believed that it is quite acceptable to use hypotheses to fill in the gaps in our historical knowledge in order to provide history with the appearance of some sort of logical development: "in the absence of any possible history, logic is the sole means of perceiving, at least by hypothetical claims, historical sequences of instruments, styles, etc." (1913a(i)(1):682).

For Durkheim, it is not sufficient to classify societies into types. One must also arrange these types in some sort of hierarchy. "A classification," he said, "is a system of which the parts are arranged according to an hierarchical order" (1912a:210).[35] The place that a social species occupies in this hierarchy is determined by its degree of composition, with more complex types taking up more "elevated" positions (1901a(i):245).[36]

The hierarchical classification of social species represents their evolutionary development, Durkheim believed. He credited Saint-Simon and Comte with being the originators of the notion of the evolutionary classification of societies (1928a:153–54; 1900b:117–18; 1915a:6). Of course, Comte's and Saint-Simon's ideas of evolution were pre-Darwinian. Comte and Saint-Simon modeled their sociological classification on Henri de Blainville's concept of an animal series, in which all animal species are arranged in a linear order from the simplest to the most complex (Comte 1830–42:1.702f., 2.127). For Comte, the differences among societies are merely apparent, representing but different stages along a single path of development (1895a:94–96; see also 1888a:89; 1900b:119; 1906a(2):140; 1915a:7). This notion of a linear ordering of societies was the legacy of a long tradition in French thought according to which the concept of a mathematical series could serve as a heuristic model for the other sciences. In the *Rules for the Direction of the Mind* Descartes proposed that through a linear arrangement of a set of phenomena according to their degree of composition, one should be able to discover the relationship governing the first few terms and then use it to derive the rest (1628: rule 6).[37] Saint-Simon's "sociological" or "historical series" is in this Cartesian tradition. Quoting from Saint-Simon, Durkheim wrote that "the future is composed of the last terms of a series of which the first constitute the past. When one has studied the first terms of a series well, it is easy to put down the following ones: thus, from the well-observed past, one can easily deduce the future" (1928a:145).

Although Durkheim saw his sociological investigations as following in the tradition of Comte and Saint-Simon, he rejected their concept of a linear arrangement of societies. He attributed Comte's belief in a single, linear path of human development at least in part to the "imperfect" state of ethnography at that time. Also, he recognized that this notion of a linear series had since become obsolete even in biology: "Already in zoology one has renounced the serial classification that had formerly seduced scholars, thanks to its extreme simplicity" (1888a:90). In a criticism of Wundt for also having conceived historical development as unilinear, Durkheim asserted that sociologists since Spencer have discredited this linear model (1913a(ii)(8):695).

Instead of a linear model, Durkheim adopted the analogy of a branching tree for the evolution of human societies as well as for animal species (1888a:90; cf. 1960b:58; 1914b:28; 1913a(ii)(8):695). To replace Pascal's comparison of humanity with a man who lives forever and is continuously learning, Durkheim suggested a comparison with "an immense family of which the different branches, increasingly divergent from one another, little by little detach themselves from a common stem in order to live their own lives" (1888a:90). Durkheim endorsed this arboreal model and insisted that his "hierarchy of social species" does not imply a linear, serial arrangement of societies (1901a(i):245–46).

Gehlke, however, objected that Durkheim cannot avoid the implication of a linear ordering, given Durkheim's method of classification of social types according to their degree of composition (1915:141). Durkheim did in fact distinguish among varieties of a social type in accordance with the degree to which their constituent parts have or have not retained their individual character (1895a:105). Thus, for example, he regarded Amerindians as "higher" than Australians (1898a(ii):57; 1903a(i):435–36; 1912a:128, 135–36, 255). However, even with this added complication it is not clear that Durkheim could avoid the implication of a linear arrangement of social types. Perhaps what Durkheim meant is that each branching point of the tree represents the adding of a degree of composition. One branch would then represent this new, more advanced type of society, while the other would represent the continuation of the older, simpler type of society. But then Durkheim's model would not be truly arboreal, as it would seem that it permits no subsequent ramification in the older branch.

Induction

The third stage of the method that Durkheim recommended, as we saw above, was to try to find by "methodical inductions" the causes of the variations of the phenomena under study (1895a:33). Here Durkheim was not referring to a simple method of inductive generalization from an enumeration of observed instances, as I have mentioned above. He was talking about John Stuart Mill's methods of eliminative induction. These methods take their name from the fact that they proceed through the elimination of various conditions in order to discover which of these conditions is the cause of the phenomenon under investigation. For Durkheim and Mill alike these methods seem to capture the logic of crucial experiments. That is, they are designed to choose among a group of competing hypotheses, each of which states a causal connection between one of these conditions and the effect to be explained.[38]

In order to determine whether one phenomenon is the cause of another, Durkheim affirmed, the only way to proceed is to compare cases where they are both present or absent, and to see how they vary. When the phenomena can be artificially produced, he said, we have the method of experiment (1895a:153; 1909e:156). When, on the other hand, "nature" presents these cases for comparison ready-made, he stated, we have the method of indirect experiment or the comparative method (1888c:262–63; 1888a:100; 1895a:153). From a strictly logical point of view, he believed, these methods are one and the same. Like Claude Bernard, Durkheim believed that it is not the production of artificial phenomena that is essential to the experimental method, but the use of comparison (1888c:262). Although one cannot perform experiments in sociology, he claimed, the comparative method will "meet the need for experiment in social science" (1960b:51). These ready-made experiments, he seems to have thought, are provided by statistical and historical comparisons.[39]

Due to the complexity of its phenomena and the "impossibility of any artificial experiment," however, Durkheim found Mill's methods of agreement, difference, and residues less applicable to sociology than his method of concomitant variations. The method of residues is not useful in a new science like sociology, he argued, because it relies on a number of laws already being known. Furthermore, he said, social phenomena are so complex that it is not possible to subtract from them the effects of all causes save one. For the same reason he believed the methods of agreement and difference are not very useful in sociology. The complexity of social facts makes it difficult to find two cases that either agree or disagree in exactly one respect, he claimed. We can never be certain that we have not overlooked some antecedent that is always either present or absent in the same way as the consequent (1895a:158–59).[40] Induction, according to Durkheim, "is not easy to practice in social matters" (1885a:86). Nor is it possible to compensate for this complexity in sociology through the repetition of experiments, he realized, as is done in the natural sciences. For these reasons the methods of agreement and difference can yield only conjectures in social science (1895a:159).

Mill made similar criticisms of the method of agreement. This method, he thought, can at best show that one phenomenon is the invariable antecedent of another. However, it can not show that one is the *cause* or "unconditional" antecedent of the other. In other words, from the fact that phenomenon *a* is always accompanied by the antecedent phenomenon *A* we can conclude nothing about whether *a* also depends on other antecedents. We may have overlooked other conditions besides *A* that are always present when *a* is present. The only way in which we can be sure that *A* is the cause of *a*, Mill said, is to find an instance where *a* is produced with no other changes besides *A*. But to do so, he pointed out, is to use the method of difference and not the method of agreement. For this reason, he concluded, only the method of difference can yield knowledge of causes. That method, however, requires the comparison of two cases that are exactly similar in all but one respect. Such conditions, he believed, are rarely met with outside of the laboratory (1843: bk. 3, chap. viii, sec. 3). Similarly, Durkheim realized that without the resource of artificial experiments the method of difference in sociology loses its advantage over the method of agreement.

The Method of Concomitant Variations

The complexity of social phenomena does not present a problem for the method of concomitant variations, however, in Durkheim's opinion. Unlike the method of difference, the method of concomitant variations does not depend upon the possibility of finding for comparison two instances of some phenomenon that are alike in all but one respect. In order for this method to be "demonstrative," he said, "it is not necessary that all of the variations different from those that one is comparing have been rigorously excluded. The mere parallelism of the values through which the two phenomena pass, provided that they have been established in a sufficient number and variety of cases, is the proof that there exists a relation between them" (1895a:159–60).

Another reason Durkheim gave for the superiority of the method of concomitant variations is that its efficacy does not depend on the comparison of a large number of cases. Since it is impossible to find two societies that are alike or different in only one respect, the other methods do require a large number of cases for comparison in order to show that the two phenomena are at least very often both present or both absent.[41] But it is difficult to collect so many facts with exactness, Durkheim pointed out. In fact, he argued, sociology has fallen into discredit precisely because sociologists, using the methods of agreement and difference, have been more intent on accumulating facts than on criticizing or selecting them. Thus, according to Durkheim, they have given the same weight to travelers' reports as to precise historical texts. The use of the method of concomitant variations, on the other hand, will allow sociologists to be more critical of their sources (1895a:162–63).

Durkheim presented a final argument for the superiority of the method of concomitant variations. Unlike the previous methods, he said, it does not show us merely "two facts that accompany or that exclude each other externally, in such a way that nothing directly proves that they are united by an internal bond; on the contrary, it shows them to us participating in one another in a continuous manner, at least with regard to that which concerns their quantity" (1895a:160). Some of the terms he used, of course, would be unintelligible if we did not keep his realism in mind. What he appears to have been trying to say is that there may be instances in which two phenomena are always present or absent together where we would nevertheless not want to say that one is the cause of the other. That two properties may have the same extension, Durkheim seems to have realized, in no way establishes that there is a causal relationship between them, either direct or indirect, or even that they share a common cause. The fact that they are always present or absent together may be purely accidental.

On the other hand, it could never be an accident that two phenomena vary concomitantly. Because "the manner in which a phenomenon develops expresses its nature," he said, a correspondence between the ways in which two phenomena "develop" must be explained in terms of their underlying natures. The very fact of concomitance, he argued, is a law (1895a:160). As Berthelot (1989:31, 37) points out, Durkheim was unconcerned with discovering a precise mathematical expression of the way in which one phenomenon varies as a function of the other. The fact that two phenomena vary in direct or inverse proportion would be sufficient to indicate the presence of a causal relationship.

Durkheim, in fact, thought that the method of concomitant variations is so powerful that its results can override those of the methods of agreement and difference. The presence of one of the phenomena without the other, he believed, can not refute the law established by their concomitant variation. In such cases, he argued, the cause may simply have been prevented from producing its effect by the action of a contrary cause or be present in a different form than that in which it has been previously observed (1895a:160–61). To make this argument, however, would seem to contradict Durkheim's principle that every cause has one and only one effect. Durkheim would have replied to this objection by pointing out that this latter principle is applicable only in the "normal" state of affairs. Causal laws discovered through the method of concomitant variation, in other words, may be defeasible in what Durkheim regarded as "pathological" cases.

However, Durkheim recognized that the concomitant variation of two phenomena does not necessarily establish that one of them is the cause of the other. "It is true that the laws established by this procedure do not always present themselves at first under the form of relationships of causality," he said (1895a:161). They may both be the effects of some third phenomenon or this third phenomenon may be "intercalated" between them, that is, the effect of one and the cause of the other. No experimental procedure can "mechanically" yield knowledge of the causal relations among things. Such knowledge can be obtained, he thought, only through what looks very much like the hypothetico-deductive method:

One shall first seek, with the aid of deduction, how one of the two terms has been able to produce the other; then one shall endeavor to verify the result of this deduction with the aid of experiments, that is to say with new comparisons. If the deduction is possible and if the verification succeeds, one will be allowed to regard the proof as done. If, on the contrary, one does not perceive any direct connection between these facts, above all if the hypothesis of such a connection contradicts laws already demonstrated, one shall apply oneself to the search for a third phenomenon upon which the other two depend equally or which was able to serve as an intermediary between them. (1895a:161)[42]

In other words, when two phenomena vary in a parallel manner, we hypothesize one as the cause of the other and try to derive the latter from the former. Just as we saw in the analysis of Durkheim's concept of explanation in chapter 4, in the passage quoted above we see that Durkheim believed that a causal relationship is a *logical* relationship that allows one to deduce effects from causes.

The hypothesis of a causal linkage is tested not only against the facts but against our background knowledge as well. Durkheim provided the example that the suicide rate varies with the level of education. Clearly, he believed, there is no way to argue that one *causes* the other. Instead, he attributed the rise of both phenomena to the weakening of the strength of religious traditions (1895a:161–62).

The Method of Comparison

Durkheim also acknowledged a further limitation on the method of concomitant variations that requires that it be supplemented with the method of comparison. Just as a few confirmatory examples chosen at random prove nothing, he believed, a number of isolated concomitances alone is not sufficient to establish the existence of a causal relationship:

That which is necessary is to compare not isolated variations, but series of variations, regularly constituted, of which the terms are connected to one another by a gradation as continuous as possible, and which, moreover, is of a sufficient extension. For the variations of a phenomenon permit the induction of a law only if they clearly express the manner in which it develops in the given circumstances. (1895a:165–66)

In other words, in order to determine whether a concomitance between two phenomena is merely an accident or provides evidence of either a causal relationship between them or perhaps a common cause, we need to make some further comparisons. These comparisons make use of our previous classifications of social facts into sociological series. Durkheim distinguished among three kinds of such comparisons: those which compare facts from a single society, those which compare facts from several societies of the same species, and those which compare facts from several social species. The first method, he believed, may suffice when we have extensive and varied data from a single society. This would be the case, for example, if we had suicide statistics for a good number of years, broken down by religious affiliations, age, sex, economic class, and so on. Nevertheless, Durkheim said, it would still be preferable to confirm our results with suicide data from another society of the same type.

When we are studying something like an institution, a legal or moral rule, or a custom which is the same for the entire society and varies only with time, however, we cannot limit our investigations to a single society, he asserted. If we did, he reasoned, we would have but a single concomitance of the conjectured cause and its effect as they vary over time. Nor would it be sufficient to compare several societies of the same type, he said, because the social institution under investigation may be the legacy of an earlier form of society (1895a:167–68). Durkheim encouraged sociologists to make use of the largest possible field of comparison (1895a:167–69; 1909e:156). A sociologist ought not to limit himself or herself to the consideration of a single people or a single epoch, he believed:

> he ought to compare societies of the same type and also of different types, in order that the variations that the institution presents there, the practice of which he wants to give an account, compared with the parallel variations that are established in the social milieu, the state of ideas, etc., allow for the perception of the relations that unite these two groups of facts and to establish between them a causal connection. (1909e:156)

The logic of these comparisons, however, would seem to be that which is captured by Mill's methods of agreement and difference. Durkheim would then appear to have agreed with Mill's position that in order "that we may be warranted in inferring causation from concomitance of variations, the concomitance itself must be proved by the Method of Difference" (Mill 1843: bk. 3, chap. 8, sec. 6). On the face of it this would seem to contradict what Durkheim said about the method of concomitant variations overriding the other methods of eliminative induction.

What Durkheim was trying to do, however, was to propose a method that combines the method of concomitant variations with the methods of agreement and difference. That is, Durkheim believed that we must not only look merely at one case in which variations in A are accompanied by variations in a but compare a number of such cases that agree in this way. This group of cases should then be compared to another that agrees with regard to a *lack* of variation in both A and a in order to confirm the hypothesis that there is a causal connection between them. The hypothesis that A causes a would be falsified if we found a case in which A varied but a did not, or vice versa. We could also describe this as a method of crucial experiment. In effect, what Durkheim was proposing was that we test the hypothesis of a causal relationship between two phenomena that vary concomitantly against the hypothesis that this concomitance is due to some conditions that are specific to some particular society or type of society. The comparisons made with other societies of the same and other types are designed to rule out these latter possibilities if they are in fact not true.

These comparisons among different societies appear to have taken on a greater importance for Durkheim's research than the method of concomitant variations itself. In fact, only *Suicide* and a couple of his articles seem to have made use of the method of concomitant variations.[43] Even in these works Durkheim did not establish the existence of concomitant variations for continuous quantities, lacking the mathematics in order to do so, as I have argued in chapter 2. In his other works, Durkheim appears to have relied largely on comparisons among different societies in order to test his explanatory hypotheses.

Comparison and Classification

I have mentioned that Durkheim believed that in order to account for a social institution found in a certain society we must compare the different forms it takes not only in societies of the same species, but in other species as well. These comparisons, he seems to have thought, presuppose that societies have already been classified into species and arranged in a hierarchy: "One cannot explain a social fact of any complexity save on condition that one follows its entire development throughout all social species. Comparative sociology is not a special branch of sociology; it is sociology itself, to the extent that it ceases to be purely descriptive and aspires to give an account of the facts" (1895a:169).

In *The Elementary Forms of the Religious Life* Durkheim faulted anthropologists for failing to recognize that classification is a precondition for carrying out comparisons among different societies. According to Durkheim, instead of classifying primitive societies into different types, anthropologists seem to consider them all under the category of "savage." He criticized Frazer in particular for having lumped together all the examples of totemism he could discover in history and ethnography, overlooking the type of social environment in which they are found (1912a:132–34):[44]

Social facts are a function of the social system of which they form a part; one cannot understand them when one detaches them from it. This is why two facts that come from two different societies cannot be fruitfully compared merely because they seem to resemble each other; but it is also necessary that these societies themselves resemble each other, that is to say that they be only varieties of the same species. The comparative method would be impossible if social types did not exist, and it can be usefully applied only within the interior of a single type. (1912a:133)

In other words, if I read Durkheim correctly, it is only by restricting our comparisons to societies of the same type that we can hope to approximate the conditions required for a "controlled experiment" or the method of differences. That is, by comparing only societies of the same kind we can try to make certain that there are no differences between the societies we are comparing except those in which we are interested.

Durkheim's admonishment that we restrict our comparisons to societies of the same type, however, seems to contradict what he said elsewhere about seeking the broadest possible field for comparison. He went on to talk in terms of the practical necessity for specialized research in the social sciences, arguing that there is a limit to the number of societies with which one can be familiar and that "one can usefully compare only those facts that one knows well" (1912a:134).

There is a difference, however, between a practical necessity and a methodological rule. As an ideal, Durkheim still found it desirable to extend one's comparisons even to societies of different types. In an early lecture on the sociology of the family, he said, "it is not through comparing two or three facts of the same kind to one another that one can make a theory" (1888c:277). In this early essay Durkheim had in mind the comparison of social facts drawn from societies at different stages of historical development. Beginning around the time of his encounter with ethnographic studies in 1894–95, he began to think that we could study the earliest stages of the development of a social institution in so-called primitive societies. Sometimes, however, the facts presented by ethnographies are ambiguous and one can understand the true meaning of a social institution in a primitive society only by tracing its subsequent development. This can involve comparisons with ethnographies written about other societies of the same type or even with other types of societies recorded in history (1912a:137–38; 1895d:609). At other times, comparisons with simpler forms of society can help us understand those that are more complex. "On this point," he stated, "there is no rule of method that can be automatically applied to all possible cases" (1912a:137–38 n. 1).

The Method of Historical Analysis

Durkheim, in fact, regarded comparisons with simpler forms of societies as an important method for taking our sociological knowledge beyond that which can be observed in our present social institutions and discovering their causal origins. As I showed in my discussion of his method of sociological classification, he believed that these simple societies represent the earliest stages of the historical development of our present form of society. A society inherits much of its forms of organization, institutions, and concepts from the society that preceded it, which may have been of another social type. Thus to explain what has been transmitted to it, he thought, we must turn to comparisons with societies of other species or types: "In order to give an account of our present family, political or economic organization, for example, it is necessary to take into account the entire historical development of humanity that has preceded us" (1914b:35).

These historical comparisons, Durkheim claimed, will allow us to see how the organization of society has "progressively grown more complex," permitting at once the "analysis and synthesis" of social phenomena (1895a:168–69). History, he suggested, can be an analysis of the present, "since it is in the past that one finds the elements of which the present is composed" (1938a:21). "In order to understand the present," he said, "it is necessary to leave it" (1899a(i):139). In fact, it is only through the method of historical analysis that we can even distinguish the elements of which society is composed. We can no more directly perceive the distinctions among these various elements, he thought, than we can perceive cells or molecules with the naked eye. For this reason, "history plays, in the order of social realities, a role analogous to that of the microscope in the order of physical realities" (1909e:154; 1908f:131–32). Just as "scientific analysis" may reveal to us the "clever organization" of what appears to be homogeneous matter, "historical analysis" may reveal a complexity in things that we never suspected (1938a:22–23). Even the nature of man, he asserted, can be explained only through historical analysis "because it is only in the course of history that it has been formed" (1914a:315):

The present is formed from innumerable elements, so intimately entangled with one another that it is not easy for us to perceive where one begins, where another finishes, what each one is, and what their relations are; thus from immediate observation we have only a cloudy and confused impression of it. The only way to distinguish them, to disassociate them, consequently to introduce a bit of clarity into this confusion, is to search in history into how they have progressively come to be superadded onto one another, to combine themselves and to organize themselves. (1938a:22)

Durkheim cautioned that the study of history can yield sociological explanations only when it abandons the simple narrative of events and turns its attention to the more "fundamental" and "permanent" institutions that lie behind these events (1903c:486–87).[45] Yet for Durkheim such social institutions as the family and religion are complex wholes that we can understand only by resolving them into their parts (1888c:264–65; 1912a:4). Through the method of historical analysis, he believed, we can observe the elements of an institution "being born . . . one after the other" as it becomes progressively more complex (1912a:4). It is only in this way that we can arrive at causal explanations of social facts, he thought. "How can we discover these causes," he asked, "if not by going back to the moment in which they were operating?" (1909e:154–55):

The principal problems of sociology consist in investigating the manner in which a political, juridical, moral, economic or religious institution, a belief, etc. has been constituted, what causes have given rise to it, to what useful purposes it corresponds. Comparative history, extended in the manner that we are going to try to make precise, is the only instrument that the sociologist has at his command for resolving these sorts of questions. (1909e:153)

The method of historical analysis, Durkheim believed, will allow us to arrive at a "general formula" or "definition" that expresses the real essence of the social fact in question (1886a:202). By peeling away its secondary or accidental properties, he thought, its essential properties will be revealed. According to Durkheim, "it is a Cartesian principle that, in the chain of scientific truths, the first link plays a preponderant role" (1912a:5). This principle suggests to him that in order to explain some phenomenon, one must study it in its earliest, most primitive form, the form in which it is the most "simple" and contains no borrowed elements (1912a:1). The essential properties of social institutions, he said, can be most easily observed in the simplest, most primitive societies (1912a:10–11).[46] He believed that primitive societies serve as "privileged cases," which are spontaneously provided by "nature," that he found analogous to the simplified phenomena that physicists attempt to produce under experimental conditions (1912a:8).

Durkheim claimed that in order to understand religion, he has turned to the study of just such a simple case (1919b:142–43). For Durkheim, the simplest religion is that of primitive Australian societies:

Every time that we undertake to explain something human, taken at a given moment in history—be it a religious belief, a moral precept, a legal principle, an aesthetic style or an economic system—it is necessary to commence by going back to its most primitive and simple form, to try to account for the characteristics by which it was marked at that time, and then to show how it became that which it is at the moment in question. (1912a:4–5)

He gave similar reasons for studying primitive classifications (1903a(i):400) and the origins of socialism (1928a:8), the origins of the family (1888c:265), the origins of the incest taboo (1898a(ii):37–38), and the origins of the French educational system (1938a:21f.; 1911c(3):75). In his course on the history of education, he said that "it is necessary to begin by knowing the constitution of the initial germ that is at the point of departure of its entire evolution" (1938a:25). Durkheim, in a way that would be consistent with his essentialism, seems to have assumed that the cause of the subsequent development of a social institution is contained in this germ.

In *The Elementary Forms of the Religious Life* Durkheim pointed out that even those religions revealed to us through history and ethnography are fairly complex. It is therefore necessary, he argued, to take our analysis beyond these "observable" cases and to try to find some original form of religious life from which the others are derived. He then proceeded to discuss the relative merits of the animistic and naturistic hypotheses of the origin of religion (1912a:67). The question of which of these was the earlier form of religion is not something that can be answered merely by comparing observable cases, he believed. Instead, we need to choose the hypothesis that best explained the subsequent development of religion. Durkheim's procedure here can be compared with that of Rousseau, whose state of nature hypothesis, Durkheim said, is not a sentimental dream over a golden age but a methodological device—*un procédé de méthode*—employed to explain the present form of society (1918b:119–20).

Durkheim adopted the method of historical analysis, along with the linear arrangement of societies that it seems to presuppose, from Comte and Saint-Simon, but he used it for different purposes. The latter were trying to discover historical trends that would allow them to predict the future. According to Durkheim, they believed that "it is only by observing very extended series of facts, consequently by descending very deep into the past, that we will be able to distinguish among the different elements of the present those that are pregnant with the future and those that are no more than monuments of a past that outlive themselves. . . . It would be easy to establish that the first belong to an ascending series, the others to a series that regresses" (1928a:145). Durkheim, on the other hand, said he was less interested in the future course of history than he was in the discovery of causal relations among social phenomena (1895a:154). Indeed, he opposed sociological speculation about the future (1928a:4).

However, as Delvolvé (1932:299) argued, in *The Division of Labor* Durkheim, much like Comte, appears to have been seeking a general law of historical development. In this work Durkheim did in fact suggest a trend away from "mechanical" towards "organic" solidarity (1902b:149). He also advocated the use of the method of historical analysis for determining the future of property rights (1902b:xl). Elsewhere he postulated trends toward greater and greater emphasis on the individual (1897a:381–82) and on free thought (1897a:430) and finally a trend away from very intense punishments, with deprivation of freedom alone becoming the normal type (1901a(i):245f.). These trends appear to be as orthogenetic as anything one finds in nineteenth-century social theorists like Comte or Saint-Simon.

In defense of Durkheim one might argue that he was interested in the discovery of trends more for the purpose of explaining the present than for forecasting the future. Furthermore, he saw these trends as governing the development of particular aspects or institutions within society rather than society as a whole. For Comte, on the other hand, at least according to Durkheim, sociology reduced to the single problem of discovering a general law governing the historical development of human society as a whole, which Comte believed to have found in his famous law of three states.[47] Durkheim criticized this law on the grounds that one stage of civilization cannot be the cause of the next: "It is not the whole that produces the whole, but each part has its own genesis which demands to be established separately" (1903c:475; 1895a:144). Comte's three-state law, for Durkheim, expresses no causal relations but only a "summary glance" of the history of mankind (1895a:145) and that only since the tenth century (1928a:327–28).

Concluding Remarks

In chapters 1 and 2 I mentioned that Durkheim had methodological reasons for the shift of his attention to primitive societies. The best, indeed the only, way to analyze the collective representations in which our social institutions exist, Durkheim appears to have thought, is to trace their formation back to their very beginnings. Paul Vogt, however, has argued that Durkheim's interest in primitive societies cannot be attributed to his use of the method of comparison. After all, Durkheim could have restricted his comparisons to contemporary societies (Vogt 1976:36–37). To raise this objection, however, is to overlook Durkheim's need to analyze collective representations into their simple, uncompounded elements. Durkheim's belief that one needed to go back to primitive societies in order to provide an exhaustive analysis of our collective representations is best seen in his sociology of knowledge. In his sociology of knowledge, as I will discuss in chapter 8, Durkheim tried to trace the origin of concepts like causation to sensations or feelings in the minds of primitive tribespeople.

Before I turn to Durkheim's sociology of knowledge and religion, however, I want to take a look at his major early works, *The Division of Labor in Society* and *Suicide*. I have two reasons. First, I want to analyze the arguments in these works in order to discover the extent to which his modes of persuasion conform to his explicit methodology or perhaps develop it further. By finding forms of argument that go beyond what is licensed by his methodology, we may also discover what he believed to be acceptable styles of argument among his academic audience that he felt no need to defend. My second reason for looking at these earlier works is that I want to demonstrate that he did in fact meet with serious conceptual difficulties in them that he thought he could best solve by turning to the study of the formation of collective representations in primitive societies.

III

An Analysis of Durkheim's Major Empirical Works

6   The Division of Labor in Society

In *The Division of Labor in Society* Durkheim argued that increasing specialization is creating a new type of social relation among people. This new type of social relation is characteristic of what he called the "organic" form of social solidarity, which he believed to be gradually replacing the older "mechanical" solidarity. The increase in specialization, he believed, is due to an increase in what he called the moral or social density of society. Social density, for Durkheim, is a function of the number of social relations in a society, which are formed through a process involving mental representations. Durkheim also explained the difference between mechanical and organic solidarity in terms of a difference in the mechanisms involving mental representations that are causally responsible for the type of social relations characteristic of each.

Durkheim's interpreters, however, have tended to read *The Division of Labor* as arguing that specialization is due to economic competition resulting from such purely physical causes as the increase in size and density of the population. Among the early reviewers of this book, Léon Brunschvicg and Élie Halévy charged Durkheim with "mechanism" and "materialism" for attempting to explain social facts in terms of the mass and density of the population (1894:565, 567). A generation later, Paul Bureau (1924:113) also saw *The Division of Labor* as a work of "rigid mechanism," offering explanations in terms of social volume and density and the struggle for existence. Talcott Parsons (1937), as I mentioned in chapter 1, regarded this work as "positivist" in character, interpreting Durkheim as having suggested that social actors choose to specialize as a rational response to population pressures. Although Parsons's interpretation of Durkheim has been challenged, it has lately found a defender in Jeffrey Alexander (1982). Similarly, Kenneth Thompson maintains that Durkheim holds to a materialist model of causation in *The Division of Labor* (1982:67, 83; 1985:15, 17). Göran Therborn considers the work "utterly naturalistic," interpreting Durkheim as saying that the division of labor "derives from the increase in the number of human animals in contact with each other within a given area" (1976:254). Finally, Jonathan Turner reads Durkheim as offering multiple causes for the division of labor, including not only the size and concentration of the population but the number and rapidity of the means of communication and transportation (1990:1094).[1]

Durkheim's conception of causality is at least partly to blame for these materialist interpretations. Because he held that there is only one cause for each effect and only one effect for each cause, his readers have confused effects with causes and thus misunderstood the causal claims Durkheim made. Specifically, they have confused social facts with their signs. For Durkheim, the population density is only the sign and not the cause of the social density. The social density of a society is neither an economic nor a demographic phenomenon but a property of its collective consciousness. Durkheim tried to clarify his concept of the causal relationship in *The Rules of Sociological Method,* which was first published the following year as a series of articles in the *Revue philosophique.* Although Durkheim's concept of causality was not new, his causal claims could easily have been misinterpreted as there was no consensus in the academic community at the time regarding the concept of causality. Durkheim wrote *The Rules* not only to clarify but to develop further the methodology of the introduction to *The Division of Labor.* In explicating his conception of the goals and methods of sociology more carefully, *The Rules* made this introduction unnecessary, he thought, and he dropped it from the second edition.

The Division of Labor can be read as one long argument in defense of Durkheim's hypothesis that specialization is giving rise to a new form of social solidarity. He defended this theory by comparing it with alternative theories of specialization in a way that agrees with the methodology of crucial facts and experiments endorsed in *The Rules of Sociological Method.* He also appears to have departed from his explicit methodology to the extent that he eliminated rival hypotheses for reasons that have to do with their conceptual rather than their empirical inadequacies. In fact, his most important reasons for rejecting theories are usually that these theories either beg the question, make unwarranted assumptions, or contain ambiguities and inconsistencies. But as I argued in chapter 1, these apparent departures from his professed methodology should not be seen as violations of his rules of method. Durkheim was simply employing patterns of persuasive argument that he believed to be generally accepted in the academic community and therefore to require no explicit defense.

Durkheim's account of the division of labor, however, is not entirely consistent with his concept of a sociological explanation. Specifically, he provided at best only a functional account of the division of labor. Although he hypothesized an increase in social density as the cause of specialization, the cause of this increase itself was not explained. Furthermore, he did not always succeed in finding genuinely sociological explanations of social facts. His account of the causes of mechanical solidarity appeals to a mental process governed by the psychological principle of association through resemblance and his account of organic solidarity appears to rely on a psychological process of habit formation. Finally, he had trouble providing a sociological account of the formation of the shared beliefs that constitute the collective consciousness, especially for societies characterized by organic solidarity. In order to solve this last problem and to try to find sociological laws governing the formation of collective representations, he set aside the study of specialization and created a new empirical focus for his sociological research program.

Goals and Methods

Durkheim conceived *The Division of Labor* as an attempt to apply the scientific method to the moral question as to whether the increasing specialization of labor ought to be resisted (1902b:xxxvii, 4). Whether the division of labor is morally desirable, according to Durkheim, is a question that can be answered only after determining its causes and effects and comparing them with the causes and effects of other morally acceptable social phenomena. A knowledge of the causes and effects of social phenomena is a subject for scientific investigation, he believed.

In the introduction to the first edition, Durkheim therefore felt the need to do two things: to explain what he believed the scientific method to be and to defend its application to moral questions. By drawing analogies between the scientific and philosophical methods, he tried to persuade philosophers that they had been applying the scientific method to moral problems all along, even if they had been doing so unwittingly. Therefore, he suggested, what he was trying to do was not all that new. At the same time, however, he tried to present his own work as bringing philosophical method into even closer agreement with scientific method through his conscious recognition of the analogies between these two methods.

Questions of morality, Durkheim argued, are usually answered by determining whether a rule of conduct is consistent with a previously established general formula. In order for moral philosophers to be able to persuade others to accept their general moral principles, he asserted, they must use something like the scientific method of hypothesis and test, although they may pretend otherwise. General moral principles, according to Durkheim, are evaluated in accordance with their ability to account for the relevant facts. Quoting from Paul Janet, one of the philosophers on his dissertation committee, Durkheim said that the facts in this case are "duties generally

admitted or at least admitted by those with whom one is arguing"
(1893b:4–5). Philosophers, that is, must try to persuade others of the value
of their general moral principles by showing that what others take to be
their moral duties can be derived from these general principles. These principles are criticized when they fail to account for such duties. For example,
Durkheim pointed out that Kant's categorical imperative had been criticized
for its failure to explain the moral rules governing the marriage relationship
(1893b:6–7). Similarly, the utilitarian principle had been criticized for its
failure to explain our duties regarding the caring for or the honoring of the
dead (1893b:12–14). One might wish to object that the fact that people
accept something as a duty is not like the sort of fact with which the natural
sciences are concerned. But this objection would be beside the point, for all
that is at issue here for Durkheim is the *logical* relationship between facts and
general principles.

In sum, according to Durkheim the fate of general moral principles, like
that of scientific theories, is ultimately decided in terms of their ability to
account for the facts. He believed that the endeavor to find a general principle for deciding moral questions should therefore begin with a scientific
study of the relevant moral facts:

In effect, since the general law of morality has scientific value only if it can account for
the diversity of moral facts, it is necessary to commence by studying the latter in
order to succeed in discovering the law. Before knowing what is the formula that
summarizes them, it is necessary to have analyzed them, to have described their characteristics, determined their functions, investigated their causes, and it is only by
comparing the results of all of these special studies that one will be able to extract the
properties common to all moral rules, that is the constitutive characteristics of morality. (1893b:15)

Note that in the first sentence Durkheim expressed the concern that a general moral principle be able to account for a variety and not merely a great
quantity of moral facts. Durkheim appears to have endorsed a method of
accepting the hypothesis that accounts for the greatest variety of moral facts.
Of course, this passage could also be read as an endorsement of an inductive
method in moral philosophy. But for Durkheim the methodological battle
in ethics is not between inductivism and a form of hypothetico-
deductivism. Rather, I believe, he saw these two methodologies as allies
against the common enemy of a priori principles and introspective methods
(1893b:15–16).[2]

Philosophers of Durkheim's day may have objected that what makes a
moral principle acceptable is that it can be shown to have a foundation in a
theory of the nature of man and not that it can account for particular duties.
Anticipating this objection, Durkheim argued that even if the nature of man

could be investigated, "the conclusions that one may draw from it by means of deduction would be, in any case, only conjectural." He explained this point by making an analogy between engineering and moral philosophy. The engineer, he said, cannot be certain of the practical consequences he has drawn from theoretical principles until these consequences have been verified by experience. Similarly, moral rules drawn from a theory of the nature of man "are only hypotheses as long as they have not undergone the proof of facts" (1893b:19–20).

Durkheim also argued that science has not yet reached the point at which it would be able to tell us what the nature of man is in the first place. Nor did he believe that one can avoid this objection simply by attempting to derive one's general moral principles from a concept of society rather than from a concept of human nature. In order to derive a principle of morality from a concept of society, he believed, one would have to know the function of morality in society. For Durkheim, the question of the function of morality is one that can be answered only by the scientific study of moral facts (1893b:19–20). Thus we would return to the problem of finding a general theory that can account for a variety of moral facts.

A moral fact for Durkheim is typically a fact that a society sanctions a certain rule of conduct. One may rightfully wonder how any theory, let alone a theory that is supposed to help us decide moral questions for ourselves, could possibly account for the fact that one can find different and even opposing moral rules in different societies. Durkheim anticipated this objection and argued that the science of morality is supposed to explain how different moral rules are appropriate to different types of society. One could, of course, press this line of objection further and point out that one can find very different moral rules in what would otherwise appear to be very similar societies. But Durkheim then introduced the additional qualification that a rule is truly moral for a given type of society only when it is "normal" for that type, with the concept of "normal" understood as a statistical mean or average among the societies of that type. Why we should consider the statistical norm as "moral," however, he did not say. If Durkheim expected his science of moral facts to help us decide the moral acceptability of the division of labor, he would have had to answer this question. In spite of this problem Durkheim, granting that a moral rule may even be present in a society without its usual accompanying sign of a social sanction, proceeded to define moral facts as follows:

One terms a normal moral fact for a given social species, considered at a determinate phase of its development, every rule of conduct to which a repressive diffuse sanction is attached in the average of the societies of this species, considered at the same period of their evolution; secondarily, the same qualification is appropriate for every rule that, without clearly presenting this criterion, is nevertheless analogous to certain of the preceding rules, that is to say serves the same ends and depends upon the same causes. (1893b:37–38)[3]

Durkheim appears to have thought that the division of labor is something that one ought to resist only if specialization were a violation of a moral rule appropriate to our sort of society. Because there is no unambiguous moral sanction associated with the division of labor and because the presence or absence of a sanction is not always a reliable indicator of the presence or absence of a moral rule, Durkheim needed to find some other means of determining what our duties are in this regard. He said that the division of labor must be studied in a "speculative" fashion, namely, in terms of a theory of its causes and functions. Once these causes and functions are known, the moral status of the division of labor can be determined by comparing its causes and functions with those of other known moral facts. What he had hoped to show was that the division of labor is morally desirable because, like other moral facts, it produces social solidarity. Durkheim then found it necessary to show that cases in which the division of labor does not have this effect are pathological. The structure of his book thus follows his plan of argument: book 1 seeks to determine the function of the division of labor, book 2 inquires into the causes and conditions on which it depends, and finally book 3 analyzes its abnormal forms in order to show that they should not be confused with its normal form.[4]

The Function of the Division of Labor

Durkheim structured the argument of book 1 of *The Division of Labor* as follows. He began with an analysis of the concept of function. Then he compared his hypothesis that the function of specialization is to produce social solidarity with what he took to be the prevailing explanation of the function of the division of labor. He rejected this competing hypothesis for its conceptual rather than its empirical inadequacies and then argued for his own hypothesis from premises grounded in his theory of mental representations. Only subsequent to these theoretical considerations did he adduce empirical evidence in the form of a concomitant variation that he believed his hypothesis could explain. Finally, he ended the first book by adducing some additional confirming evidence and removing some anticipated objections.

Let me now put some flesh on this skeletal outline. Durkheim defined "function" as a relationship of correspondence between an organism's actions and its needs. For Durkheim the needs to which the division of labor corresponds are affective and not economic. The function of the division of labor, he believed, is to create the social attachments that constitute social solidarity. According to the alternative, economic account, the division of labor augments "the productive force and ability of the worker" and is thereby "the necessary condition of the material and intellectual development of societies" and the "source of civilization." In rejecting the economic hypothesis, Durkheim was not disputing that the division of labor can have economic effects but only the claim that producing these effects is the function that it serves. If the division of labor had no other effect, he asserted, there would be no reason to attribute any moral value to it as there is nothing morally obligatory about the advance of civilization. More importantly, Durkheim also argued that if the effect of the division of labor were purely economic, there would be no reason for it to exist, as the economic needs that it satisfies are only those that it creates itself (1902b:11–17).

In other words, Durkheim seems to have said that the economic account of the function of the division of labor begs the question as to what brought it about in the first place. However, it is not at first clear that an account of the function of specialization must also provide its original cause. Durkheim appears to have overlooked the possibility, for which, as we have seen, he allows in *The Rules*, that the function of some social phenomenon may help to maintain it in existence without having caused it in the first place. In *The Rules* he clarifies his position, suggesting that the problem with the economic account is that it can at best provide *only* the function and not the cause of the division of labor.

Durkheim took the sexual division of labor in conjugal relationships as his paradigm of specialization and then argued that the function of the division of labor is to satisfy certain affective needs (1902b:19–25). His argument rests ultimately upon the theory that one's mental representations of others who are different from oneself give rise to feelings of dependency that serve as the basis of social relationships. The causal processes that explain these feelings appear to be psychological, involving the habitual association of mental images:

The image of the one who completes us becomes in ourselves inseparable from our own, not only because it is frequently associated with it, but above all because it is the natural complement of it: it thus becomes an integral and permanent part of our consciousness, to such a degree that we are not able to do without it and that we seek out everything that will increase the energy of it. This is why we love the society of the one that it represents, because the presence of the object that it expresses, through making the representation pass into a state of present perception, gives it more emphasis. On the other hand, we will suffer from every circumstance, such as physical separation or death, that can have for effect to prevent the return or to diminish the vivacity of it. (1902b:25)

Durkheim then suggested that his theory of the function of the division of labor can be generalized to include social groups larger than the family (1902b:26–27). There are, of course, obvious and important differences between conjugal relationships and other human relationships based on the division of labor. Given Durkheim's assumption that every type of effect has a different type of cause, one would expect to find these differences in relationships reflected in differences in their causes. Even if the mental representations of men and women served as "natural complements" of one another, it is not at all certain that all relationships of mutual dependency could be explained in this manner. Nor did Durkheim ever explain what it means to say that one image is the "natural complement" of another.

Be that as it may, Durkheim proceeded to contrast the "mechanism of images" underlying relationships based on differences with a different mechanism which he thought underlies relationships among people who are similar to one another. The mechanism involved in relationships based on differences, he said,

is not identical to that which serves as a basis for sentiments of sympathy of which resemblance is the source. Undoubtedly, there can never be solidarity between an other and us unless the image of the other unites with our own. But when the union results from the resemblance of two images, it consists in an agglutination. The two representations become solidary because being indistinct, totally or in part, they are confounded and form but one, and they are solidary only to the extent to which they are confounded. On the other hand, in the case of the division of labor, they are outside one another, and they are linked only because they are distinct. The sentiments thus cannot be the same in the two cases, nor can the social relations that derive from them. (1902b:25–26)[5]

Durkheim associated relationships based on similarity with what he called the mechanical solidarity of so-called primitive tribes. In one of his early essays he characterized mechanical solidarity in terms of "the similarity of consciousness, . . . the community of ideas and sentiments" (1888c:258). He associated relationships based on differences, on the other hand, with "organic solidarity," which he believed to be characteristic of contemporary society. According to the two accounts just given, then, the two kinds of social solidarity for Durkheim appear to have their ultimate causes in reactions among mental representations in individual minds. The mental process that gives rise to mechanical solidarity appears to be governed by the principle of the association of ideas through resemblance and the mental process that gives rise to organic solidarity seems to be one of habit formation. Not only do these mechanisms appear to be psychological, but they cry out for further explanation. What makes the similarity of mental representations cause them to join together, or, as he said elsewhere, to "fuse"? What explains the association of representations that are opposites or, as he called them, "natural complements"? Also, what makes two representations alike or different? In sum, Durkheim's argument that the function of the division of labor is to create a new type of social relationship would seem to raise more questions than it answers.

To be fair, Durkheim did not believe that his theory of the function of the division of labor could be justified solely through speculating about mental processes. His theory stood in need of empirical confirmation. Durkheim's next move was to provide such evidence. The remainder of book 1 of *The Division of Labor* was devoted to providing empirical verification of Durkheim's hypothesis that the function of specialization is to create the social relations characteristic of organic solidarity. According to Durkheim, such relations exist only in mental representations and thus cannot be readily observed. In order to submit his hypothesis to empirical test, then, Durkheim had to derive from it a test implication concerning some observable state of affairs.

Drawing the Test Implication

Durkheim expressed the need to draw such a test implication in somewhat different terms than we would use today. As he put it, to refer again to passages that I have discussed in chapter 3, social solidarity does not lend itself to observation and measurement and can be known only through the intermediary of its social effects. Therefore, he argued, we must substitute for this "internal fact" of social solidarity an "external" one that symbolizes it. "This visible symbol is the law," he said (1902b:28). The law for Durkheim is nothing but the most fixed and stable expression of a society's social relationships. He believed that the proportion of different types of law within a society's body of law should reflect the relative importance of different types of social relationships and thus of different types of social solidarity in that society.

In chapters 2 and 3 of book 1, Durkheim argued that mechanical solidarity corresponds to repressive or punitive law and that organic solidarity corresponds to restitutive law. He then drew the test implication that the relative proportion of restitutive law for a society should increase concomitantly with its degree of specialization of labor. Empirical evidence that corroborates this test implication was presented in his chapters 4 and 5. An analysis of the way in which he drew out and confirmed this test implication will reveal some of his underlying assumptions about explanation and method in sociology.

First, Durkheim tried to establish through an analysis of the concept of crime that repressive law is the expression of mechanical solidarity. Crime, according to Durkheim, is the breaking of the bond of social solidarity to which repressive law corresponds. To inquire into the nature of this bond, he reasoned, is then to investigate the cause of punishment or "that in which crime essentially consists" (1902b:35). He claimed to have established his definition of crime "inductively" (1902b:73) but the method of induction he used was not one of enumeration. Indeed, he argued that the method of defining crime cannot be one of "enumerating all the acts that have been at all times and in all places designated as crimes, in order to observe the characteristics that they present" because the number of such universally recognized crimes is too small (1902b:36). The method he used was one of considering a series of objections to a provisional definition of crime and then either refining the provisional definition or removing these objections for conceptual or empirical reasons. Durkheim treated his definition of crime as a *hypothesis* that he sought to *verify* by showing how it could account for the characteristics of punishment (1902b:52). The assumption underlying this method is his model of explanation through essences, which in this case allowed him to conflate a causal explanation of punishment with a definition of crime.

Durkheim began with the provisional definition that a crime is an act that harms the larger interests of society. He rejected this definition for several reasons, beginning with the conceptual problem that by ascribing too large a part to calculation and reflection it contradicts what he said elsewhere about the role of individual psychology in sociological explanation. Then he raised the empirical problem that there are many acts that are considered criminal but that are not harmful, and that even when they are harmful, the degree of harm is not proportional to the severity of the punishment (1902b:37–38).

Modifying his original definition Durkheim then stipulated that a crime is an act that merely *seems* harmful to society. This definition was immediately criticized on conceptual grounds for being "a veritable truism" or tautology. However, Durkheim also cited empirical evidence in its favor. He said that the definition is partly correct in recognizing that what all crimes have in common is that they are reproved by the members of society. Furthermore, he believed that this definition could explain the fact that while civil codes spell out both our obligations and the sanctions for violating them, criminal codes spell out only sanctions. The duties that are violated by crimes, he stated, are already well understood. Finally, he thought this definition of crime could account for the fact that the functioning of repressive law remains diffuse in certain types of society (1902b:38–42).

However, Durkheim objected, to define crime merely as an offense against collective sentiments is not to distinguish acts that are criminal from those that are only immoral. Thus he felt the need to qualify this definition by adding that the sentiments offended by criminal actions must have a greater average "intensity" than those offended by merely immoral actions. But even this definition, he thought, could be refuted by the fact that there are some very intense sentiments, such as filial piety and compassion for those in misery, that are protected only by moral sanctions. He therefore added the condition that the sentiments offended by crime must be "precise," that is, "relative to a well-defined practice." Penal rules, he said, are characterized by greater precision and clarity than moral rules (1902b:43, 45).

Durkheim was then ready to state his final definition of crime: "an act is criminal when it offends strong and well-defined states of the collective consciousness." He considered the objection that there are criminal acts that are punished with a severity that is out of proportion with the degree of indignation to which they give rise or which do not directly offend any collective sentiment. In reply, he said that all the crimes of which this is true are those that are in some way directed against a governing body, which comes to symbolize the collective consciousness. Because "the same fact cannot have two causes," he added, punishment must always result from an offense to the collective consciousness (1902b:47, 49–50).

Having arrived at a definition of crime, Durkheim sought to "verify" this definition by demonstrating that it could account for the characteristics of punishment. First, however, he needed to determine what these characteristics are. He arrived at a definition of punishment in a manner similar to that in which he defined crime, that is, by making successive refinements in the light of objections to a provisional definition. His derivation of the definition of punishment from that of crime then proceeded in terms of the mechanism of images responsible for mechanical solidarity.[6] He began his argument that collective indignation could be useful to society by making an analogy with the way in which anger can be useful to an individual in a dangerous situation (1902b:52–64, 66). Then he explained what he considered to be the "well-known" mechanism responsible for collective indignation.[7] Strong emotions can arise in a crowd, he said, when each individual's mental representation fuses with his or her representation that others share the same representation, resulting in a representation with greater energy. For Durkheim, this mechanism explains the strength of society's reaction against offenses to the collective consciousness (1902b:67–73).[8]

His definition of crime, Durkheim believed, is "verified" by its ability to explain the strong emotions associated with punishment. He then proceeded to argue that the proportion of penal as opposed to restitutive law will measure the relative strength of mechanical as opposed to organic solidarity. The true function of punishment, he said, is to maintain social cohesion among those who are not being punished. This kind of social cohesion or solidarity, he believed, arises from the fact that there are a number of states of consciousness common to all the members of a society. Repressive law represents these shared states of consciousness for Durkheim. Thus, he concluded, the proportion of law that is penal is a measure of the importance of this kind of social solidarity (1902b:73–78).

So far, however, Durkheim's job is only half done. From his hypothesis that the function of specialization is to create the social relations characteristic of organic solidarity he was trying to draw the test implication that the relative proportion of restitutive law should increase concomitantly with the degree of specialization of labor. In order to draw this test implication, he needed to demonstrate not only that the proportion of penal law measures the strength of mechanical solidarity in a society, but that the proportion of restitutive laws expresses the strength of organic solidarity.

Social relations characteristic of organic solidarity, we recall, are grounded in individuals' perceptions of their differences from one another. For Durkheim the rules of restitutive law presuppose these differences among individuals, spelling out their obligations that arise as a result of the division of labor. In order to reach this conclusion Durkheim had to distinguish two types of restitutive law: the law of property on the one hand and domestic, contract, procedural, and administrative law on the other. Only the latter category, which he called cooperative law, is concerned with the regulation of specialized social functions. The relative proportion of a society's body of law that consists of cooperative law expresses for him the degree to which labor is specialized in that society. Durkheim called the corresponding form of social solidarity organic because of the resemblance of this type of society to a living organism composed of organs that exercise specialized functions (1902b:84–101).

Because rules of cooperative law cover only specialized tasks that do not concern everyone, Durkheim said, they are outside of the collective consciousness (1902b:79–83, 97). Presumably, however, rules of cooperative law are nevertheless represented in the social consciousness (cf. 1902b:46). That is, each and every rule is not contained in a collective representation in the minds of all the members of society but only of some of its members. To say that he believed these rules are outside the collective consciousness is not to suggest that he thought they are simply imposed on people by some legislative body. Written codes of law are again but the observable "signs" of social relationships that exist for Durkheim first and foremost in mental representations. At least in normal cases, he believed, the rules of cooperative law merely express relationships of mutual dependency and obligation that arise spontaneously through a process of repetition and habit formation among a society of people with specialized functions. He contrasted the normal with pathological cases in which there is insufficient contact for these relationships to develop among people performing their specialized functions, resulting in what he called the anomic division of labor (1902b:357–60).

Finally, Durkheim was prepared to draw his test implication. If mechanical and organic solidarity have the legal expressions that he claimed for them, he reasoned, then repressive law should have a greater preponderance over cooperative law when mechanical solidarity is more pronounced. Inversely, he said, as the division of labor increases, the proportions of these two kinds of law should be reversed (1902b:103).

Empirical Verification

In order to compare his test implication with the facts, Durkheim felt the need to introduce some additional assumptions. Although he found relatively straightforward the task of comparing the proportions of cooperative and repressive law in the legal codes of various societies, he had trouble finding a way to assess the degree of division of labor. Instead of providing a way to measure a society's degree of specialization, he substituted for it a society's place in a linear arrangement of social types. He then argued that the individual members of a society resemble one another more the closer that society is to the primitive end of his linear classification. The evidence of resemblance he cited, however, has little or nothing to do with the division of labor. It includes not only the fact that all the members of so-called primitive societies practice the same religion, but also the craniological measurements made by Lebon and Broca and some rather dubious evidence that these anatomical similarities are the expression of psychological similarities (1902b:103–5).

Although Durkheim insisted that he was using an arboreal and not a linear model of the classification of social types, he nevertheless believed he was justified in considering those social types higher on the tree to be more advanced (1902b:112 n. 2). In accordance with this model he assumed that one can compare ancient Jewish law, the Hindu laws of Manu, and the Roman laws of the Twelve Tables as representing legal codes of successively more advanced societies. In a similar fashion he compared the legal codes of the Salian Franks, the Burgundians, and the Visigoths, ordering these Germanic tribes according to what he believed to be their degree of development. Not surprisingly, he found that the proportion of penal law progressively decreases in more advanced societies and declared his test implication confirmed by the facts (1902b:108–17). Durkheim appears to have been in danger of committing some sort of tautology here. That is, it seems to be true almost by definition that more advanced societies will have a greater relative proportion of restitutive law. One should keep in mind that for Durkheim, the degree to which a society is "advanced" is not determined by the time at which it appeared in history. For instance, he considered the Salian Franks at the time that their laws were codified in the fifth century to be less advanced than the Romans in the fourth (1902b:116). Instead, the position that a society occupies in Durkheim's hierarchy is determined by its degree of articulation into social subgroups. One would expect that the more a society is divided into parts, the greater would be its need for laws regulating the harmonious cooperation of these parts. Minimally, Durkheim needed to provide some measure of the degree to which a society is divided into parts that is independent of its code of law. Better still, Durkheim should have provided an independent way of measuring a society's degree of specialization.

Durkheim rejected an alternative explanation of the trend away from punitive law and toward restitutive law according to which legislators merely responded to the degree of violence they observed in their societies and found punitive laws less important as society became less violent. Durkheim said that this explanation makes the law an artificial creation of the legislator instituted to "contradict the public mores and react against them." Assuming his own views of social causation, he argued that the law is not opposed to such mores but is the very expression of them (1902b:117).

It could be argued that Durkheim had his facts wrong and that in actuality restitutive law developed before punitive law. One might point to pre-literate societies in which even murder is treated as some sort of primitive tort, with the defendant forced to pay damages to the family of his or her victim. Such evidence, however, would not necessarily disprove Durkheim's theory of the function of the division of labor. Even if it were true that restitutive law is the older form of law, this fact would undermine his theory only if the assumptions he made in deriving his test implication and comparing it with the facts were unimpeachable. Ironically, any criticism one could make of the rather tenuous connections he assumed between types of social solidarity and types of law could actually serve to protect his theory against this empirical refutation. That is, if it could be shown that Durkheim's theory that the function of the division of labor is to create organic solidarity does *not* imply that penal law is the earliest form of law, then the fact that restitutive law is the earlier form of law would not refute his theory.

Durkheim then presented another type of evidence in support of his belief that mechanical solidarity is giving way to organic solidarity. First he argued that the social bonds characteristic of mechanical solidarity are weaker than those characteristic of organic solidarity to begin with, citing what he took to be the relative ease with which members of more primitive societies can leave their society and be assimilated into another one (1902b:120–24). He then adduced evidence that is supposed to show that the collective consciousness associated with mechanical solidarity is becoming even weaker and less well-defined.

The "force" of the bonds of mechanical solidarity depends on three factors, Durkheim asserted: the relative "volume" or "extension" of the collective and individual consciousnesses, the average intensity of the states of the collective consciousness, and the degree to which these states of the collective consciousness are well-defined. The latter two factors are decreasing, he argued. Even though the relative proportions of the individual mind occupied by the collective and the individual consciousnesses have remained the same, he believed, the proportion of the collective consciousness that corresponds to penal law has diminished (1902b:124–25). What he appears to have meant by all of this is that states of the collective consciousness occupy just as much of the mind of modern as of primitive human beings except that in the modern mind a smaller proportion of these collective representations correspond to rules of penal law. As evidence for this claim, he appealed to a decrease in the number of *types* of crime. He cited a number of aspects of social life that are either no longer regulated by penal sanctions or regulated by sanctions that have become less severe. His evidence concerns relations between parent and child and between the sexes, regulations concerning food, dress, and economic activities, and the disappearance of religious crimes (1902b:126–37). Many of these types of social actions and relations, of course, are now governed only by customs or moral rules. Collective representations that correspond to morals and customs are vaguer and have less energy than those that correspond to laws, Durkheim seems to have believed.

Furthermore, Durkheim added, the number of different types of crime has diminished without compensation by new varieties being added. He then considered an apparent exception to this claim and tried to convert this exception into a confirming instance of his general theory. Specifically, he took into account the objection that crimes against the person, such as murder and robbery, are taken more seriously today than they have been in times past. Durkheim interpreted this fact as evidence that individual rights have been extended to a larger segment of society and argued that this extension of individual rights only attests to the diminution of the strength of the collective relative to that of the individual consciousness (1902b:138–42).

In conclusion Durkheim argued that if mechanical solidarity is weakening, another type of social solidarity must take its place if society is to hold together and that this new type of social solidarity results from the division of labor (1902b:148). In support of his conclusion that the collective consciousness associated with mechanical solidarity is growing weaker, Durkheim cited additional evidence regarding the diminishing social role of religion and the disappearing of proverbs (1902b:142–45).

Because the function he assigned to the division of labor is very different from that which the economists had assigned to it, Durkheim felt compelled to remove the objection that specialization is a purely economic arrangement of no particular moral value. His argument began by distinguishing his concept of organic solidarity from Spencer's concept of a modern industrial society based on contracts formed among self-interested individuals. He then attacked Spencer's theory not only on evidential grounds but by arguing that the possibility of individuals entering into contracts presupposes the prior existence of a society (1902b: bk. 1, chap. 7; see also 79–82). Durkheim believed that the division of labor is something that actually ties the individual more closely to society and is not a mere economic arrangement resulting from self-interest. Having rejected the economists' explanation of the function of specialization, Durkheim then turned to the question of the causes of the division of labor.

The Causes of the Division of Labor

Like his account of the function of the division of labor, Durkheim's explanation of the causes of the division of labor is rooted in his theory of mental representations. His causal explanation appears quite ambiguous, however, due at least in part to his failure to clarify his concept of causality in the careful way that he tried to explicate what he means by "function." This ambiguity has led his critics to mistake what he said are the signs or effects of social facts for their causes. As a result, these critics have attacked Durkheim for holding positions that are nearly the opposite of what he actually maintained and have overlooked the real problems with his theories. Specifically, they have failed to notice that he never really provided a causal account but at best only a functional account of the division of labor. His critics have also failed to notice that Durkheim never came up with a genuine sociological account of the shared ideas that constitute the collective consciousness of a society.

Durkheim began book 2 by rejecting the classical explanation of specialization provided by political economists. According to the classical account, he said, the division of labor is due to the desire man has of ever increasing his happiness. Durkheim rejected this account for the reason that political economists were trying to explain a social fact in terms of psychological causes, that is, in terms of individuals acting with certain ends in view. The cause of the division of labor could not consist in "an anticipated representation of the effects that the division of labor produces" (1902b:211). He then argued that if the division of labor progressed in order to increase our happiness, that the division of labor would have reached its limit and ceased developing long ago. We cannot assume, Durkheim reasoned, that primitive man would have been less happy than we are because he lacked the things that make us happy (1902b:215–19). Civilization is not a goal that motivates people, he said, but the necessary effect of a cause. Durkheim qualified this remark by stating that he was not suggesting that civilization could not come to be perceived as useful or even as a goal in itself. He meant to say only that the utility of the division of labor and the civilization that it brings about could not be the original cause of the division of labor (1902b:327–30). For something to be perceived as useful or for a need to be felt, he believed, is a social fact that requires further explanation in terms of social causes (1902b:48–49). Appealing to the social suicide rate as an "objective" measure of the general level of happiness in a society, he also argued that the evidence points towards increasing unhappiness with the advance of civilization (1902b:226–30).[9] Finally, Durkheim rejected a variant of the classical explanation according to which people specialize not in order to increase their happiness but merely to relieve their boredom (1902b:232–36).

The Social Environment

Having rejected the economists' individualistic, psychological explanation of the division of labor, Durkheim proposed to explain it in terms of changes within what he called the social environment (1902b:237). One cannot appeal to the physical environment in order to explain changes in society, he believed, because the physical environment remains relatively stable (1902b:232).[10] Furthermore, he argued, there is insufficient variation in the physical environment to account for the great diversity of human social functions (1902b:245–46). What he meant by the social environment, however, was left implicit in what follows and was not explicitly clarified until he wrote *The Rules of Sociological Method*.

In *The Rules* Durkheim prescribed that social facts are to be explained in terms of the social environment (1895a:138, 147–48). "This conception of the social environment as the determining factor of collective evolution is of the highest importance," he explained, for without it, sociology could provide no causal explanations at all (1895a:143). He proceeded to distinguish two aspects of the social environment, which he labeled "external" and "internal." The "external" environment of a society, for Durkheim, is constituted by its neighboring societies and thus, he thought, affects only its functions of attack and defense. To try to explain social facts in terms of the external social environment, he reasoned, would then be to try to explain the present in terms of the past alone. Durkheim's line of reasoning, of course, not only seems to assume that all wars were in the past but overlooks the possibility of economic and other peaceful relationships among societies. Nevertheless, Durkheim drew the conclusion that social facts are to be explained only in terms of the "inner" social environment, by which he meant "the determinate whole formed by the union of elements of all kinds that enter into the composition of a society" (1895a:138). Elsewhere he affirmed that in order to explain social facts, "it is necessary to attach them to a determinate social environment, to a well-defined type of society, and it is in the constitutive characteristics of this type that it is necessary to go to search for the determining causes of the phenomenon under consideration" (1899a(i):136).

The constitutive elements of a society, according to Durkheim, are of two kinds: persons and things. Things for Durkheim include laws, customs, works of art and literature, and material objects of various manufactures. He asserted that these things have no "motive power" to initiate changes in a society and that things can affect only the rate and direction of change. Therefore the sociologist will be principally concerned with the "human environment." He then distinguished what he considered to be the two important characteristics of this human environment: the "social volume," by which he meant the number of social units, and the social, moral, or "dynamic" density. *Dynamic density* refers to the number of social relationships entered into by the members of a society, he explained, and is not to be confused with the mere physical density of the population (1895a:138–39).

In chapter 2 of book 2 of *The Division of Labor* Durkheim presented his theory that the division of labor increases with the moral or dynamic density of society. He claimed to be able to "induce" this cause immediately from his discussion of the function of the division of labor in book 1. The argument that Durkheim provided, however, is inductive in the eliminative sense. Referring to book 1, he said that we saw there that the division of labor "develops regularly" to the extent to which what he called the "segmentary" type of society disappears (1902b:237). The segmentary type of social structure is one that consists of the repetition of similar social units, in contrast with the type that consists of social units having specialized functions (1902b: bk. 1, chap. 6). Durkheim then considered two different hypotheses: that the disappearance of segmentary social structure causes the development of the division of labor and that the division of labor causes the disappearance of segmentary society. He rejected the second hypothesis because, he said, "we know" that segmentary organization is an "insurmountable obstacle" for the division of labor. To reject the latter hypothesis, he qualified, is not to deny that the division of labor may contribute to the further disappearance of the segmentary type of society once it has already begun to fade. However, he concluded, the vanishing of the older segmentary type of society is primarily the cause and not the effect of specialization (1902b:237).

One might wish to object that Durkheim has rejected the alternative to his hypothesis not for empirical reasons but because the alternative hypothesis contradicts his theory of social structure and social solidarity. Given Durkheim's definition of the segmentary type of society, it is nearly a tautology that the division of labor is due to the disappearance of this type of social structure. To avoid this objection, he would have to provide the causes for what he referred to as the "natural erosion" of the segmentary type of society (1902b:330 n. 1) and then show that these causes also explain the division of labor. He attributed this erosion to the members of society, who were formerly separated, coming into sufficient contact in order to act and react upon one another, increasing the number of social relationships. "This drawing together and the active commerce that results from it" is what Durkheim called the "moral or dynamic density" of a society (1902b:237–38). Thus for Durkheim, the progress of the division of labor is in direct proportion to the degree of moral or dynamic density.

The moral density of a society is a function of the number of its social relations. For Durkheim, for a social relation to exist between any two individuals depends upon their sharing mental representations. These mental representations and thus the moral density to which they give rise, however, cannot be directly observed.[11] The moral density of a society must be measured in terms of a visible sign. For Durkheim, this sign would appear to be the physical population density. The way in which he expressed this point, unfortunately, is highly ambiguous. According to Durkheim, "this moral drawing together cannot produce its effect unless the actual distance among individuals has itself diminished. . . . The moral density thus cannot increase without the material density increasing at the same time, and the latter can serve to measure the former" (1902b:238). So far he seems to be saying that the physical density of population is a necessary condition of the moral density of a society. But a necessary condition can be either an effect or a part of a cause. He then went on to say that it is "useless to investigate which of the two has determined the other"; the moral and the material or physical density of the population, he said, are "inseparable" (1902b:238). Instead of clarifying his position, these words only increase the ambiguity of Durkheim's account, suggesting that an increase in population density is both a necessary and a sufficient condition for an increase in the moral density of a society. Because sufficient conditions are easily mistaken for causes, it is not hard to see how this passage could have been interpreted as saying that the physical density is the cause of the moral density of a society. Not until *The Rules of Sociological Method* did he make clear that he denies the plurality of causes for a given effect and the plurality of effects from a given cause and that for him the cause and effect are each necessary and sufficient for the other under normal conditions.

In Durkheim's opinion, the physical population density of a society is a sufficient condition for its moral density only in the sense that the degree of the former is a sufficient indicator of the degree of the latter. The context from which the quoted passages are drawn reveals that Durkheim regarded the moral density as more fundamental than the physical density of a society.

Durkheim was not suggesting that the density of population may be the cause of the degree to which people enter into social relationships with one another. Rather, he would have said that the social relationships among the members of a society are what caused them to concentrate in an area in the first place. Indeed, he attributed the rise of cities to the needs of individuals to be in constant intimate contact with one another (1902b:239). Once they are settled, the physical proximity of the members of the society can then react upon its cause, strengthening it and maintaining it in existence.

Far from believing that the physical density is the cause of the moral density, Durkheim said that the physical density is only one of three different signs by which the moral density is revealed.[12] In addition to population density, he mentioned the formation of cities as a "another symptom" and the number and speed of means of communication as a "visible and measurable symbol [that] reflects . . . the moral density" (1902b:239, 241). If signs and symptoms are the effects of their causes, however, it would seem that moral density has at least three different effects. Durkheim would thus have violated his rule against the plurality of effects from the same cause, unless he could argue that all three of these signs are all part of the same effect.

However, if the physical population density is only a way of measuring and not the cause of the moral density of a society, the question then arises as to what in fact is the cause of moral density. The fact that Durkheim cited the difference between settled agricultural and nomadic societies as an example of the greater population density of more "advanced" societies (1902b:239) might suggest to some that the ultimate causes of social density are economic. In a footnote, however, he cautioned that "we do not wish to say that the progress in density results from economic changes. The two facts are mutually conditioning and that suffices for the presence of the one to attest to that of the other" (1902b:239 n. 4).[13] In another footnote he explained that there can even be a kind of superficial economic development that allows for a high degree of division of labor and population density in a society without a high degree of moral density. Durkheim provided the case of England relative to the Continent as an example. In spite of England's high degree of development of the economic division of labor, he believed, the segmentary type of social structure is still very pronounced there (1902b:266–67 n. 4).

Citing this same example of England again in *The Rules of Sociological Method* he added that England also has a higher degree of population density than France (1895a:141). He also said that he had made a mistake in *The Division of Labor* in presenting the physical population density as the "exact expression" of the dynamic density (1895a:140 n. 1). The dynamic density, he explained, is "a function of the number of individuals who are actually in relationships that are not only commercial but moral" (1895a:139). Because such physical factors as the number of people per unit area and the means of communication and transportation affect commercial more than moral relations, he argued, they cannot provide an exact measure of the moral density (1895a:140–41, 140 n. 1).

In addition to the moral density of a society, Durkheim seems to have considered the social volume or the number of members in a society to be another factor that can result in an increase in the division of labor. An increase in the social volume, he said, generally produces an increase in the number of social relationships in a society and thus can have the same effect as an increase in the social or moral density. Again, however, he pointed out exceptions to the rule, in this case societies of great volume but little social density. Although "the number of social relations generally increases with that of individuals," he claimed, this is not always true. "The increase in volume is not necessarily a mark of superiority" in a society, he said, "if the density does not increase at the same time and in the same proportions" (1902b:241–43). In his dissertation on Montesquieu, Durkheim described a situation in which a large number of people may be subject to the same authority or leader but be widely dispersed and have few relations among themselves (1960b:38–39). An increase in social volume, then, is not a sufficient condition for an increase in social density. Nor does it appear to be a necessary condition. Even if the population of a society were to remain stable, Durkheim argued, social relations would continue to change through population movements and the division of labor would continue to increase (1902b:332–33).[14]

In sum, Durkheim believed that the division of labor within a society results from an increase in its moral or social density. The moral density of a society is a function of the number of moral relations among its individual members. Durkheim never explained what causes the moral density of a society to increase, however. In his defense, one could argue that for Durkheim the function of some social phenomenon is supposed to react upon its cause, maintaining it in existence. Thus, if the function of specialization is to create new kinds of social relations, as specialization increases the number of social relations and thus the social density increase. What Durkheim left unexplained, however, is what gets the whole process started. He thereby left himself open to the interpretation that the division of labor is due to an increase in the physical population density.

Competition and Adaptation

Durkheim's explanation of the way in which an increase in the moral density brings about an increase in the division of labor is based on an analogy with Darwin's account of the divergence of characteristics under natural selection.[15] The greater the moral density, Durkheim stated, the "more ardent" becomes the struggle for existence (1902b:248). Specialization, he believed, is a response to this struggle that allows former rivals that competed for the same resources to coexist by exploiting different niches within the same environment. We specialize not in order to produce more, he maintained, "but in order to be able to live in new conditions of existence that we have made." The increase in productivity that economists cite as the cause of the division of labor, he said, is merely an unintended side effect (1902b:253, 259).

The competition for survival is not simply the result of conditions in the physical environment, according to Durkheim. Above all, people must *perceive* themselves to be in competition with one another, he asserted, and in order to do so they must sense that they are members of the same society. Although he did not explain why people must be members of the same society in order to compete, he did give reasons why they must feel themselves to be members of the same society in order for competition to give rise to the division of labor. If the bonds of social solidarity are too weak, he argued, people may simply emigrate instead of specialize. If specialization is chosen, he believed, it is only because it offers the path of least resistance. Other possible solutions to the struggle for existence besides specialization, such as emigration or even suicide, are resisted by our love of life and home (1902b:270–71; 1895a:115). The fact that he allowed for other possible solutions to the struggle for existence may help to explain why he thought that the division of labor increases with the number of moral rather than merely commercial relations. People united only by the latter may choose one of the alternatives to specialization.

For Durkheim, then, specialization is an adaptive response to conditions in the social as much as in the physical environment. Parsons was then mistaken in his interpretation of Durkheim's appeal to the struggle for existence as an attempt to explain sociological facts in terms of biological factors (1937:323, 350). Durkheim was not attempting anything like a reduction of sociology to biology. Instead, he saw biology merely as a source of analogies on which to base provisional sociological hypotheses that nevertheless require further verification. He found Darwinian evolutionary theory to be an especially attractive source of hypotheses because it suggested to him a way to avoid the kind of explanations that political economists had proposed that are expressed in terms of the agent's ends-in-view. Ultimately, however, Durkheim's explanation of the division of labor also had recourse to representation of its beneficial effects in the minds of the members of society. He merely substituted the desires to live and not to leave home for the political economists' explanatory appeal to the desire to increase one's happiness.

Furthermore, as Durkheim soon came to realize, to explain specialization as an adaptive response is to provide only a functional and not a causal account of it. In *The Rules of Sociological Method* he conceded that the tendency towards self-preservation is not a sufficient explanation of the division of labor because "it is necessary that the division of labor have already begun to exist in order for the utility of it to be perceived and for the need for it to make itself felt" (1895a:115). In other words, the struggle for existence is not a cause of the division of labor but rather the division of labor has for one of its effects to make this struggle less arduous. To the extent Durkheim relied upon this struggle to explain the division of labor, he thus failed to provide what he regarded as a genuinely causal account of it.

Be that as it may, Durkheim concluded that the division of labor presupposes the prior existence of a society. He added that people must already be in communication with one another in order for their different specialties to harmonize with one another (1902b:259–61). The present conclusion and that of book 1, he said, "verify" and "confirm" one another (1902b:261). That is, he believed, his investigations of both the function and the cause of the division of labor show that social life cannot be explained in terms of the actions of self-interested individuals alone (1902b:262–64).

The Division of Labor and the Collective Consciousness

We have seen that the division of labor for Durkheim is the result of the pressure of competition within an already existing society. But the pressure of competition, he believed, can be "neutralized" by a pressure in the opposite direction exerted by the collective consciousness. The pressure of competition pushes us to specialize while the collective consciousness pushes us to conform. For the division of labor to take place, then, the collective consciousness must become weak enough to allow for individual variation without becoming so weak that everyone decides simply to leave (1902b:268). The collective consciousness must become not only weaker but less well-defined, he added.

Durkheim regarded what he called the "regression" and "progressive indetermination" of the collective consciousness as only a "secondary factor" in bringing about specialization and not as its cause (1902b:270–72; 1895a:114). The appeal to secondary factors would seem to indicate that his principle of explaining a fact in terms of a single cause or essence does not rule out the possibility that this single cause may be complex, consisting of a number of contributing conditions.[16] In fact, as he went on to explain, the progressive weakening and indetermination of the collective consciousness is itself the result of the increase in social volume and density that is causally responsible for specialization (1902b:272–75). Durkheim provided indirect evidence of the weakening of the collective consciousness in the form of population statistics attesting to the relative youth of the urban as opposed to the rural population. He argued that traditions weaken as young people in urban areas experience greater personal liberty and come into contact with other people besides their elders (1902b:278–85).

When Durkheim tried to explain how an increase in social volume and density makes the collective consciousness weaker and less well-defined, he again appealed to mental processes. Members of a society, Durkheim believed, will share similar mental representations only to the extent to which they share similar environments, as they do in small, primitive societies. In such societies, he said, everyone lives under the same conditions and the states of consciousness that represent these conditions "thus have the same character," giving rise to a collective consciousness that is well-defined. The collective consciousness of societies characterized by organic solidarity, however, is vague. These societies tend to be much larger than primitive societies, he affirmed, with their members distributed over a larger area. As a result, Durkheim tried to argue, members of such a society will share collective representations only in the form of abstract general ideas. When society spreads over a larger area, collective representations become more "abstract" because "there are hardly any except general things that can be common to all these various environments." Members of a society no longer share a representation of a particular animal, spring, or forest, but only of a species of animal, or springs and forests "*in abstracto*" (1902b:272).

So far there appears to be nothing particularly sociological about the process by which collective representations are formed in the minds of individuals. The examples he gave seem to involve mental representations forming as the result of each individual's perceptions of the physical environment. However, Durkheim provided the additional example of the development of the notion of divinity from animism, through polytheism, to monotheism: "The notion of the divinity becomes more general and more abstract, because it is formed not from sensations, as in the beginning, but from ideas." The same can be said for rules of law and morality, which, he asserted, become "universalized" and lose clarity (1902b:274). Elsewhere he said that such rules in contemporary society are nothing but "combinations of abstract ideas" (1925a:104). Of course, he still needed to provide sociological laws of collective ideation that would explain the processes by which these religious and moral collective representations are formed.

There is a second problem with Durkheim's explanation of the way in which social volume and density purportedly bring about a weakening of the collective consciousness. Not only did he lack the laws of collective ideation; he postulated the existence of a kind of mental representation referred to as an abstract general idea. As I mentioned in chapter 1, Durkheim realized that Locke's notion of an abstract general idea had been criticized and rejected long before by George Berkeley. Berkeley was the first to argue that no idea or mental image could adequately represent all of the things that fell within the extension of a general concept. The clearest example is provided by the concept of a triangle. For Berkeley it is impossible to imagine a triangle that is neither obtuse, right, nor acute angled, nor equilateral, isosceles nor scalene, but all and none of these at once.[17]

The fact that the collective consciousness of modern society is supposed to consist of abstract general ideas left Durkheim's research program in a rather awkward or difficult situation. As we saw in chapter 1, Durkheim understood the necessity of language for general concepts at least by the time he began lecturing on Rousseau around 1901–2. Lacking a sociological theory of language at the time of *The Division of Labor,* however, Durkheim was unable to explain the formation of the collective representations that constitute the collective consciousness of contemporary society. Similar difficulties, perhaps, did not arise for Durkheim with respect to giving an account of the representations shared among the members of a primitive society just to the extent to which these representations are supposed to be concrete and particular. One of his reasons for having turned his attention to primitive societies may very well have been that he was seeking a way to account for the shared ideas that hold a society together without having to rely upon the problematic notion of abstract general ideas.

Abnormal Forms of the Division of Labor

Although normally the division of labor produces social solidarity, Durkheim said, sometimes it produces different and even opposite results. It is important to investigate these pathological forms, he believed, because as long as it has not been established that they are in fact deviant, exceptional cases, "the division of labor could be suspected of implying them logically" (1902b:343). In contemporary terms, the purpose of book 3 is to remove anomalies to Durkheim's theory of the function of the division of labor. The study of deviant forms, Durkheim added, will also allow one better to determine the conditions of existence for the normal state. That is, when we study the circumstances in which specialization does not result in organic solidarity, we will know better what is necessary to produce this effect.

Durkheim's method of comparing normal with pathological cases can be compared to Mill's method of difference. Both methods seem to involve the comparison of cases that are similar in all but one respect in order to determine whether this difference can be attributed to a difference in antecedent conditions. Durkheim wanted not only to remove pathological cases as anomalies to his theory of the function of the division of labor but to be able to explain them in terms of his theory of the causes of the division of labor and thereby increase the explanatory scope of his theory.

As I mentioned above, Durkheim explained the anomic division of labor as an unregulated form of specialization resulting from a lack of communication among the various specialized functions. He found this abnormal form to be characteristic both of large industries and of scientific research, especially in the social sciences (1902b:346–47, 362). When there is a lack of sufficient communication in industry, he argued, the government is not able to regulate the economy and to bring into harmony functions that are not already in accord with each other. Durkheim argued for this conclusion by drawing an analogy with the brain, reasoning that the brain does not create but merely "expresses" the unity of the organism. Similarly, just as government cannot create economic harmony out of discord, the philosophy of science cannot assure the unity of the sciences (1902b:351, 353). In fact, it could be argued that one of Durkheim's reasons for collaborating on the journal *L'Année sociologique* was his belief that it would bring greater unity to sociology by facilitating communication among social scientists.

The constrained division of labor occurs when regulation does not arise spontaneously but is imposed by force and is exemplified by the class struggle that results. A third and final abnormal form arises when individuals performing specialized tasks do not have sufficient work to bring them into repeated contact with one another often enough for the relationships characteristic of organic solidarity to form (1902b: bk. 3, chaps. 2–3).

In sum, Durkheim tried to explain all three of these abnormal forms of the division of labor in terms of his theory of social solidarity. The division of labor is able to produce social solidarity, he thought, only when the conditions are appropriate for new types of social relationships to form through the repeated interactions of people performing specialized functions. In the normal state of affairs, he believed, these relationships become habitual or customary and are ultimately crystallized in the form of moral and legal rules of conduct. These rules of conduct spell out a system of rights and obligations that link the members of society to one another (1902b:402–3). If for one reason or another this process of repetition, which resembles habit formation in individuals, is not allowed to take place, the division of labor will not result in organic solidarity.

The Reception of *The Division of Labor in Society*

Alexander's externalist account of the development of Durkheim's intellectual career, as we saw in chapter 1, is premised upon a problematic reading of *The Division of Labor in Society*. Although Alexander presents little reason to believe that Durkheim supposedly abandoned materialism in favor of idealism in order to advance his career, Alexander does raise the important issue of how *The Division of Labor* was received. In addition to the review by Brunschvicg and Halévy that I referred to above, Alexander (1982:232, 1986a:101–2) cites Charles Andler (1896:255 n. 1), Paul Barth (1897:296), and Georges Sorel (1895:148) as having interpreted Durkheim as having suggested some form of physical or economic determinism. They appear to have read the book as advocating that specialization is due to such purely physical causes as the size and density of the population and competition for survival.

However, one should not confuse the "mechanism" and "materialism" with which Durkheim was charged by his early reviewers with Marxist economic determinism, as Alexander does. By "mechanism" Brunschvicg and Halévy seem to have meant Durkheim's proposal to subject social or moral facts to scientific explanation (1894:567–68). Above all, they objected to Durkheim's project of founding morality on science. As I mentioned in chapter 4, they also criticized Durkheim's concept of a scientific explanation for its assumption that there is a unique cause for every effect and thus for overlooking the role of contributing factors. There is a kind of ironic justice in their criticism, for Durkheim's concept of the causal relationship led to precisely the sort of misunderstanding of *The Division of Labor* to which Brunschvicg and Halévy were subject. That is, it is because Durkheim presented the moral and physical density of a society as necessary and sufficient conditions for each other that he was interpreted as proposing that the physical population density is the cause of a society's moral density.

Far from accusing Durkheim of Marxism, Sorel (1895:164–69, 179–80) and Barth criticized Durkheim for having overlooked the class struggle. Sorel even read Durkheim's attack in *The Rules of Sociological Method* on the random use of examples as a criticism of the methods of Marx's *Capital* (1895:1), even though Durkheim never mentioned Marx in this work. What Andler actually said was the Durkheim generalized Marx's economic "thingism" into a sociological "thingism" and thereby generalized Marx's mistake (1896:255 n. 1). Far from having accused Durkheim of being an economic determinist, Andler ridiculed what he took to be Durkheim's idealism, caricaturing Durkheim's concept of the collective consciousness, as I showed in chapter 3, as some sort of soul hovering over society. Andler also criticized the psychological mechanisms on which Durkheim's theories rest, arguing that there is no "fusion" of ideas among the management of a modern corporation (1896:250). Of course, Durkheim would never have used the concept of the fusion of mental images to explain what goes on in the boardroom, as this psychological mechanism is supposed to be characteristic only of mechanical solidarity. Presumably the managers of a company are not all alike. Barth, who did charge Durkheim with attempting to explain the development of morality in terms of economic conditions, may have been misled by a passage in the preface to *The Division of Labor* where Durkheim said that morality changes for "reasons of an experimental order" (1902b:xxxviii). What Durkheim meant by this expression is simply that the collective consciousness cannot be studied through introspection or the analysis of ideas but must be studied with methods modeled on the experimental method of the natural sciences.[18]

Regardless of what Durkheim actually meant, however, the misinterpretations of *The Division of Labor* must have dismayed him. His desire to avoid misunderstanding may explain the fact that he stopped using social density as an explanatory concept and that in *Suicide,* as I will argue in the next chapter, he substituted the concepts of social integration and regulation for the concepts of mechanical and organic solidarity, respectively. The latter change may have been motivated to some extent simply by a wish to avoid terminological confusion. Tönnies was also using the terms *organic* and *mechanical* but in precisely the opposite sense from that of Durkheim. In Tönnies, Durkheim recognized, the former term was associated with *Gemeinschaft,* or traditional society, and the latter with *Gesellschaft,* or modern society (1889b:419–21).

There were also conceptual problems with Durkheim's theory of organic solidarity. Not only does it appear to rely on psychological mechanisms but it draws on a problematic conception of abstract general ideas in order to account for a society's shared moral, religious, and other beliefs. Durkheim thus lacked the explanatory resources to account for modern societies composed of highly differentiated subgroups each with their own sets of beliefs and regulations. Similarly, we might add, Durkheim would have had difficulty explaining how individuals in contemporary society are able to function simultaneously in various social groups governed by different and even conflicting values. For this reason also he would have been unable to develop further his ideas about the way in which the conceptions of the goals and methods of scientific research specialties function like moral rules and govern the members of these disciplines. Durkheim's failure to develop this last idea may help to explain the fact that it has been overlooked by contemporary sociologists of scientific knowledge.

Durkheim struggled throughout his entire career with the problem of explaining how beliefs and values are shared among individual members of society. As I will show in a later chapter, Durkheim began to consider seriously the role of language in society and to think of words as more than merely the expression of ideas in the mind only at about the time of *The Elementary Forms of the Religious Life*. Before I turn to this discussion, however, I first want to analyze *Suicide* (1897a). By bringing out some similarities between *Suicide* and *The Division of Labor* that previously have been overlooked, I hope to strengthen my case that there was no radical discontinuity in Durkheim's thinking between *The Division of Labor* and his later works.

7 Suicide

Durkheim's goal in *Suicide* was to demonstrate the explanatory power of his theory of the social environment by showing how it could account for variations in the social suicide rate. There are four different types of suicide, he believed, resulting from either insufficient or excessive social integration or insufficient or excessive social regulation. *Integration* and *regulation* were his new terms for the social relations characteristic of the two types of social solidarity, mechanical and organic. The collective representations associated with the two kinds of social solidarity give rise to social forces that, when sufficiently strong, protect one from suicide but, when excessive, drive one to it.

Durkheim's conception of the social environment in *Suicide* places greater emphasis on the structure of a society than on the density of its social relations. Of course, the idea of explaining social facts in terms of social structure was not entirely new for Durkheim. The difference between mechanical and organic solidarity, after all, reflects a difference in the structure of society. The explanatory appeal to social structure was also prefigured in his dissertation on Montesquieu, where Durkheim said that "if social science is really to exist, societies must be assumed to have a certain nature which results from the nature and arrangement of the elements composing them and which is the source of social phenomena" (1960b:13).

In *Suicide* Durkheim's emphasis on social structure led him to suggest a new hypothesis concerning the relation between the social environment and states of the collective consciousness. In *The Division of Labor in Society*, where he had emphasized social density, he had said, "the major part of our states of consciousness" derive from "the way in which men once associated mutually affect one another, according as they are more or less numerous, more or less brought together" (1902b:342). In *Suicide*, on the other hand, Durkheim suggested that the collective representations associated with religion reflect the structure of society:

Religious conceptions are the products of the social environment, far from their producing it . . . People are able to represent the world to themselves only in the image

150

of the small social world in which they live. Religious pantheism is thus only the consequence and, as it were, a reflection of the pantheistic organization of society. (1897a:245)

This passage from *Suicide* contains one of Durkheim's earliest statements of his hypothesis that our basic categories of thought reproduce the structure of our society. The following year, Durkheim clarified the suggestion that religious ideas reflect the structure of society in "Individual Representations and Collective Representations" (1898b), which will be considered toward the end of this chapter along with some of his other publications from the period immediately following *Suicide*. While developing and clarifying his theories in *Suicide* and these other works, Durkheim nevertheless continued to conform to the same conception of sociology that governs *The Division of Labor* and is expressed in *The Rules of Sociological Method*.

The Argument of the Work

Unlike *The Division of Labor,* Durkheim's *Suicide* does not begin with a long introduction explaining the methods he is using in the work. Perhaps he felt that, having just published the monograph edition of *The Rules of Sociological Method* two years earlier, such an introduction would be unnecessary. The introduction to *Suicide* is limited to laying out the goals specific to that work.

Durkheim began his introduction with a conceptual analysis that results in his controversial definition of suicide. As I argued in chapter 5, he saw no reason that a scientist must adhere to common sense concepts. According to Durkheim, social actions should be categorized into types in accordance with scientific theories and not common sense. Durkheim was proposing a theory of a kind of self-imposed death that includes as a subset what is commonly understood as suicide. He also had additional methodological reasons for defining suicide as any case of death that results from an action that the agent knows will cause his or her death instead of defining suicide in terms of intention. Concepts, as he explained in the *Rules,* should be defined in terms of readily observable characteristics. A person's intentions can not be observed by others. Of course, what a person knows can also be regarded as consisting in unobservable mental states. Nevertheless, it is easier to impute foreknowledge than intent to agents on the basis of observation. By defining suicide in terms of foreknowledge Durkheim would then be in keeping at least with the spirit if not the letter of his rules governing the definition of concepts.

Having defined his concept of suicide, Durkheim then proceeded in the introduction to make clear that he was interested not in individual acts of suicide but in the suicidal tendencies of social groups. He provided statistical evidence that the rate of suicide for each European country remains fairly constant over time and argued that the stability of these rates attests to the existence of suicidal tendencies characteristic of each society. Next, he distinguished two possible types of explanation of these collective tendencies: a society's suicidal tendency is either "a *sui generis* state of the collective soul" or merely the sum of individual states (1897a:14–15).

The argument of *Suicide* is then divided into three parts: book 1 rejects what Durkheim regarded as nonsociological, individualistic, putative explanations of collective suicidal tendencies. Book 2 defends Durkheim's own sociological theory, according to which there are four different types of suicide that result from four different social causes rooted in variations in the strength of the prevailing forms of social solidarity. Books 1 and 2 taken in conjunction can then be thought of as an extended crucial experiment between the individualist and the collectivist approach to explaining social suicide rates. Book 3 concludes in favor of the collectivist approach and explains collective suicidal tendencies in terms of Durkheim's theory of the social forces generated by collective representations. It also removes some objections to his explanation of suicide, provides reasons for rejecting a few alternative collectivist accounts, and offers a suggestion for bringing the rate of suicide down. Durkheim's remedy is based on his theory of the causes of suicide. Having diagnosed what he regarded as the current abnormally high rate of suicide as due to insufficient social integration, he then proposed the formation of occupational societies in order to increase the level of social integration or solidarity.

The fact that Durkheim provided his theory of the causes of suicide only after having presented the reader with a great deal of empirical data has misled some of his interpreters into thinking that Durkheim was attempting to *derive* his theory from the data. To read Durkheim in this way, however, is to confuse the chronological order in which he presented his data and conclusions with the logical nature of the evidential relations between them. Durkheim appealed to suicide statistics in order to corroborate a theory of suicide that he deduced from his more general theories about how the collective consciousness binds individuals to their societies. His theory of suicide is not an inductive generalization from these statistics. Thus to demonstrate the unreliability of the statistics he was using, as many of his critics have done, is not to refute his theory but only to leave it as an unconfirmed hypothesis.

As I explained in chapter 5, what Durkheim meant by "induction" is captured by Mill's methods of eliminative induction and not the naive method of induction by enumeration. Even those among Durkheim's critics who seem to understand what he meant tend to forget that Durkheim considered Mill's to be methods of "proof" for testing hypotheses and not for generating hypotheses in the first place. Peter Manicas provides a prime example of this error when he criticizes Durkheim on the grounds that the method of concomitant variation can reveal a knowledge only of correlations and not of causes (Manicas 1987:161–65). Durkheim, however, never regarded concomitant variation as a method for discovering causal relations. For Durkheim, as I will demonstrate below, relations of concomitant variation serve as crucial facts for deciding between competing causal hypotheses. In the method he used, each hypothesized cause of suicide entails the test implication that the rate of suicide should vary concomitantly with that putative cause. If the evidence does not agree with this prediction but agrees with the test implication of another hypothesis, then the latter hypothesis is to be preferred. In sum, Durkheim's argument consists largely in showing that the available evidence agrees best with predictions based on his theory of the social causes of suicide.

Extrasocial Factors

In book 1 of *Suicide* Durkheim's concern was to demonstrate that a society's suicidal tendency cannot be explained by summing up the factors that cause individuals to commit suicide. He considered and rejected a number of attempts at explaining collective suicidal tendencies in nonsociological terms. These putative explanations include theories that attribute these tendencies to psychological causes such as insanity, neurasthenia, and alcoholism, biological factors such as race and heredity, and environmental factors such as climate and temperature. Many of these putative causes can be found in Enrico Morselli's book on suicide, in which he defended a natural selection hypothesis. According to Morselli, those individuals with morbid states of the brain resulting from biological and environmental factors will find that they are unable to compete in the struggle for existence and will commit suicide (1882:353–68). Durkheim also devoted a final chapter of book 1 to refuting Tarde's theory of suicide, in which Tarde explained suicide rates in terms of his theory of imitation.

Empirical Arguments

Durkheim eliminated some of the putative causes of suicide simply because of a lack of concomitant variation with the social suicide rate. For example, in rejecting alcoholism as a cause, Durkheim presented evidence that, contrary to what this hypothesis would seem to imply, there is no concomitant variation between the geographic distribution of suicide and that of prosecutions for alcohol abuse, mental illnesses caused by alcohol, and alcohol consumption (1897a:46–51). Explanations in terms of climate and seasonal temperature are also discredited for reasons having to do with the method of concomitant variation. Durkheim refuted Morselli's hypothesis that a high suicide rate is associated with the temperate zone of Europe by pointing to evidence that the suicide rate has varied historically for various regions of Europe while their climate has remained constant. Morselli, Enrico Ferri, and Cesare Lombroso had attempted to explain the seasonal variation in suicide rate in terms of the seasonal variation in temperature. Durkheim, however, argued that for various European countries there is a closer fit between seasonal variations in the suicide rate and day length than there is between seasonal variations in the suicide rate and temperature. He then proposed the hypothesis that the seasonal variation in suicide rate is to be explained in terms of the intensity of social life, which is greater during daylight hours. The hypothesis that the intensity of social life is related to the causes of suicide receives additional corroborating evidence in the form of variations of the suicide rate with time of day, day of the week, and urban versus rural patterns of living (1897a:83–106).

Durkheim hence used empirical evidence concerning concomitant variations in at least three ways: he rejected a hypothesis if its test implication was contradicted by the absence of a concomitant variation, if the hypothesis could not explain or says nothing at all about the presence of a concomitant variation, and if an alternative hypothesis had a test implication that provided a more accurate fit with the data. So far Durkheim seems to have been using a method of hypothesis and test that combines the method of concomitant variation with the method of comparison in the way I described in chapter 5. That is, in order for X to be the cause of Y, according to Durkheim, variations in X and Y must in all cases be either present or absent together.

In the case of some other hypotheses regarding the causes of suicide, Durkheim analyzed the meaning of the terms contained in the hypotheses before he investigated whether the factors in question vary concomitantly with the suicide rate. As we saw in chapter 5, Durkheim was very explicit about the need to clarify the meaning of a hypothesis in order to determine whether it in fact agrees with the data. Durkheim's analyses of the concepts of race, heredity, and imitation, which I shall discuss below, are then in keeping with the methodology of *The Rules*.

The hypothesis that each race has a suicidal tendency proper to it had been suggested by Adolf Wagner, Morselli, and Oettingen. Durkheim pointed out that anthropologists at the time were not in agreement as to what the concept of race meant, defining it either in terms of heritable resemblances or relationships of filiation. He chose to consider the racial hypothesis under the first interpretation, arguing that filiations are difficult to establish. In other words, Durkheim preferred to define the concept of race in terms of the observable characteristics of races. However, he pointed out, to define race in terms of heritable resemblances would allow for a group of people as limited as a single European nation to count as a separate race. Morselli's hypothesis provided a case in point, as he distinguished four European "races," the Germanic, Celto-Roman, Slavic, and Ural-Altaic. Morselli further divided each "race" into "varieties," such as the Celtic and Cymric varieties of Celto-Romans in France. Defining race so narrowly, Durkheim pointed out, makes it difficult in the case of Europe to separate the effects of racial differences from those of differences in religion. To the extent he could make Morselli's racial hypothesis clear, Durkheim found it refuted by variations in the suicide rate among nations that Morselli would have considered as belonging to the same race. Although Morselli claimed that his hypothesis was corroborated by an association between the average height and the suicide rate for the northern and southern regions of France, Durkheim argued that this association did not hold for any finer division of France into regions or for Europe generally. Furthermore, Durkheim reasoned, regional differences in the suicide rate admit of alternative explanations in terms of cultural differences (1897a:54–68).

The theory that explains suicidal tendencies in terms of race, Durkheim argued, implies that the propensity toward suicide is inherited. In order to test this implication against the facts, he said, it is necessary to clarify precisely what it is that is supposed to be inherited. If what would be inherited were the sort of temperament that would merely predispose one toward suicide in the appropriate circumstances but would not necessitate it, this hypothesis would have no explanatory scope [*portée*] (1897a:69). Durkheim's argument here appears to presuppose his model of causation, according to which a cause is a necessary and sufficient condition for its effect and not merely a factor that increases the probability of the occurrence of an effect. He then considered the alternative interpretation of the hereditarian hypothesis. According to this alternative interpretation, some sort of monomaniacal tendency toward suicide is inherited. Although this hypothesis conforms to his concept of causality, the cases that seem to support this hypothesis, which consist of two families with a marked tendency toward suicide, are not however sufficient in number to confirm it (1897a:69–71).

At this point in his argument Durkheim laid down three conditions upon the empirical confirmation of hypotheses: First, he said the facts must be in sufficient number so as not to be attributable to accidental circumstances. Exactly how many facts will suffice he did not make clear, but the two cases he cited that agree with the heritable monomania hypothesis are obviously not enough. Second, the facts must not allow for some other explanation. From a strictly logical point of view, of course, this condition cannot be satisfied. In principle, it is always possible that another hypothesis will explain the same set of facts. In practice, however, it is not always easy dream up alternative explanations of the facts. In fairness to Durkheim, we should then interpret this second condition as stipulating merely that there is no available alternative explanation of the facts. This interpretation is consistent with what I have said about Durkheim's methods of argument presupposing a sociology of science. For Durkheim, it seems, to confirm one's hypothesis involves showing how it is superior to the hypotheses of one's rivals. One way to do this is to adduce facts that one's own hypothesis explains but that are not explained by the hypotheses of one's competitors. Third, the facts must not be contradicted by other facts. The context makes clear that what he meant by the third condition on hypotheses is that in order for a set of facts to confirm a hypothesis, the *hypothesis* must not be contradicted by other facts. What Durkheim actually did in this context was the following: he drew further test implications from the hypothesis that a tendency towards suicide is inherited and then cited facts that disagreed with these predictions. Making some rather erroneous assumptions about the mechanisms of heredity, he inferred that the hereditarian hypothesis would imply that the suicide rate should be equally distributed between the sexes and between adults and children. Durkheim found these implications refuted by the facts (1897a:71–80).

Conceptual Arguments

The other hypotheses Durkheim considered in book 1 were rejected for a combination of empirical and conceptual reasons. These hypotheses include Tarde's explanation in terms of imitation and various hypotheses postulating psychopathological causes. The empirical methods Durkheim used are similar to those he used to refute the hypotheses of alcoholism, climate, temperature, race, and heredity. Durkheim's conceptual arguments fall into two categories: some are grounded in his assumptions concerning sociological explanation and others involve considerations of logic and method that were not specific to Durkheim's research program but, one can assume, were shared with the wider research community.

Like his arguments against the racial and hereditarian hypotheses, Durkheim's argument against Tarde's imitation hypothesis begins with an attempt to clarify what the hypothesis entails. He distinguished among three different meanings that have been given to the term *imitation:* it has been used to refer to the process of reciprocal interactions among individuals that results in their all thinking alike, the need that compels individuals to conform to the society of which they are a part, and "apelike" imitation, involving the automatic reproduction of an action perceived by the agent without further deliberation. The contagious spread of yawning provides his best example of simian imitation. For Durkheim the third sense is the only proper sense of the term *imitation* (1897a:108–15). Interpreting Tarde's hypothesis accordingly, he drew a test implication regarding the geographic distribution of suicide rates. Durkheim argued that Tarde's hypothesis would imply a pattern of distribution of suicide rates in which the rate would be highest in important urban centers and would gradually decrease in the surrounding regions. Finding no such pattern, Durkheim rejected Tarde's hypothesis (1897a:120–29).

The connection between Tarde's hypothesis and the test implication that Durkheim drew from it is rather obviously tenuous. One assumes that Durkheim had other, more important reasons for rejecting Tarde's hypothesis than the fact that it fails this empirical test. Durkheim, I think, rejected Tarde's theory of imitation because he believed the goals and methods of Tarde's research program to be so utterly different from his own. According to Durkheim, Tarde provided psychological explanations of social facts, violating one of Durkheim's important rules governing sociological explanation. Durkheim even went so far as to accuse Tarde of being metaphysical and unscientific (1897a:136–38). In book 3 of *Suicide* Durkheim criticized Tarde's theory again, this time for failing to provide a causal mechanism by which imitation would yield a constant suicide rate year after year for each society. For good measure, Durkheim also attacked Tarde for relying on the method of introspection (1897a:346–51).

Durkheim seems to have employed the widest range of argument strategies, however, against explanations of collective suicidal tendencies that postulate psychopathological causes. The theory that psychopathic states affect the social suicide rate takes pride of place, forming the subject matter of the first chapter of book 1. Durkheim began by distinguishing two "ways of defending" the theory that suicide results from psychopathological causes. In effect, he considered two different hypotheses: that suicide is a specific kind of mental illness or monomania and that it is an episode in one or several different kinds of mental illness. He rejected the first hypothesis by questioning the existence of monomanias in general on the grounds that the theory of monomania has been linked historically to a kind of faculty psychology that no longer found many supporters (1897a:21–25). In effect Durkheim was providing a reason grounded in an implicit sociology of science for rejecting an explanation of suicide. That is, he associated theories in science with research communities organized around their defense. Whether he needed to take a theory seriously, he thought, depended upon the strength of the relevant community.

Distinguishing insanity from neurasthenia, Durkheim next considered two versions of the hypothesis that suicide is merely an episode in mental illness. He criticized on methodological grounds the hypothesis that suicide eventuates from one or several types of insanity. First, he said that this hypothesis is not falsifiable, not allowing for an *expérience negative*. This hy-

pothesis is not falsifiable, according to Durkheim, because we cannot investigate every single case of suicide in order to determine the role of mental
illness in each. Durkheim's way of formulating this objection would suggest
that he found that the insanity hypothesis would be difficult to falsify in
practice rather than that it is unfalsifiable in principle. He added that one
cannot prove this hypothesis merely by citing positive instances, as the fact
that one has observed only positive instances does not rule out the possibility of negative instances (1897a:21, 26).

Durkheim concluded that "the only methodical way of proceeding" is to
classify the reported cases of insane suicides and to see whether voluntary
self-imposed deaths fit into any of the categories. He classified these cases
according to the type of mental illness involved and then showed that the
definitions of each of these mental illnesses and thus of insanity in general
are inconsistent with the very concept of suicide. Durkheim reviewed four
major categories of insane suicides employed by his contemporaries in medicine, including maniacal, melancholy, obsessive, and impulsive suicides,
and found them all either lacking in motive or resulting from imaginary motives. On the other hand, he argued, the majority of voluntary deaths stem
from motives with some foundation in reality (1897a:26–31). In sum,
Durkheim believed that in order for one to commit suicide one must know
what one is doing, which would not be the case if one were insane. For example, someone who kills herself jumping off a tall building in the deluded
belief that she can fly does not thereby commit suicide.

Furthermore, to link suicide with insanity would be to restrict arbitrarily
the meaning of the world *suicide* in such a way as to exclude cases of self-
sacrifice committed for religious or political reasons (1897a:31–32). On
the face of it, Durkheim appears to have begged the question in favor of his
own theory of suicide, according to which self-sacrifice is attributed to excess social integration, one of Durkheim's four causes of suicide. Whitney
Pope (1976:185) has criticized Durkheim for having employed a double
standard, having rejected the insanity hypothesis because it cannot explain
every case of suicide yet having provided no hypothesis of his own that can
explain every case. However, I think Pope has misread Durkheim's standards. Durkheim never said that a putative cause must explain *all* cases of
suicide. It seems to me that all Durkheim was suggesting is that we ought to
seek a hypothesis that explains *more* cases than the insanity hypothesis does.
Unlike the insanity hypothesis, his own hypothesis, which postulates four
different causes for four different kinds of suicide, includes self-sacrifice
within its explanatory scope. Durkheim made the valid point that insanity
can explain at best only a very limited number of self-imposed deaths,
which, given his previous argument, perhaps should not be regarded as suicides in the first place.

Unlike the insane, neurasthenics could be said to act on the basis of knowledge. Durkheim then proceeded to test the hypothesis that neurasthenia plays a causal role in suicide. It would seem, however, that the neurasthenia hypothesis would be unfalsifiable for the same sort of practical reasons that he considered the insanity hypothesis to be unfalsifiable. Nevertheless, Durkheim did not raise this objection again here. Instead, he proceeded to test whether suicide varies concomitantly with neurasthenia. Unfortunately, however, there were no statistics regarding the incidence of neurasthenia and Durkheim had to introduce some additional assumptions in order to derive a test implication that could be compared with empirical data. Assuming that the incidences of neurasthenia and insanity should vary together as these diseases differ only in degree of severity, he concluded that the rate of admissions to insane asylums could serve as a reasonable indicator of the distribution of both diseases. We can thus test the hypothesis of psychological causes, he believed, by seeing whether the rate of admissions to mental asylums varies concomitantly with the rate of suicide (1897a:36).

First, Durkheim considered the evidence that is favorable to the hypothesis of psychological causes. He found that suicide and insanity were both more prevalent in the city than in rural areas. However, he argued that this evidence admits of an alternative explanation in terms of social causes connected with city life that are responsible for both suicide and mental disease. This mode of argument is in agreement with the rules he prescribed for the method of concomitant variation, in which he cautioned that a concomitance between two factors may be evidence that both are caused by a third. Next, Durkheim cited a great deal of negative evidence to the effect that insanity and suicide do not vary in the same way with regard to sex, religion, age, country, and stage of historical development. Durkheim criticized the evidence that Morselli adduced that shows that the rate of suicide is high in countries where the rate of admissions to mental asylums is high, faulting this data for combining the retarded with the mentally ill (1897a:36–44). Even if we overlook this problem, Durkheim argued, Morselli's statistics do not reveal a concomitant variation: there is no "consistent correspondence [*correspondance suivie*] . . . between the two orders of phenomena. . . . The social rate of suicides therefore does not sustain any definite relation with the tendency to madness, nor, by way of induction, with the tendency to different forms of neurasthenia" (1897a:43–44).

Durkheim's use of the term *induction* here refers to a method of generalizing from what is true of insanity to what may be true of all mental illness. One wonders why Durkheim did not use the data concerning admissions to insane asylums in order to falsify the insanity hypothesis but used it only to falsify the neurasthenia hypothesis. I think he was being devious with a purpose. To have turned directly to the data on admissions to asylums in order

to falsify the insanity hypothesis would have deprived him of the opportunity to argue that the insanity hypothesis is incoherent. For Durkheim the
more persuasive reason for rejecting the insanity hypothesis is not its lack of
unequivocal supporting data but the argument that the insane by definition
are incapable of the kind of premeditation entailed by the concept of suicide. In presenting this conceptual argument first, Durkheim revealed his
belief that the academic community finds conceptual problems with a hypothesis to be more serious than empirical refutations of it.

Social Causes and Social Types

Summarizing the results of book 1 on the first page of book 2, Durkheim
said that there is a suicidal tendency for each social group that can be explained neither by "the organic-psychic constitution" of individuals nor by
factors in the physical environment. By elimination, he argued, this suicidal
tendency must depend upon social causes (1897a:139). Book 2 is then concerned with the investigation of these causes.

Method

Durkheim began book 2 with a short chapter defending his use of the
method of hypothesis and test both for classifying suicides into types and
for investigating the causes of each type. This chapter employs a subtle rhe-
torical strategy that continues to dupe his interpreters to this day. Durkheim
recognized that some of his readers may be opposed to the method of hypothesis and test and prefer a method of induction by enumeration. For example, Morselli, a leading member of the Italian school of criminology,
defended a method of enumerative induction in his book on suicide
(1882:12), citing Newton's famous attack on hypotheses in query 31 of the
Opticks (Newton 1730:404). Durkheim pretended to want to use the
method of enumerative induction but cited practical difficulties that necessi-
tated his use of hypotheses. Of course, Durkheim had intended all along to
employ the method of hypothesis but he realized that to defend his choice of
method by attacking the alternative method of enumerative induction
would serve only to alarm his opponents unnecessarily. Using the strategy
that he does, Durkheim hopes to gain allies among both hypotheticalists
and inductivists.

Before we can investigate the social causes of suicide, he said, we must
first inquire as to whether in fact the social tendency toward suicide is
"simple or indecomposable" and results from a single cause or whether it
consists in "a plurality of different tendencies," each with its own cause. The
method for answering this question that he initially appears to have endorsed is the very sort of naive empirical method that I have argued that he

disowned. He seems to have recommended that we begin with the observation of the greatest possible number of individual cases, grouping them all into a single category if they display "the same essential characteristics." If they do not, he said, we should constitute "a certain number of species according to their resemblances and differences" (1897a:139–40).

However, Durkheim then went on to reject this method of classification in accordance with empirical resemblances and differences, disingenuously citing a lack of documentation as his reason. He suggested that we instead "reverse" the order of our studies and classify suicides according to their causes: "we are able to constitute the social types of suicide, not by classifying them directly according to their previously described characteristics, but by classifying the causes that produce them" (1897a:141). Such a classification, Durkheim said, will be "etiological" rather than "morphological." This "reverse method," he added, is the "only fitting one for the problem at hand" (142).

Durkheim attempted to justify his etiological method of classification through an appeal to his concept of the causal relation, according to which there is no plurality of causes for any given effect and no plurality of effects from any given cause. At the same time, however, he seems to have recognized that his taxonomy of causes is only an hypothesis, which must be verified by deducing from it a taxonomy of types of suicide that can then be tested against the facts:

> Once the nature of the causes is recognized [*connue*], we can try to deduce from it the nature of the effects which thus will be found to be characterized and classified by the same action for the sole reason that they will be attached to their respective sources. It is true that, if this deduction were not in any way guided by the facts, it would risk losing itself in contrivances of pure fantasy. . . . [The facts] will show us the direction in which the deduction ought to be directed, and by the examples that they will furnish us, we will be assured that the species thus deductively constituted are not imaginary. Thus, from the causes we will descend again to the effects and our etiological classification will be completed by a morphological classification that will be able to serve to verify the former, and *vice versa*. (1897a:142)

Toby Huff sees in this passage a certain amount of confusion on Durkheim's part between the concepts of deduction and abduction and interprets Durkheim's "reverse" method as a method of abductive inference. Specifically, Huff thinks, Durkheim meant to say that the species of suicide were "abductively" and not "deductively" constituted (1975:249–51). The concept of an abductive or retroductive inference is due to Charles Sanders Peirce. According to Peirce, in an abductive inference one begins with the observation of some surprising fact C. One then reasons that if the hypothesis A were true, then C would follow as a matter of course. Finally, one concludes that there is good reason to suspect that A may be true (Peirce

1940: chap. 11).[1] Abductive inference, in other words, is a method for arriving at a theory rather than for testing one.

Judging by Durkheim's lectures on pragmatism (1955a), however, he seems to have been unfamiliar with Peirce's writings on scientific method, identifying pragmatism largely with William James's theory of truth. Of course, Durkheim could have arrived at a method of abduction without reading Peirce. However, in the passage quoted above, Durkheim seems to have been describing a process of verification and not an abductive inference. The types of suicide are "deductively constituted" in the sense that they are deduced from a theory of the causes of suicide. When Durkheim wrote about deduction being "guided" or "directed" by the facts, I think he meant that out of the many possible test implications that one could derive from a theory, he wished to derive only those to which the facts at hand are relevant.

To be sure, by going almost immediately to the data in each of the following chapters Durkheim at least gave the *appearance* of proceeding abductively. That is, it seems as if in each chapter he began by considering a group of suicide statistics and then sought an explanation for them. However, that Durkheim was not proceeding abductively becomes clear once we consider how the suicide statistics are distributed among chapters 2 through 5 of book 2. The breakdown of suicide statistics by religion is considered in the chapter on egoistic suicide, military suicides are considered in the chapter on altruistic suicide, and so on. That is, the data is distributed among the chapters in accordance with his theory of the causes and the test implications he has drawn from this theory. Rather than proceeding abductively from the data in each chapter, Durkheim was using the data to support conclusions he had drawn from his theories of social integration and regulation.

As I mentioned in chapter 5, Lukes objects that Durkheim has begged the question by presupposing his theory of the causes of suicide in his classification of suicide into types (1973:201–2). Like Huff, Lukes seems to be confused about the purposes for which Durkheim was using the statistical data. Durkheim classified suicides into types and distributed the statistical data into the appropriate chapters accordingly not in order to arrive at an explanation but in order to corroborate his theory. As I mentioned earlier, Durkheim was interested only in those classifications that are relevant to testing causal hypotheses. It was not at all clear to him that one could arrive at such a classification through exhaustive descriptions of individual cases. Furthermore, he argued, suicides cannot be classified according to the motives reported in individual cases, as these reports merely reflect the opinions of the recording officials. As proof, he cited the fact that the proportions assigned to various motives remain the same even when the suicide rate changes. Also, these proportions are the same even for entirely different occupational groups, which seemed highly implausible to Durkheim (1897a: 142–47).

Egoistic Suicide

Durkheim's two chapters on egoistic suicide provide the first illustrations of his methods of corroborating his theory of the causes of suicide. Egoistic suicide, he thought, is caused by a lack of social integration. Religion, the family, and political association are all sources of social integration, he believed. He therefore inferred that the social rate of suicide should vary inversely with the degrees of religious, domestic, and political integration.

In order for Durkheim to have compared his conclusions with the facts, he would have needed to find a way to measure the degrees of social integration, but he never provided one. Instead he proceeded in the following manner. First, he drew from the statistical data a low-level empirical generalization concerning the putative existence of a concomitant variation between the social suicide rate and some religious, domestic, or political factor. This generalization was usually one that had been accepted by other writers on suicide. Second, he showed that either this empirical generalization or the explanations that have been given for it were inconsistent with other data. Sometimes when the empirical generalization was inconsistent with the data, he proposed a corrected version of the generalization. Third, he introduced additional assumptions that would allow him to explain either this empirical concomitance or its corrected version in terms of his theory of social integration. Finally, Durkheim attempted to show that his theory of the causes of suicide is corroborated just to the extent that it accounts for more of the data, explaining either what were anomalies for earlier theories or some additional data that other theorists had overlooked.

Durkheim's procedure here can be compared to Mill's inverse deductive method. Like Mill, Durkheim seems to have regarded empirical generalizations as unreliable unless they could be explained by an acceptable scientific theory (Mill 1843: bk. 3, chap. 16). Mill's inverse deductive method begins with historical generalizations and then seeks to explain them in terms of a higher-level sociological theory (1843: bk. 6, chap. 10). Durkheim's method, on the other hand, begins with statistical generalizations.[2]

Egoistic Suicide and Religion

Durkheim devoted the first of his two chapters on egoistic suicide entirely to religious factors. This chapter has usually been read as arguing *for* the "law" that Catholics have lower suicide rates than Protestants. Far from trying to establish such a "law," however, Durkheim was actually trying to persuade the reader that his theory of social integration provided a better explanation of the suicide data than the hypothesis that religion is a cause of suicide. The case for empirical correlations between religion and suicide rate had already been presented by Morselli (1882:119f.). Durkheim argued that his own theory of the causes of suicide not only yields a higher-level explanation of the fact that Catholicism *seems* to provide better protection than Protestantism from suicide but accounts for the exceptions to this generalization as well.

Durkheim first raised doubts about the hypothesis of religious causes by suggesting that the effects of religion may have been confused with those of "levels of civilization." He cited the very low rate of suicide among the Greek Orthodox and insinuated that it may be due to their lower level of civilization rather than to their religion. In order to avoid confusing the effects of religion with those of other cultural factors, he believed, one must compare the effects of religion within the same country (1897a:149–50). In other words, Durkheim recommended a use of Mill's method of difference in which one holds the country constant and varies only the religion.

By implication Durkheim was also suggesting that the level of civilization is causally related to the suicide rate. As we saw in the previous chapter, Durkheim had argued in *The Division of Labor* that mechanical solidarity gives way to organic solidarity with the advance of civilization. He believed, I think, that suicide rates are to be explained ultimately in terms of the changing composition of social solidarity. Social integration was simply his new term for mechanical solidarity. As civilization advances, the degree of social integration decreases and as a result the rate of suicide increases unless it is counteracted by other social forces.

The theory that the rate of suicide increases with the rise of civilization and the decline of social integration, he seems to have thought, can account for the anomalies to an empirical law relating religion and suicide rate. The first anomaly is that presented by Norway and Sweden, which, although Protestant, have a lower rate of suicide than the Protestant countries of central Europe. Durkheim explained the lower rate in Scandinavia in terms of the supposed fact that these countries were at a lower "intellectual level" or level of civilization than central Europe (1897a:153). England presented a more difficult anomaly, being a Protestant country with a high level of civilization and a low suicide rate. Durkheim explained that England is only nominally Protestant and that the Anglican church functions in much the same way as the Roman Catholic church, resulting in the same level of protection against suicide (1897a:157–62).[3] In other words, Durkheim introduced the rather dubious assumption that the Anglican church produces a higher degree of social integration than other Protestant religions.

Believing that he had explained the anomalies to the generalization that Protestants have higher suicide rates than Catholics, Durkheim then sought to explain the generalization itself. Durkheim explained both the generalization and its exceptions in terms of his theories of social solidarity and the collective consciousness. He argued that the lower suicide rates among Catholics is not a question of moral precepts, which are the same for all Christians. Furthermore, he added later, suicide is not proscribed for Jews, who have the lowest rate of suicide. The difference between Catholic and Protestant suicide rates, he said, is to be attributed to the Protestants' "spirit of free inquiry," which itself is nothing but the effect of "the shaking of traditional beliefs" (1897a:154, 172, 157). In other words, Durkheim believed that the collective consciousness is much weaker among Protestants than among Catholics.

Durkheim then drew out two test implications from his theory of egoistic suicide. Assuming that the taste for "free inquiry" is associated with a taste for education and that people aspire to education only to the extent to which they are free of the yoke of tradition, he argued,

if the progressive weakening of collective prejudices and customs inclines toward suicide and if it is thence that the special predisposition of Protestantism comes, one ought to be able to establish the two following facts: (1) the taste for instruction ought to be more keen among Protestants than among Catholics; (2) to the extent that it denotes a shaking of common beliefs, it must, in a general manner, vary with suicide. (1897a:162)

If anything in Durkheim is clear, it is that the hypothesis under test here is that the suicide rate is causally dependent upon the weakening of the collective consciousness and not that the rate depends upon religion.

First Durkheim considered the test implication that Protestants should be more interested than Catholics in education. He looked at rates of school attendance and found that it was in fact higher in Protestant countries, including Norway and Sweden, than it was in Catholic countries. The fact that these Scandinavian countries had a high rate of school attendance but a relatively low rate of suicide would seem to present an anomaly for Durkheim that required an explanation. Such an explanation, however, he did not provide. He then turned to the test implication that suicide should vary with level of education. Durkheim cited corroborating statistics that show that the suicide rate varies with the rate of literacy *within each religion*. He was using the method of difference again, holding religion constant in order to see whether the suicide rate varies with the rate of school attendance. Holding both religion and country constant, he also used the method of difference to show that the suicide rate was higher for the professions requiring more education and that it was higher for men, who are generally better educated than women. The sole exception to the latter claim is the case of American Negroes, among whom the women were both better educated and had a higher suicide rate than men (1897a:163–68).

A more serious anomaly for the second test implication is presented by the case of European Jews, who had a higher level of education yet generally a lower rate of suicide than either Catholics or Protestants. Durkheim attempted to explain this anomaly in terms of a "general law" that religious minorities seek education as a form of self-protection or in order to emulate the surrounding society (1897a:168–70). This "law" was corroborated by data concerning school attendance among Protestant minorities in Catholic countries and Catholic minorities in Protestant countries (169–70 n. 2). Thus he said "the exception is merely apparent; it even confirms the law" that the propensity for suicide is aggravated by a weakening of traditional beliefs (170). In other words, the level of education is not always a reliable indicator of the strength or weakness of collective opinions.[4] According to Durkheim, Jews had both a lower rate of suicide and a higher degree of education than Christians *because* they were more socially integrated. However, he noted that Bavaria, where Jews had a higher rate of suicide than Catholics did, appeared to be an exception. He explained this exception by pointing out that Bavarian Jews tended to be urban and professional and thus by implication less well socially integrated than Catholics (153–54).[5]

Durkheim concluded that religion preserves one from suicide through the social integration of the religious community. Suicide and the quest for knowledge are both effects of the same cause, he believed—that is, they result from the loss of "cohesion" in the religious community (1897a:173, 171). On my interpretation of Durkheim, many of the attempts to "refute" him have been misguided. These attempts at refutation fail because they do not touch Durkheim's theory that it is the degree of social integration and not religion that is the important factor in explaining suicide. Some of his critics question the reliability of his data, arguing that the lower reported rates of suicide in Catholic countries can be explained by a greater desire to conceal suicides.[6] Others have adduced evidence that shows that there is no difference between the suicide rates of Catholics and Protestants.[7] However, if Durkheim had been confronted with statistics that established, say, that Protestants have *lower* suicide rates than Catholics, he would have sought a way to argue that Protestants are more socially integrated than Catholics.[8] His theory was not derived from the data; it was used to explain the data. To refute Durkheim effectively one must attack his theories of social integration and not the low-level empirical generalizations he tried to explain in terms of his theories.

Egoistic Suicide and the Family

In the second chapter devoted to egoistic suicide, Durkheim drew the additional test implication that if religion protects one from suicide "because and to the extent it is a society, it is probable that other societies produce the same effect" (1897a:174). He then considered two forms of society: the family and political society.

Durkheim began his discussion of domestic society by eliminating on empirical grounds Wagner's hypothesis that the married have higher suicide rates than the unmarried because of the greater burdens and responsibilities of married life. Durkheim suggested that Wagner had looked at only the absolute numbers and not the rate of suicide and failed to correct the data for the fact that a large proportion of the unmarried population consisted of children under the age of sixteen. Durkheim argued that when investigating the effects of marriage on the suicide rate one must hold the age constant. Thus again he recommended the method of difference. Unfortunately, he said, none of the published statistics on suicide present the data on age and marital status together in such a way to permit the appropriate comparisons (1897a:174–75, 179).

For the first and only time in his work, Durkheim presented previously unpublished data. He cited approximately 25,000 cases of suicide that had been reported to the Ministry of Justice that he said "we" classified by age and marital status in order to separate the effects of each (1897a:179–80, 183, 204).[9] From this data Durkheim drew the following empirical generalizations: that early marriage aggravates the propensity to suicide among men; that after twenty years of age, marriage protects one from suicide at a rate that varies with age; that the degree to which married persons are protected varies with sex; and that widowhood diminishes but does not completely suppress the effect of marriage on suicide rates (1897a:182–85). He then sought to explain these empirical generalizations in this chapter and in the chapter on anomic suicide.

Durkheim tried to persuade the reader that his theory of social integration provides the best explanation of the relative immunity to suicide enjoyed by married people. He set up a crucial experiment between the hypothesis that this immunity is due to the influence of the domestic environment itself and the hypothesis that it can be explained by matrimonial selection. According to the matrimonial selection hypothesis, which had been suggested by Bertillon and Letourneau, people tend not to choose as marriage partners people who are unfit and hence more likely to commit suicide. Durkheim questioned on what he called a priori grounds Bertillon's and Letourneau's assumption about the way in which people choose marriage partners, criticizing them for equivocating with regard to the concept of fitness. Then from the marital selection hypothesis Durkheim inferred test implications concerning how the suicide rate should vary with age and sex and showed these test implications to disagree with the facts (1897a: 186–89). He conceded that the marital selection hypothesis does explain the fact that childless widowers are less likely to commit suicide than unmarried men. However, he tried to provide an alternative explanation of this fact in terms of a "salutary influence" of marriage that survives the death of the spouse (202–7).

By elimination Durkheim was left with the family environment hypothesis. He then cited evidence that it is the presence of children more so than spouses that protects one from suicide, showing that the rate of suicide varies inversely with the number of children in the family (1897a:191–212). The more children one has, apparently, the greater the extent to which one is socially integrated.

Egoistic Suicide and Political Society

Durkheim then cited statistics to the effect that the suicide rate actually falls during wars and revolutions such as the crises of 1830, 1848, and 1870. He explained that these crises gave rise to collective sentiments that increased the level of social integration. An alternative explanation of the data that he considered is that poor records are kept during times of crisis. He rejected this alternative for empirical reasons: the effect is the same for victor and vanquished; the effect persists after the crisis is over; the effect is greater in cities than in rural areas, as people in cites are more caught up in political events; and national wars have a greater effect than dynastic struggles (1897a:216–22).

Concluding his discussion of egoistic suicide, Durkheim said that the suicide rate varies inversely with the degree of religious, domestic, and political integration. He called this type of suicide egoistic because, when society disintegrates, the individual considers his personal interests above those of the larger society. According to Durkheim, this state of affairs is unhealthy and the individual actually has a need to be integrated into a larger whole.

Altruistic Suicide

Where egoistic suicide is caused by a lack of social integration, Durkheim believed, altruistic suicide is caused by excess social integration. Altruistic suicides are found primarily in less advanced societies characterized by mechanical solidarity: "inferior societies are, pre-eminently, the terrain of altruistic suicide" (1897a:246). In *The Division of Labor* Durkheim had asserted that suicide appears only with civilization and excluded such acts of self-sacrifice as the Hindu widow throwing herself on the funeral pyre as not true suicides "in the vulgar sense of the word" (1902b:226). In *Suicide* he rejected the "vulgar" concept and broadened the definition of suicide in order to include these acts of self-sacrifice (1897a:261–63). Not all altruistic suicides occur in less advanced societies, however, according to Durkheim. The sacrifices of religious martyrs and soldiers also exemplify this type of suicide for Durkheim.

Of all the parts of modern society, he said, the army "is that which best recalls the structure of inferior societies." The empirical data in his chapter on altruistic suicide is then largely concerned with suicides among the military. Among military personnel, the rate of suicide is higher than that among the general population, a fact that Durkheim explained in terms of soldiers being too strongly integrated into the service. He rejected alternative explanations, including bachelorhood, alcoholism, and disgust with military life. These hypotheses, he said, either are limited cases of more general hypotheses that he has already rejected or have test implications that do not agree with the statistical data. Durkheim argued that his theory of excess social integration can explain certain facts that would seem to refute the hypothesis of disgust with military service. The suicide rate, he showed us, increases with years of service and is higher for volunteers, reenlists, non-commissioned officers, and elite troops than for draftees and simple soldiers. These facts seem to indicate for Durkheim that suicide is caused by a strong attachment to the service rather than disgust with it. He believed that his theory could also explain the fact that the military suicide rate varies inversely with the civilian suicide rate and the fact that the military rate was currently in decline. Civilian and military suicides result from opposite causes, he believed. The larger society's strong sense of individualism that is responsible for civilian suicides tends to weaken the military spirit that causes altruistic suicides (1897a:248–59).

Anomic and Fatalistic Suicide

Anomic suicide, according to Durkheim, is caused by insufficient social regulation and fatalistic suicide is caused by excess social regulation. Society regulates individuals, he believed, to the extent that it sets limits to their desires, whether economic or sexual. Durkheim achieved a kind of symmetry in his theory of the causes of suicide. That is, there are two types of social solidarity for Durkheim and an insufficient or excess amount of either type can result in a high suicide rate, yielding a total of four causes and thus four kinds of suicide.

There is a considerable literature, however, objecting that Durkheim did not clearly distinguish social regulation from integration and has thus discovered only one or two causes of suicide.[10] In fact, the relationship between his theory of the social causes of suicide and his theories of social solidarity in *The Division of Labor* is for the most part only implicit. Only the connection between altruistic suicide and mechanical solidarity was made explicit. The connections of the other causes with his theories of social solidarity, however, can be inferred. For example, since he opposed egoistic and altruistic suicide, one can assume that egoistic suicide results from a lack of mechanical solidarity. The connection between anomic suicide and organic solidarity can be inferred from the concept of the anomic division of labor. The anomic division of labor, as we saw in the last chapter, occurs when people performing specialized functions are not in sufficient contact for the relationships characteristic of organic solidarity to develop. These relationships are supposed to regulate these specialized functions in a way that allows for their harmonious cooperation. In other words, the anomic division of labor is unregulated specialization and by analogy anomic suicide is then due to insufficient regulation of the sort characteristic of organic solidarity. The chapter on anomic suicide is largely concerned with comparisons between suicide rates and the incidence of domestic and economic crises because such crises disrupt the social relationships that arise from the sexual and economic division of labor.

Anomic is opposed to fatalistic suicide. The excess regulation associated with fatalistic suicide must then result from too much contact and relationships that are too confining. Durkheim's examples of the fatalistic suicides of very young husbands and childless married women are cases of people who feel too restricted by their relationships with others. So is his example of the suicide of the slave (1897a:311 n. 1). The forced division of labor found in slave societies is an abnormal form according to Durkheim. Only regulation and not integration can occur through force (279). A slave can be highly regulated by society without being highly integrated into it. The very fact that fatalistic suicide received but a footnote and not an entire chapter could be explained in terms of Durkheim's belief that the division of labor and organic solidarity had not yet evolved to the point where many people would be excessively regulated. For Durkheim, the contemporary malaise resulted from the fact that society was in a transitional period in which traditional forms of social solidarity were withering away and new forms were still evolving. Thus egoistic and anomic suicides would be the dominant types in contemporary European society, with egoistic suicide due to the decline of mechanical solidarity and anomic suicide due to the still incipient state of organic solidarity. "Mixed" cases of egoistic-anomic suicide were also possible (324).

Durkheim began his discussion of anomic suicide with the anomie brought about by economic crises. During times of economic crisis, he believed, society fails to exercise its function of imposing limits on the satisfactions of desires. The failure to impose limits leads to a kind of unhappiness that results in an increase in the social rate of suicide (271–81). He rejected the alternative explanation in terms of increased misery due to poverty during economic crises because further test implications of this hypothesis did not accord with the facts. Durkheim showed that prosperity has the same effect as adversity on the suicide rate and that the suicide rate varies geographically with the rate of wealth rather than poverty. That is, the suicide rate is higher where there are more wealthy people (266–71). Poverty actually protects one from suicide, Durkheim argued, because it lowers one's economic expectations (282, 287).

Domestic Anomie

Durkheim explained anomic suicides resulting from domestic crises in a manner analogous to the way in which he accounted for suicides resulting from economic crises. In the case of domestic anomie, however, the suicide victim is tormented by unregulated sexual rather than economic desires.

The fact that the suicide rate varies concomitantly with the divorce rate had already been established by Bertillon and explained in terms of psychopathological causes responsible for both suicide and divorce. Durkheim criticized Bertillon, however, for having failed to provide independent evidence that the incidence of psychopathology varies from country to country in the same way as the rates of suicide and divorce. Durkheim tried to show that his theory of domestic anomie explains a wider variety of the statistical data than the hypothesis of psychopathology. For example, he cited the fact that in countries with a high divorce rate, married men had a higher rate of suicide than unmarried men, while in countries where divorce is outlawed, married men had a lower rate of suicide than bachelors. Durkheim then explained this fact in terms of the effect that the institution of divorce has on the marriage relationship. For men at least, according to Durkheim, marriage serves to regulate sexual passions. Bachelors, on the other hand, he believed, are tormented by the desires aroused by unlimited sexual possibilities. Where divorce is possible, he argued, husbands do not feel the beneficial restraint of marriage and are more like bachelors (292–98, 303–5).

Suicide rates among women, however, present an anomaly for Durkheim's account of anomic suicide. Unlike the suicide rate of husbands, the suicide rates of wives vary *inversely* with the divorce rates. That is, the suicide rate of married women is *lower* in countries where divorce is permitted and women seem to benefit from the institution of divorce (298–302). Overlooking what would appear to today's reader to be the rather obvious explanation of this fact in terms of spousal abuse, Durkheim attempted a rather ad hoc explanation of this anomaly. He asserted that women's sexual needs have less of a mental character than men's and are regulated by their instincts rather than by society (306). A less ad hoc explanation would have been premised on differences in social integration and regulation between men and women.

Durkheim then ended his discussion of sexual anomie by adducing some additional supporting evidence for his theory of anomic suicide. He cited the fact that divorce has a greater effect than mere legal separation on the suicide rate and explained that separation does not grant the same sexual liberty as divorce. Also, he argued that his theory of anomie can explain the fact that the suicide rate among bachelors grows most rapidly during the years of life when sexual activity would be at its peak. Finally, he stated that his theory of anomic suicide can explain some of the residual facts left unaccounted for in his chapters on egoistic suicide, such as the fact that early marriage aggravates the suicidal tendency among men. The passions of young men are so strong, Durkheim argued, that the regulation imposed by marriage appears excessive to them (308–9). In other words, Durkheim seems to have believed that his theory of anomic suicide can explain fatalistic suicide as well.[11]

Suicide as a Social Phenomenon

Durkheim drew his conclusions and explained them in book 3 of *Suicide*. Books 1 and 2, we recall, can be characterized as an extended crucial experiment between two alternative types of explanation of the suicidal tendencies of various societies: this tendency is either a state of society itself or the sum of the states of individual members of society. Durkheim concluded in favor of the first alternative: the social suicidal tendency, he said, is "a distinctive trait of each collective personality" (1897a:364).[12] He explained the collective suicidal tendency in terms of the social forces that arise from collective representations (345–63), as we saw in chapter three.

Durkheim then anticipated and tried to remove an objection to his account of suicidal tendencies. Specifically, he saw that he needed to explain why only certain individuals within a group of individuals who all share the same description will commit suicide. Although he believed that such individual factors as mental illness and alcoholism cannot explain the social suicide rate, he hypothesized that, due to such individual factors, the "mental constitution" of suicide victims may offer less resistance to "suicidogenic currents" (366). In other words, the explanation of an individual case of suicide for Durkheim was not a question of narrowing down the class to which the individual belongs by using only sociological factors. An individual suicide for Durkheim is a psychological fact that requires a psychological explanation. The social suicide rate, on the other hand, is a sociological fact that requires a sociological explanation.

Of course, it was not enough for Durkheim simply to reject individualistic explanations of the social suicidal tendency and to explain and defend his own collectivistic account. Other explanations in terms of collective properties were still possible and Durkheim had to eliminate them as well. Thus we find Durkheim arguing against Adolphe Quetelet's account of suicide rates in terms of his concept of the "average man." The stronger criticism he raised against Quetelet is that Quetelet failed to provide a mechanism by which the average type, which is but an arithmetic mean, can affect individuals. Durkheim also said that because only a tiny minority within any society kill themselves, in computing the average type the tendency toward self-destruction will be overwhelmed by the opposite tendency toward self-preservation (337–42).

Durkheim then attributed to Moritz Drobisch another explanation of suicide rates in terms of collective properties. According to Durkheim's understanding of Drobisch, the constancy of the social suicide rates is due to the stability of the rates of unfortunate circumstances that lead to suicide, such as bad marriages and bankruptcies. Durkheim criticized Drobisch's hypothesis for begging the question, arguing that the unchanging quality of the rates of misfortune would then cry out for an explanation (343–44).

Finally, Durkheim considered the hypothesis put forward by Ferri and Morselli. According to these Italian criminologists, the rate of suicide varies inversely with the rate of homicide because suicide and homicide are but two outlets for the same organic-psychological tendency, with the choice of outlet dependent on social conditions. Durkheim criticized this hypothesis by drawing test implications from it and comparing these implications with the facts. First, he reasoned that if suicide and homicide were expressions of the same organic-psychological tendencies, then the rates of homicide and suicide would vary in the same way with respect to such conditions as sex, age, and time of year. He then showed that the statistics do not bear this prediction out. In a second test implication, he reasoned that if Ferri and Morselli's hypothesis were correct, then one would expect a higher rate of suicide for incarcerated murderers due to the frustration of their preferred outlet for their homicidal tendencies. Although Ferri and Morselli adduced statistics that would seem to support this conclusion, Durkheim criticized their data for failing to distinguish murderers from other convicts. He also cited additional information that contradicts this test implication. Finally, Durkheim argued that if Ferri and Morselli's hypothesis were true, then one would expect the suicide and homicide rates to vary in opposite ways with social conditions. He then showed that this test implication is not borne out for the variations of these rates with respect to historical events, geographical distribution, religion, and family life (386–405).

Theory and Method in *Suicide:* A Summary

As I mentioned toward the beginning of this chapter, the most significant theoretical difference between *Suicide* and *The Division of Labor* concerns Durkheim's conception of the social environment. Durkheim no longer used the notion of social density in his sociological explanations but emphasized social structure instead. He tried to explain the overall rise in European suicide rates in terms of a trend away from one very general type of social structure toward another. Having put aside the notion of social density, *Suicide* is free of much of the ambiguity that plagued *The Division of Labor* and was never misconstrued as offering "materialist" explanations of suicide rates. For this reason, *Suicide* is often regarded as marking a major turning point in Durkheim's thinking. "Individual Representations and Collective Representations" helped to foster the impression that he had moved toward a more idealist conception of sociology.

Durkheim's new emphasis on social structure, however, did not represent a change in his conception of either a sociological explanation or the subject matter of sociology. Far from his theory of the social causes of suicide representing a turning point in his intellectual development, it appears that these causes were derived from his more general theories of social solidarity in *The Division of Labor*. The difference between his theories of social solidarity in his first two major works appears to be largely terminological, with *Suicide* adopting "social integration" and "regulation" as names for the social relations characteristic of "mechanical" and "organic solidarity." Similarly, I have argued, his substitution of "collective representations" for "states of the collective consciousness" was only a terminological change motivated by his desire to avoid misunderstanding. The close theoretical relationship between *Suicide* and *The Division of Labor* is hardly surprising if we assume that he must have been working on the problem of suicide rates while writing *The Division of Labor*. Although *Suicide* was published in 1897, four years after *The Division of Labor*, Durkheim had published an article on suicide as early as 1888 and appears to have been collecting data for *Suicide* even while composing *The Division of Labor*.[13]

Not only did Durkheim's conceptions of sociological explanation and subject matter stay the same, but so, too, did his methods for the comparative evaluation of theories. *Suicide* conforms to a method that for the most part is provided by *The Rules of Sociological Method*, in which competing causal hypotheses are evaluated in terms of their ability to account for empirical relationships of concomitant variations among observable factors. According to this method, from any hypothesis about the causes of suicide one draws the logical implication that the suicide rate will vary concomitantly with this putative cause from society to society. Hypotheses are then eliminated when the facts disagree with the test implication or when another hypothesis provides a better, more precise explanation of the same set of facts. Durkheim tried to persuade the reader to accept his theory of the social causes of suicide largely by arguing that it had greater explanatory scope and precision than the available competitors.

Durkheim's empirical procedures can be characterized in somewhat more detail as follows. First, he offered explanations of the empirical relationships of concomitant variation that his competitors' theories could also explain. Sometimes he tried to show that his theory provided a more precise explanation of the same facts, as in his account of the variation of the suicide rate with the time of year. Sometimes, too, as in his account of the relationship between suicide rates and religion, he argued that his theory can explain both the fact that a generalization holds for the majority of cases and the fact that other cases deviate from this generalization. He then attempted to account for facts that either refute the alternative hypotheses or that these alternatives leave unexplained. These refutations include the empirical absence of relations of concomitance that are implied by the hypotheses in question. By explaining additional concomitances, he implicitly challenged his rivals to come up with explanations of these facts as well.

In our discussion of his reasons for rejecting the hypothesis of hereditary suicidal tendencies, we saw that he laid out three conditions on the empirical verification of hypotheses. First he said that the confirming facts must be sufficient in number so as not to be attributable to accidental circumstances. This condition agrees with the stipulation in *The Rules* that one must appeal not to isolated concomitances but to *series* of concomitant variations. The second condition is that the facts adduced must not allow for some other explanation. I interpreted this condition earlier as stipulating that the facts are not explained by any actual alternative hypothesis that currently finds supporters among one's rivals. This interpretation is consistent with a concern that a concomitant variation between two factors may be explained in terms of a third, as he said in *The Rules*. As such a third factor is again always *logically* possible, Durkheim seems to have meant that there must be no *actual* hypothesis that explains this concomitance in terms of some third factor. In *Suicide,* as we have seen, Durkheim tried to rule out the possibility of alternative explanations by attempting to hold constant the factors that were not under investigation. For example, in order to control for the influence of nonreligious factors such as race on suicide, Durkheim considered the effect of religion on the suicide rate separately for each country.

The third condition on the verification of hypotheses that he stipulated in *Suicide* is that one's hypothesis must not be contradicted by other facts. This last stipulation hardly needed to be defended in his methodological treatise. Durkheim attempted to satisfy this condition by anticipating and explaining what were empirical anomalies for his theory. Sometimes, however, especially in his discussion of the relationship between suicide rates and divorce rates among women, the way in which he attempted to defend his theories against empirical refutation appears to be ad hoc. Although defending one's theory in this manner is not sanctioned by his explicit methodology—nor by that of any other methodological writer for that matter—it is nevertheless a common but discreditable practice among scientists. Thus although Durkheim may deserve Douglas's (1967) charge of engaging in casuistry, to accuse him of being unscientific on this basis is problematic. Casuistry is the art of reasoning about ambiguous cases and usually involves showing that a case is an exception to a rule because it falls under another rule. In science, casuistic argument consists in showing that an apparent empirical refutation of one's theory does not in fact falsify it because this fact can be explained by some other part of one's theory. Only when a scientist explains away exceptions in terms that are not a part of his or her theory is he or she engaging in casuistry in the pejorative sense.

There are other ways in which the procedures Durkheim used in *Suicide* are in agreement with his explicit methodology. As we saw in chapter five, Durkheim stressed the need to clarify the meaning of terms in order to determine the scope of the hypotheses in which they occur. In *Suicide* Durkheim analyzed the meanings of the terms used in hypotheses before he subjected these hypotheses to empirical test. Sometimes Durkheim criticized the empirical data adduced in support of his rivals' hypotheses, as when he faulted Ferri and Morselli for having failed to distinguish convicted murderers from other prisoners and to separate the retarded from the mentally ill. Although criticizing data for its ambiguity is not a practice prescribed in *The Rules,* it does not however appear to violate any of his methodological stipulations.

One way in which Durkheim appears to have departed from the explicit methodology of *The Rules* is by having rejected hypotheses not only on empirical but on methodological and conceptual grounds. Sometimes his reasons had to do with the concepts of the methods and explanatory goals of sociology that he explicated in *The Rules.* I showed this to be the case with his rejection of Tarde's program of introspective research into the laws of imitation as well as with Durkheim's criticism of Quetelet for having failed to provide a causal mechanism. Durkheim also appealed to these basic assumptions about goals and methods in defense of his procedure for classifying suicides into types in accordance with his theory of their causes.

The nonempirical reasons that Durkheim provided for rejecting hypotheses, however, were not always grounded in his philosophy of sociology. Among the conceptual reasons he gave were that the theories of his rivals suffered from problems of ambiguity, inconsistency, and question begging. Some of these arguments would seem to be unexceptionable, such as his criticism of the insanity hypothesis for being conceptually incoherent and his criticism of Drobisch's account of the stability of suicide rates for begging the question. Durkheim apparently believed that the norms of the academic and scientific profession justified his rejection of theories on such grounds. More surprising is his argument that the monomania hypothesis need not be taken seriously because it belonged to a theory of faculty psychology that had few defenders. That Durkheim would have used such an argument would seem to indicate a belief that his readers shared a perception of academic research as divided into competing communities, with each community organized around a set of theoretical assumptions.

Durkheim's overall strategy in *Suicide* of eliminating alternative hypotheses then conforms with what he said in *The Rules* about the need for crucial experiments. However, *Suicide* has been criticized on two grounds. First, his critics have charged that the alternative hypotheses he considered are not exhaustive and that there may be other possible explanations of the social suicide rates that Durkheim has not taken into account. One can claim to have proven one's hypothesis through the method of elimination, this objection runs, only when one has eliminated every hypothesis that is logically possible. Second, Durkheim's critics have charged that the alternatives that he eliminated may not be mutually exclusive. That is, the social suicide rates may be due not to one but to a combination of factors.[14]

Stephen Turner (1984:59–60; 1986:132) defends Durkheim against the accusation that his eliminative arguments do not exhaust the alternatives. To raise this objection against Durkheim, he argues, is to fail to understand that his goal was not so much to achieve certainty as to persuade his readers of the relative strength of his sociological research program. He was not trying to demonstrate that he had the only possible explanation of suicide rates but only trying to convince his readers that his explanation was the best one available. For this purpose, he did not need to eliminate every logically possible explanation other than his own, but only those explanations that were actually put forth by serious rivals. In other words, Durkheim used eliminative arguments in *Suicide* for polemical purposes as much as for purposes of proof. Ultimately, of course, a scientific theory must stand on its own evidence and not just the evidence that is used to eliminate its competitors. Scientific research, however, as Durkheim recognized, is a collective enterprise. In order to obtain further evidence to support his sociological theories, he thought, it is important to attract others to one's program of research. However, Durkheim did not entirely succeed in his endeavor to convince his readers of the explanatory power of his sociological theories. As I indicated in earlier chapters, not even his collaborators on *L'Année sociologique* were impressed with *Suicide*. Simiand dismissed Durkheim's realist conception of social forces and Lapie thought that Durkheim's methods of statistical analysis were rapidly becoming obsolete. Although one of his collaborators, Maurice Halbwachs, took up the study of suicide, he was also critical of Durkheim's methods and theories in *Suicide*.

The objection that the alternatives Durkheim eliminated may not be mutually exclusive has been phrased in different ways. Pope (1976:185) and Gustavo Tosti (1898:474–75) have complained that Durkheim did not consider the possibility of "interaction" among nonsocial factors. Hanan Selvin (1976:41) is more precise. He says that each of the putative causes Durkheim eliminated may have made some statistical contribution to the suicide rate. That is, the statistical correlation between the suicide rate and each of these factors is not zero. However, as Turner (1986:134–35) argues, to raise this objection is to presuppose an entirely different concept of causality than Durkheim's. A cause is not a mere statistical correlation for Durkheim, but, as we saw in chapter 4, a necessary and sufficient condition for its effect in the normal state of affairs. Turner points out that Durkheim was using statistics not to establish a probabilistic relationship between causes and their effects but to measure the strength of "suicidogenic currents."

The notion of explanation by statistical correlation was totally foreign to the philosophical tradition in which Durkheim was educated and within which he sought to legitimize sociology. To point this out, however, is only to raise the question as to why Durkheim should have been satisfied with a traditional concept of explanation. As Selvin points out, in the years 1895–99 Pearson and Yule had taken up and refined Galton's work on the coefficient of correlation and regression analysis. Meanwhile, Durkheim continued to appeal to concomitant variations that were not of continuous quantities and did not attempt to apply even Quetelet's methods of error analysis. Durkheim seems to have associated Quetelet's concept of a normal distribution with his practice of explaining social phenomena in terms of the mathematical idealization of an average type, a practice that did not conform to Durkheim's conception of a causal explanation in sociology.

Selvin then asks why neither Durkheim "nor any of his disciples ever came to be aware of the defects in the methodology of his quantitative analysis" or even to "acknowledge the existence of these analytic techniques that would have been so useful" to them. He proposes to look at the conditions under which these new statistical techniques "would have diffused" and then asks "what went wrong" (1976:41–44). To answer his own question Selvin provides an analysis of what he takes to be four requirements for the intellectual diffusion of an idea or method to take place. These requirements include (1) a channel or path of communication, which he believes was present in this case; (2) comprehensibility on the part of the potential adopters, which he argues was not a problem for the Durkheimians; (3) a social structure that would support and reward those who adopt it; and (4) what he calls value consonance (1976:45–47). By the latter he means that "the potential adopters must perceive the message as legitimate and useful" (45). Overlooking the possibility of a close relationship between (3) and (4), Selvin follows the externalist fashion and attributes Durkheim's "failure" to adopt these new statistical methods largely to the lack of a supportive social structure. Selvin concludes with the admonishment that Durkheim should have been more open to new procedures (49).

There is a rather obvious presentist bias to the very question Selvin asks. He assumes that it was some sort of *mistake* for Durkheim not to have adopted the new statistical methods because they are in general use among social scientists today. More recent sociologists of scientific knowledge would no doubt criticize Selvin for trying to find external causes only for what he believes to have been an "error" on the part of Durkheim. The sociology of scientific knowledge, they would say, should not be limited to explaining scientists' mistakes.

I would make the somewhat different argument that Durkheim's decision not to adopt the methods of Pearson, Yule, Galton, or Quetelet was not necessarily a mistake. What sorts of things count as rational and what are mistakes depend upon the cognitive norms and values to which a scientist subscribes. Selvin never makes clear that the methods of Galton, Pearson, and Yule would have been "useful" to Durkheim given the explanatory goals of his research program, especially his concept of a cause. With regard to the question of the consonance of Durkheim's values with the statistical methodology of Pearson and Yule, Selvin does concede that "there is a bit of solid ground in what is known of his desire to make sociology a discipline independent of all others, especially mathematics. Hence, it seems unlikely that he would have even considered seriously any claim that Yule's statistical formulas might have helped him" (1976:48). However, Selvin is confusing substantive with methodological issues as well as empirical with nonempirical sciences. For Durkheim, sociology was supposed to be independent only of other empirical sciences and only in terms of its substantive theories. The serious problems of value consonance lay elsewhere. Pearson, for example, was an extreme nominalist for whom there could be no knowledge of causes. All we could hope to know, Pearson believed, were correlations among sense impressions (Porter 1986:305; Stigler 1986:340). Donald MacKenzie (1978, 1981) points out that Pearson's and Yule's conceptions of statistical methodology were not all that compatible with each other. Thus Selvin should have addressed the issue of value consonance between Durkheim's sociology and the new statistical methodologies separately for each.

Nevertheless, Durkheim's misfortune to publish *Suicide* at a time when the field of statistics was undergoing some significant development did present him with a dilemma. On the one hand, he could not continue to use his dated statistical methods without eventually making his research appear unoriginal and irrelevant. On the other hand, it was not clear that learning the new methods would help him to achieve his goals of discovering the laws governing collective ideation. The statistical methods of *Suicide*, after all, allowed him to demonstrate only the strength of the social forces arising from the collective representations associated with the different types of social structure. They did not help him to find a way to relate specific collective representations to specific social structures, as he did in his later works. His turn to ethnographic studies, far from indicating a new conception of sociology, indicates a decision to study primitive societies in order to develop his ideas about the relationship between collective representations and the structure of the social environment.

Subsequent Theoretical Developments in Durkheim's Sociology

In the year following the publication of *Suicide,* Durkheim published the first of several essays in *L'Année sociologique* that drew upon ethnographic data to defend hypotheses in the sociology of knowledge and the sociology of religion. These included his papers on incest (1898a(ii)) and totemism (1902a(i)), and the paper on primitive classifications written with Marcel Mauss (1903a(i)). In these papers Durkheim attempted to explain the beliefs and practices of a society in terms its social structure. The prohibition against incest among Australian tribesmen, for example, is explained in terms of the way in which their tribes are divided into clans, phratries, and matrimonial classes. Durkheim said that "one cannot comprehend the meaning and the scope of this system of regulation if one does not know the manner in which these classes are composed" (1898a(ii):47).

Durkheim's explanatory appeal to social structure has misled those of his critics who interpret Durkheim as identifying this social structure with some sort of physical arrangement of people. These critics overlook the fact that Durkheim conceived the structure of society as consisting of social relations that depend for their existence upon mental representations. Complaining that he was still being misinterpreted as a materialist (1898b:49), Durkheim attempted to clarify his concept of the relationship between social structure and collective representations in a paper published in the *Revue de métaphysique et de morale* in 1898. In this paper, "Individual Representations and Collective Representations," he said:

Society has for its substratum the whole of associated individuals. The system that they form through uniting and that varies according to their disposition over the surface of the territory, the nature and the number of the means of communication, constitutes the base upon which social life is raised. The representations that are the web of it are disengaged from the relations that are established among the individuals thus combined or among the secondary groups that are intercalated between the individual and the total society. (38–39)

The relationship between these representations and their substratum, he went on to explain, is like that between a function and its organ: the former is both dependent on and distinct from the latter (45). He tried to clarify this point further by drawing an analogy between the case of individual representations in psychology on the one hand and that of collective representations in sociology on the other. For Durkheim, individual representations are not the mere epiphenomena of neural states but form a natural kind distinct from neural states. He further distinguished individual representations into two kinds: sensations and images. Sensations, he asserted, are the result of a kind of chemical synthesis among the neural states. This synthesis gives rise to new properties that cannot be reduced to physiological phenomena. Mental images in turn result from the synthesis of these sensations (14–16, 46–47).[15] These representations, he affirmed, exist in a memory that is purely mental (17–23). Concepts or collective representations, he argued, are formed in turn from the synthesis of mental images and bear the same relationship to their substratum that individual representations bear to theirs (14, 47).[16] In other words, I take it Durkheim was trying to say, just as one's mental life cannot be explained in terms of neurophysiology alone, social life cannot be explained in terms of its substratum of individual human beings associated in groups.

Durkheim then proceeded to distinguish between two kinds of collective representations: those that arise immediately from the social substratum of associated individuals and those that are subsequently formed from other collective representations. From what he said in the passage quoted below, it appears that only the first group of collective representations represents the structure or morphology of the society. According to Durkheim,

the forms that [the collective life] takes on at the moment that it arises from [the collective substratum] . . . bear the mark of their origin. This is why the prime matter of every social consciousness is in close relation with the number of social elements, the manner in which they are grouped and distributed, etc., that is to say, with the nature of the substratum. But once a first foundation of representations is thus constituted, they become, for the reasons that we have stated, partially autonomous realities that live a life of their own. They have the power to attract and repel each other, to form among themselves syntheses of all kinds, that are determined by their natural affinities and not by the state of the environment in the midst of which they evolve. Consequently, the new representations, which are the product of these syntheses, are of one and the same nature: they have for proximate causes other collective representations, not such and such a character of the social structure. (45–46)[17]

To say that these new collective representations are not caused by the structure of society would seem to imply that the older ones are caused by the social structure. However, if Durkheim also said that the older representations result from individual representations, it would seem that for Durkheim the social structure would exist first in individual representations. I think that what Durkheim meant to say, however, is that the first type of collective representation forms through the synthesis of individual representations as individuals form social relations with one another. These social relations and thus the structure of the society exist in this primary category of collective representation.

"Individual Representations and Collective Representations" does not mark any major break with his original concept of a sociological explanation, as some of his critics have argued.[18] In fact, far from indicating a turning point in his career, this essay simply explicates a fairly traditional theory of mental states that Durkheim appears to have held all along. As Charles Elmer Gehlke has already pointed out, Durkheim has simply adopted what William James referred to as the "mind-stuff theory" of the compounding of mental states and carried it over to sociology. The idea that sensations combine to form images and images combine to form concepts, Gehlke said, can be found in the psychological theories of Spencer, Tyndall, and Wundt (1915:19–25, 31, 94).[19] All Durkheim did was to identify concepts with collective representations and then allow these collective representations to combine to give rise to yet further collective representations.

This distinction between two kinds of collective representations allowed Durkheim to show how individual representations could be necessary for the existence of collective representations without being sufficient to explain them. As Gehlke explained, individual representations must be included among the necessary conditions of the type of collective representations that are formed first (1915:67–68). Although individual representations are thus indirectly necessary for the second type of collective representations as well, they are not sufficient to explain the formation of this second type, which depends upon the emergent properties of collective representations. Durkheim thus clarified his distinction between psychological and sociological explanations and his goal of finding genuinely sociological laws governing collective representations.

Durkheim's essay of 1898, however, left several problems unsolved. By his own admission he could provide no account of how collective representations give rise to other collective representations. "There is a whole part of sociology that should investigate the laws of collective ideation and that still entirely remains to be done," he said (1898b:47 n. 1). Nor did he shed much light even on the problem of how collective representations form from individual representations. For example, as I mentioned above, Durkheim assumed in *The Division of Labor* that the resemblance of two representations will cause their fusion. Instead of explaining in the essay of 1898 how resemblance leads to fusion, he argued merely that it would not be impossible for resemblance to be a cause of the association of ideas (28–32).

Three years later, in the preface to the second edition of *The Rules of Socio-*

logical Method, Durkheim conceded that the laws of collective ideation were still unknown. He now said that the distinction between individual and collective representations does not necessarily rule out the possibility that the laws of collective ideation may "resemble" the laws of the association of ideas. He admitted to a lack of sufficient knowledge at the time of this writing to determine whether or not in fact these two kinds of laws do resemble each other (1901c:xvii–xviii). Durkheim thus had no explanatory resources other than the so-called laws of the association of ideas, which did not themselves rest on a secure foundation.

Lacking a knowledge of the laws of collective ideation, it seems that at best Durkheim could explain only those collective representations that are formed from individual representations. Durkheim believed this kind of collective representation constitutes the collective consciousness of primitive societies, while the collective consciousness of more advanced societies also contains collective representations formed from other collective representations. He was thus unable to explain the formation of many of the collective representations that are characteristic of contemporary societies. He could not use the principles of associationist psychology to explain the formation of the second type of collective representation without violating his own rules concerning sociological explanations. If he had been in possession of the laws of collective ideation, he might have been able to provide social causes for the deterioration of the collective consciousness that he believed was making the division of labor possible and giving rise to higher rates of suicide.

Durkheim's shift to a concern with primitive societies then does not reflect a *new* interest in collective representations, as others have argued. He had always assumed that sociology was the study of collective representations. Rather, I think that he turned to the study of primitive societies in order to try to solve the theoretical difficulties to which the concept of collective representations gave rise. Durkheim's last major research article that does not deal almost exclusively with ethnographic data, "Two Laws of Penal Evolution," was published in the same year as the new preface to *The Rules of Sociological Method* (1901a(i)).

Durkheim's lack of sociological laws governing the collective representations of modern society, in conjunction with his reasons for deciding to specialize in the sociology of knowledge discussed in chapter 2, led him to concentrate on the problem of providing accounts of primitive categories of thought. His hypothesis concerning the relation between collective representations and social structure was subsequently developed into a full-blown sociology of knowledge that made its first appearance in Durkheim's and Mauss's well-known paper on primitive classifications. In this paper, the authors argued that "the classification of things reproduces the classification of men" (1903a(i):402). In *The Elementary Forms of the Religious Life* (1912a) Durkheim generalized this thesis to include the primitive categories of space, time, causality, class or genus, and so on, which he explained in terms of the social life of primitive societies. This thesis, which he continued to defend throughout the remainder of his career (for example, 1913a(ii)8: 680; 1914a:331), will be discussed in greater detail in the following chapter.

8 The Elementary Forms of the Religious Life

For Durkheim, the problem for a science of religion is to discover that part of reality that is causally responsible for religious beliefs. The hypothesis of *The Elementary Forms of the Religious Life* (1912a) is that what people take to be their religious experiences are in fact caused by an underlying, unobservable reality of social or collective forces. In the conclusion to this work Durkheim illustrated this hypothesis with an analogy from physics. To say that our experiences of light and heat have a causal basis in an independently existing reality, he pointed out, is not to say that this independent reality resembles the way it appears to the senses (1912a:597). Our subjective experience of colors, for example, exists objectively as different wavelengths of light. Similarly, he said, it does not follow that the reality behind the religious experience is that which believers claim it to be. There is no reason to believe, he argued, that people's religious experiences inform them about the causes of these experiences any better than our sensations inform us about the objects of nature (596–98).[1] Durkheim recommended that one proceed in the science of religion in the same way that one does in the physical sciences, namely, by substituting a "scientific and conceptual representation" that accounts for the representations of sense. He claimed that by following this method, he has seen that "this reality, which mythologies have represented under so many different forms, but which is the objective, universal, and eternal cause of these *sui generis* sensations of which religious experience is made, is society" (597).

A society, for Durkheim, exists in the collective representations in the minds of its members. Thus when a believer feels powerful religious forces within him or herself, according to Durkheim what he or she is really experiencing is the effect of these collective representations on the mind. In subsequent defenses of *The Elementary Forms* he complained of being misunderstood and tried to explain this thesis in terms of what he called the "duality" of human nature (1913b; 1914a). For Durkheim the traditional belief in a distinction between body and soul is not purely illusory but grounded in the distinction between "two groups of states of consciousness," one relating to the senses and the other relating to society (1914a:330). What the believer takes to be his or her soul, for Durkheim, is nothing more than the part of oneself that one owes to society. Thus even such beliefs as a belief in a god or a soul are not mere social constructions but have causes in an underlying social reality.

In *The Elementary Forms* Durkheim not only continued to adhere to the same conceptions of sociological explanation and subject matter, but employed many of the same methods of persuasion and comparative theory evaluation that he used and defended elsewhere. For the most part, Durkheim's arguments in *The Elementary Forms* are in agreement with the explicit methods of *The Rules of Sociological Method*. As in his earlier works, some of his arguments are rooted in his conception of sociological explanation rather than in his methodology. He began his argument with an analysis of the concept of religion in order to avoid being misled by common sense notions of religion and to ensure that the scope of this concept corresponded with that of a natural kind. He then formulated a provisional definition of religion in terms of what he took to be its "observable" characteristics: a religion is a shared set of beliefs and practices relative to maintaining a distinction between the sacred and the profane. The remainder of the work consists of an extended crucial experiment or series of crucial experiments between sociological and psychological accounts of religious beliefs and practices. He argued that only his collectivist, sociological account of religion explains the causal origin of the distinction between the sacred and the profane. According to Durkheim, believers regard those things that are known through individual representations alone as profane and those things that are invested with the power of collective representations as sacred. This conclusion can be reached most easily, he believed, through the analysis of the most primitive religions. Theories of the origin of religion that are grounded in individual psychology, he argued, cannot account for the distinction between the sacred and the profane.

Durkheim was actually struggling against two major traditions in the study of religion. Both of these traditions drew on the associationist tradition in individual psychology for their arguments. One was the ethnographic tradition that included people like James Frazer and Edward Tylor. Durkheim associated this tradition with the animistic hypothesis that religion began with a belief in the human soul and the subsequent attribution of anthropomorphic souls to things in nature. He also opposed the tradition associated with the naturistic hypothesis, which he credited to Max Müller, that religion owes its origins to the experience of fear and awe in the face of what were believed to be the supernatural forces responsible for unexpected events in nature. Durkheim's own collectivist account maintains that totemism with its system of taboos regarding sacred objects was the earliest form of religion. His crucial test between individualistic and collectivistic accounts of the causes or origins of religion therefore becomes a three-way battle among the theories that totemism, animism, and naturism, respectively, were the earliest forms of religion.

Durkheim's Goals: The Sociology of Religion and the Sociology of Knowledge

Durkheim began *The Elementary Forms* with a statement of his explanatory goals. As he clarified in his analytical table of contents, this work actually has two objects (1912:639). Its principal purpose is to *analyze* and to *explicate* the most primitive, simple religion, not for its own sake, as a historian or ethnographer would study it, but with a sociologist's interest in explaining the present (1912a:1–2). Its secondary purpose is to investigate the origins of the fundamental notions of thought or categories that are "at the root of our judgments" (12).

The two purposes of *The Elementary Forms* were not unrelated for Durkheim. Defining religion as a system of beliefs and practices that maintain a distinction between the sacred and profane, he made the question of the origin of religious belief a problem in the sociology of knowledge. The sacred is distinguished from the profane not merely by occupying a higher place in some sort of hierarchy, he argued, but what distinguishes them is their radical heterogeneity. The sacred and the profane are two mutually exclusive categories that are completely separated from each other by a system of interdictions. If one believes that sacred beings are distinguished from others only by the strength of their powers, he reasoned, the problem for a theory of the origin of religion would then be to explain where people acquired the idea of this force. But if, on the other hand, the sacred and the profane are of entirely different natures, he argued, the problem of the origin of religious belief then becomes one of the origin of this idea of separating things into radically different classes (51–58). Durkheim then sought the origin of religion in the origin of our ability to systematize or to classify. He explained the origin of this ability through his account of the genesis of the categories of genus and species. These categories, he believed, owe their origin to the totemic division of the tribe by phratries and clans.

Durkheim's evaluations of competing theories of the origin of religion turn on questions in the sociology of knowledge. Defenders of the naturistic hypothesis who say that religious belief began with the concept of powerful supernatural forces are obligated to explain the origins of the category of cause or force, he thought. Likewise, proponents of the animistic hypothesis who claim that religious belief began with the concept of a soul are required to account for the genesis of this concept. Durkheim thus located the field of battle for competing theories of the origin of religion within the domain of the sociology of knowledge. For Durkheim, a theory of religion must account for the origins of the ideas of sacred beings, religious forces, and radical heterogeneity. His strategy was to argue that only his hypothesis that totemism was the earliest religion could explain the origins of all of these ideas.

Durkheim framed the basic problem for the sociology of religion in nothing less than the terms of traditional epistemology: the sociology of religion concerns the sources of *ideas* that Durkheim believed are not met with in sense experience. At the same time, Durkheim thought that his theory of the categories would renew the theory of knowledge by providing a sociological point of view (12). He criticized what he took to be the two most important philosophical doctrines regarding the categories: a priorism and empiricism.

According to Durkheim, the empiricist theory says that the individual constructs the categories for him or herself from elements taken from his or her sensations. Durkheim rejected this theory for its failure to account for the universality, generality, and necessity of the categories. For Durkheim, the generality of a concept is a question of its extension while the universality of a concept is a question of its universalizability or communicability. That is, a concept is universal because it is not private and can be held in common with others (619 n. 1). Subjective mental states, including sensations, perceptions, and images, on the other hand, cannot be communicated and thus, he believed, cannot explain the origin of the categories (618–9; cf. 1914a:317; 1955a:204). That is, Durkheim criticized empiricism because it assumes the existence of only subjective mental states and is thus unable to account for the intersubjective or social nature of the categories. Nor can empiricism account for the generality of our categories, he thought. It was not plausible for him that the individual could construct such general concepts from his or her sensations, which are always of particulars. The extension of these general concepts goes far beyond the personal experience of any individual, suggesting to Durkheim that they are the work of the collectivity (1912a:620–21). Finally, he argued that empiricism cannot account for the necessity with which the categories impose themselves on our thought. Although he conceded that sensations are also imposed on us "in fact," he nevertheless granted us "by right" a kind of liberty with respect to the way we *conceive* sensations, even with regard to the order in which they are produced (19). Durkheim developed the argument from generality in his account of the categories of genus and species and the argument concerning necessity in his account of the category of causality. The idea of the communicability of our categories is implicit in his account of the origin of the concept of the soul.

Although the bulk of his arguments outside the introduction and conclusion to *The Elementary Forms* seem to be directed against empiricist theories of the origin of the categories, Durkheim was not happy with a priorist accounts either. To say that the categories are "inherent in the nature of human intelligence" is not to explain the power of the mind to transcend experience and to add things to it, Durkheim argued: "It is still necessary to give some indication of whence we hold this surprising prerogative and how we are able to see in things relationships that the spectacle of things cannot reveal to us" (20). To say that the categories make experience possible, he continued, is only to beg the question of why experience should depend on conditions that are external and antecedent to it. He argued that to propose that they "are necessary because they are indispensable to the functioning of thought" is merely to offer a tautology. He rejected as unscientific the hypothesis that the categories are an emanation of divine reason. To say that they are innate, Durkheim said, is merely to state in a positive fashion that they do not come from experience. Furthermore, the a priori philosophy cannot account for the fact that the basic categories of thought vary according to time and place (20–23).

One can avoid the difficulties faced by both empiricism and a priorism once one admits the social origin of the categories, he believed. For Durkheim, the sociological theory of knowledge is a type of rationalism that occupies an intermediate position between empiricism and a priorism and has the advantages of both without their disadvantages (27, 26 n. 2). He believed his hypothesis could preserve the "fundamental principle of *a priorism* [that] knowledge is formed from two sorts of elements that are irreducible to one another," that is, empirical and conceptual elements. The empirical elements, he said, are "individual states, that are explained entirely by the psychic nature of the individual" (21–22). Concepts and categories, on the other hand, cannot be explained in this way. "If, as we think, the categories are essentially collective representations, they translate above all states of the collectivity: they depend upon the manner in which the latter is constituted and organized, on its morphology, on its religious, moral, economic institutions, etc." (22). Thus, he affirmed, between the empirical and conceptual elements of our experience "there is all the distance that separates the individual from the social, and one can no more derive the second from the first than one can deduce society from the individual, the whole from the part, the complex from the simple" (22).

The fact that the categories are social in origin, Durkheim thought, can explain both the fact that they are communicable and the fact that their extension goes beyond the confines of any individual's experience (619–20). Their social origin will also explain the necessity with which they impose themselves on our thought (23). Durkheim explained the ability of the categories to impose themselves on our thought in terms of the power of social forces, as we will see below through an analysis of his account of the origin of the category of causality.

The fact that Durkheim included space and time among the categories provides evidence that he was not working with Kant's concept of a category. He distinguished his concept of the category of space from Kant's concept of space as a form of intuition, adopting Octave Hamelin's critique of Kant: "As Hamelin has demonstrated, space is not this vague and indeterminate milieu that Kant had imagined: purely and absolutely homogeneous, it would be of no use and would not even offer anything for thought to grasp" (15). Space must be differentiated in order for one to be able to locate sensations in it. Similarly, Durkheim argued that one cannot represent the notion of time to oneself in abstraction from ways of measuring it. Among his reasons for believing that the categories are of social origin are the facts that the units in which time is measured and the differentiations of space are taken from social life (14–15). As evidence for the latter claim he adduced Australian and Amerindian tribes for whom the number of spatial directions is taken from the number of clans in the tribe. The names these tribes give to the directions of space are taken from the names of the clans that occupy the corresponding situations at tribal gatherings (15–17; cf. 1903a(i):425–45).

For Durkheim the categories are not conditions of possible experience, as they were for Kant, but somehow in things themselves. Elsewhere he tried to clarify the difference between his concept of a category and Kant's. For the Kantians, he said, "the categories preform the real, whereas for us, they recapitulate it. According to them, they are the natural law of thought; for us, they are the product of human artifice" (1909d:757 n. 1). Far from the categories being conditions of possible experience for Durkheim, he believed that experience is possible without them. The relations that the categories express exist implicitly in individual consciousnesses, he said. The individual has a sense of time, place, resemblances, regular succession, and so on that does not depend on the categories (1912a:628–29). An individual has no need of the categories of space and time in order to orient himself and to satisfy his personal needs, as even animals can do this. Nor do we need the category of causality in order to escape danger. Similarly, he contended, the classification of things by genus and species is not necessary for the ability to recognize relationships of resemblance (632).

Durkheim's concept of a category resembles Aristotle's concept of a predicable or type of predicate more than it resembles Kant's concept of a condition of possible experience. In an autobiographical account given toward the end of his career, Marcel Mauss asserted that the Durkheimian school had been attempting to provide sociological accounts of Aristotle's, not Kant's, categories. They regarded even Aristotle's list of ten categories as nothing more than a starting point, leaving the total number of such concepts an open question (Mauss 1938:1).[2] Unlike Kant, furthermore, Durkheim believed that the categories vary with time and place and that as the categories change over time they gradually come closer to expressing relations that exist in things themselves: Durkheim was a realist, not a transcendental idealist.

Durkheim's Methods

Durkheim's appeal to primitive religion in order to understand our present modes of thought is an instance of his method of historical analysis. For Durkheim, our concepts have their origin in religious beliefs, which in turn are to be explained in terms of social causes. An analysis of primitive religion, he believed, will reveal that the distinction between the sacred and the profane can be explained in terms of the differences between collective and individual representations.

Durkheim's method of historical analysis, as I explained in chapter 5, assumes that societies can be classified according to type in serial fashion, beginning with the most simple and progressing according to their degree of composition. The degree to which societies are compounded from social subgroups is also supposed to reflect their order of historical development, with the earliest societies being wholly undivided into parts. What is assumed to be true of societies as a whole is also supposed to be true of their institutions. Thus Durkheim believed that we cannot understand the most recent religions without following through history the way in which they have become progressively *composée* (1912a:4). The essential characteristics of religion or of any other social institution are best revealed through a study of their most primitive form, he thought, as they acquire encumbering secondary properties through successive stages of the composition of society (6–7, 11). Primitive societies thereby serve as "privileged cases" (8), that is, cases for study that are analogous to the simplified phenomena produced in the physical laboratory but which are spontaneously produced by nature at the beginning of history (11).

In constructing series of progressively more complicated social institutions, Durkheim cautioned, we must choose the first link in the chain with care, as it will have an important effect on the rest of the explication. Religious facts will be explained differently according to whether we choose as our hypothesis that naturism, animism, or some other belief system was the earliest religion (5). In order to find the truly original form of religious life from which all the others derive, he argued, we must even descend by analysis below "observable" religions, as the religions known through history and ethnography are already very complex (67). The general claims that Durkheim made about totemic religion therefore cannot be refuted simply by contesting the ethnographic reports on which we relied.[3] He could always dismiss as later developments cases of totemic societies that, say, have a larger number of clans than of spatial directions or that have totems only for clans and not for phratries. He also cautioned the reader that he used terms like *primitive* and *origins* in only a relative and not an absolute sense (11 n. 1).

Durkheim justified his decision to study Australian tribes by arguing that their clan structure is the simplest known type of social organization (136). He referred to *The Division of Labor* where he described this type of social organization as segmentary: characterized by the repetition of identical units in a way analogous to the structure of annelid worms (1902b:150). He defended his comparisons of Australians with North American Indians by arguing that in order to understand an institution it is advisable to follow it through successive stages of development. North American Indians, he asserted, represent but a different variety of the same social type as Australians (136–38).

In justifying his focus on Australian totemism, he argued that because the value of facts is more important than their number, it is less useful to accumulate experiments than to have significant well-made experiments (134–35). Anticipating the objection that he has made an induction from the study of a single case, he argued in the conclusion that when a law has been proved by a well-made experiment, the proof is universally valid. Because it is not conceivable that the same effect can result from different causes, he reasoned, if religion can be shown to be due to social causes in one case then this claim will be true for all cases. An induction based upon a "well-defined" experiment, he believed, is less risky than a summary generalization that tries to get at the essence of religion without resting upon the analysis of any religion in particular (593–94). Durkheim had offered a similar defense of "well-made experiments" earlier in the book, as part of his argument that the totemistic hypothesis provides a better account than the animistic hypothesis of Arunta beliefs about conception (352–53).

Durkheim associated the notion of a well-made experiment with the concentration of one's research on a single social type. Social facts cannot be understood in isolation from the social system of which they form a part, he argued, and thus we cannot usefully compare social facts unless they are from societies of the same social types or species. If one is not careful, he cautioned, one may compare facts that do not have the same meaning and scope ["*ni la même sens ni la même portée*" (133)]. In order to compare social facts, he argued, it is not sufficient that they resemble one another. The societies from which these facts come must resemble one another as well. Durkheim seems to have been concerned that if we compare cases that disagree with respect to a number of different causal antecedents, it is not at all clear to which of these antecedents we can attribute each of the effects. By restricting our comparisons to societies of the same social type, we can limit the amount of variation among the causal antecedents. With such cases for comparison, one can attribute similarities in their effects to similarities in their causal antecedents and any differences in their effects to differences in their antecedents.

What Durkheim appears to have meant by a well-made experiment, then, is an experiment that tests a causal hypothesis in accordance with Mill's canons of induction, especially his methods of agreement and difference. As I mentioned in chapter 5, the methods of agreement and difference capture the logic of crucial experiments for Durkheim and Mill, providing a way to adjudicate among competing causal hypotheses. In *The Elementary Forms*, I think, Durkheim was using the term *well-made experiment* as a name for crucial facts or experiments, which he endorsed in *The Rules*. When he characterized his study of Australian totemism as a well-made experiment, he could then be interpreted as having put it forth as a crucial test between the totemistic hypothesis on the one hand and the animistic and naturistic hypotheses on the other. All three hypotheses must account for this Australian religion successfully. The two hypotheses that maintain that the original form of religion is something other than totemism must nevertheless account for the appearance of totemism. As I have shown in my analysis of his earlier works, Durkheim's concept of a crucial test allows for the criticism of theories for their conceptual as well as empirical inadequacies. Showing that one of these competitors has begged the question or made an ad hoc assumption in accounting for the facts, for example, will prove to be just as decisive for Durkheim as showing that his hypothesis can explain a fact that a competitor cannot.

Durkheim's Definition of Religion

After explaining the goals that he will pursue and the methods that he will use in *The Elementary Forms,* Durkheim formulated a preliminary definition of the type of phenomenon under study. In accordance with the methodology of *The Rules,* he expressed a need to define religion in terms of easily perceptible external signs that would allow one to recognize religious phenomena wherever they are met with and not to confuse them with others. In formulating such a definition, he believed, we must free our minds of preconceived ideas, prenotions, and prejudices. These socially constructed concepts for Durkheim, as I explained in chapter 5, do not necessarily correspond with real kinds. Our immediate goal should be not to capture the common-sense notion of religion but to find the observable characteristics that allow us to distinguish religious from other kinds of social phenomena. Without such a definition, he feared, we are liable to make two kinds of errors. The first type of mistake would be to regard as a religion a system of ideas and practices that is not of a religious character. The second and apparently more important type of error for Durkheim would be to overlook religious facts without perceiving their religious character.

Durkheim, as I argued earlier, treated definitions like empirical hypotheses, comparing them with regard to their empirical scope. Definitions, for Durkheim, are supposed to circumscribe a class of phenomena for a theory to explain. Ultimately, he believed, we must try to discover the underlying essence that will allow us to account for the observable characteristics in terms of which we have defined this kind of social phenomena (1912a:31–32). Thus the greater the scope of the definition, the greater will be the scope of the theory. When Durkheim criticized his competitor's definitions of religion for being insufficiently inclusive he was thereby criticizing their theories of religion for lacking empirical scope. He never faulted anyone for including too many systems of belief under the concept of religion.

The concept of definition that Durkheim endorsed here is comparable to that which he employed in his other works. Durkheim would have regarded those of his critics who would object that totemism is "not a religion" as committing the same methodological blunder as those who would object that self-sacrifice is "not suicide." In fact, Durkheim was willing to consider even certain patriotic celebrations as having a religious character (610). As Godlove argues, the question is not whether in fact they are religious but what insights are to be gained by defining religion so as to include patriotic celebrations (1989:145–46). The same argument could be made in defense of Durkheim's decision to include totemism as a religion.

Among Durkheim's contemporaries, Frazer raised the objection that totemism is not a religion.[4] According to Frazer, totemism is merely a form of magic. Frazer's argument, however, presupposes his own definition of religion, as it is expressed in terms of deities and divine beings. When the totemist performs a sacrifice, Frazer said, he does so only in order to encourage the species of the animal sacrificed to reproduce so as to ensure a plentiful supply of food. He argued that the purpose of the rite is not to achieve a kind of "mystical communion with a deity" and that the primitive does not believe the killed animal to be "divine" (Frazer 1900:202; Jones 1985:83–84).

The disagreement between Durkheim and Frazer is less a question of whether totemism is a religion than a question of how religion is to be defined in the first place. Durkheim, in fact, criticized Frazer for committing the methodological error of providing a definition that is not sufficiently inclusive (1912a:31). Frazer, along with Tylor and other proponents of the animistic hypothesis, defined religion in terms of a belief in gods or spiritual beings. Durkheim rejected this definition, citing the existence of systems of belief such as Buddhism and Jainism, in which the idea of gods and spirits is absent or plays a secondary or minor role, but which he nevertheless regarded as religions. Furthermore, he argued, even in deistic religions one can find rites that are independent of the idea of gods or spiritual beings. He adduced the Jewish rite performed during the Feast of the Tabernacles in which the believer beats the air with willow branches in order to raise the wind and to produce rain. Jews believe that if performed correctly, this rite will yield the desired results directly, without divine intervention, Durkheim reported. Thus he concluded that religion "overflows" the idea of gods and spirits and cannot be defined in these terms (40–49).

Durkheim found the definition of religion agreed upon by adherents of the naturistic hypothesis to have insufficient empirical scope. Defenders of various versions of this hypothesis, including Max Müller and Frank Byron Jevons, defined religion in terms of a belief in the supernatural. Durkheim criticized this definition by arguing that the distinction between the natural and the supernatural is of relatively recent date. In response to Jevons's anticipation of this objection, Durkheim argued that for events to be regarded as due to supernatural causes they must not only be unexpected but thought to be impossible. Thus the idea of the supernatural depends upon and therefore could not arise before the scientific notion of a deterministic natural order. To argue in this way, of course, is to conflate the issue of the logical dependence of one concept upon another with the problem of the order of their historical development. Nevertheless, Durkheim concluded that to define religious phenomena in terms of the supernatural would be to exclude all those religions that existed before the development of science (33–40).

Durkheim's own definition of religion was motivated by a desire to include within its scope what other theorists of religion have not. He began his search for a definition by distinguishing two types of religious phenomena: beliefs and rites. Religious rites can be distinguished from other prescribed modes of action such as moral rules only by the nature of their object, he said. Since the object of a rite is expressed in a belief, one can define religious rites only after having defined religious beliefs. He then asserted that all known religious beliefs suppose a classification of things into two opposed classes: the sacred and the profane. Even the four noble truths of Buddhism and the practices that derive from them have a sacred character, which is why he regarded Buddhism as a religion (50). Thus Durkheim included within the scope of his theory of the origin of religion those belief systems that are not included within the scope of the animistic hypothesis.

If Durkheim could criticize his opponents for maintaining definitions of religion that he believed were not sufficiently inclusive, he recognized, his opponents could reply that his definition was *too* inclusive. Frazer, in particular, would have criticized Durkheim for including totemism, which Frazer regarded as a form of magic. Durkheim therefore needed to draw the boundary between religion and magic in a way that would allow him to escape this objection. He then distinguished religion from magic in terms of a "moral community" or "church" that adheres to a religion (65). The relationship between a magician and his clientele is more like that between a doctor and his patient than that between a priest and his parishioners, he argued (61–62). Durkheim then defined religion not just as a system of beliefs and practices regarding the sacred and the profane, but as a system of such beliefs and practices that unites its adherents into a single moral community (65). To explain the origin of religion is thus to explain how a *group* of people came to share such a system. Framing the problem in this way, it is not at all surprising that religion will turn out to have social causes.

Totemism, Animism, and Naturism

Having criticized the definitions of religion maintained by proponents of the animistic and naturistic hypotheses, Durkheim turned next to the criticism of these hypotheses themselves. He argued that these hypotheses not only fail to explain the origin of the distinction between the sacred and the profane but give rise to problems that they cannot solve and that do not arise for his totemistic hypothesis. By elimination, he concluded that totemism must have been the original form of religion. Finally, he suggested that the other two primitive religions, animism and naturism, may be derived from totemism (124). That is, he believed that the totemistic hypothesis could account for the origins of the naturist idea of forces and the animist idea of souls as well as the totemic idea of classification. Therefore, he concluded, of all three hypotheses the totemistic hypothesis has the greatest explanatory scope and its claim that totemism is the earliest religion should be accepted.

The Animistic Hypothesis

According to the animistic hypothesis of E. B. Tylor, religion began with the belief in a soul, Durkheim reported. Durkheim then adumbrated three problems that any theory of religion built around this hypothesis must solve and criticized the proposed solutions to these problems offered by the animistic hypothesis. First, he said, since according to this hypothesis the idea of the soul is the earliest religious idea, the animistic theory must show how the idea of the soul was formed without borrowing elements from any earlier religion. Second, it is necessary to show how souls were transformed into spirits and became the object of a cult. Finally, he added that the animistic theory must explain how the cult of nature is derived from animism (69–70). With this third problem, Durkheim pitted the animistic hypothesis against the naturistic hypothesis, the other leading competitor to his totemistic hypothesis.

Tylor, according to Durkheim, tried to explain the origin of the idea of the soul by arguing that this idea was suggested to the primitive by the poorly understood spectacle of a double life of waking and dreaming. That is, it was believed that when one dreamed, one's soul left one's body and went to the place of which one dreamed and conversed with the souls of those about whom one dreamed (70). Durkheim criticized Tylor's theory by asserting that for the primitive, the soul is not radically distinct from the body. He then argued, in effect, that the primitive would have found the hypothesis of the soul unacceptable. Specifically, he said that this hypothesis is neither the only nor even the simplest explanation of dreams. Also, he added, the hypothesis that one's soul met with those of others during dreams would be easily refuted upon waking and talking with these other people. Durkheim furthermore questioned Tylor's assumption that people would have sought to explain dreaming in the first place, arguing that even today there are many possible problems that we do not recognize and try to explain. Finally, Durkheim pointed out that even if it were true that the primitive explains his dreams this way today, it does not follow that this was the *origin* of the idea of a soul (78–82). In several of these arguments Durkheim assumed that the primitive thinks the way we do, evaluating hypotheses according to criteria such as simplicity and freedom from contradiction and so on. Later in the book, in his criticism of Frazer's attempt to derive totemic beliefs from a belief in the soul, he made explicit this implicit assumption about primitive mentality.

Tylor's solution to the second problem, in which he gave an account of how the idea of a spirit derives from that of the soul, constituted for Durkheim the "very core" of the animistic theory (85). Durkheim reported that according to Tylor, souls, by analogy with the way in which they are thought to leave the body during sleep, are thought to leave the body at death and become transformed into spirits, forming an arsenal of ever-ready explanatory causes (72–73). For Durkheim, however, this analogy did not suffice to explain the cult of ancestors. Death alone does not explain how the soul, once detached from the body, transforms itself from something profane into something sacred (85). In other words, Durkheim rejected the animistic theory for its failure to account for the distinction between the sacred and the profane. He also cited Jevons as having made a similar criticism of

Tylor's theory. Although Jevons agreed with Tylor that the idea of the soul comes from dreams and is then projected onto things in nature, he argued that it does not follow that these natural things animated by souls would be considered supernatural beings. On the contrary, he said, they would be thought to resemble us to the extent that we, too, possess souls. As empirical confirmation of the claim that death alone does not import sacredness to the soul, Durkheim adduced a Melanesian cult in which only some of the dead are honored (85–88).

Durkheim then drew from Tylor's theory a test implication that he believed to be refuted by the facts. He argued that if the first sacred beings were the souls of the dead and the first cult that of ancestors, this cult should have a larger place in religious life the more "inferior" the society. Durkheim's reasoning here illustrates the methodology of *The Rules:* he drew a test implication concerning the existence of a concomitant variation. According to Durkheim, the animistic hypothesis entails a concomitance between the degree of inferiority of the society and the extent of the cult of ancestors. Of course, neither of these variables could be measured precisely. The degree of inferiority of a society for Durkheim was simply a function of its position in his classification of societies according to their degree of composition. Nevertheless, he found this test implication refuted by the facts regarding the type of society he believed to be the most primitive. The Australians have funeral rites, he said, but no true cult of ancestors. The ancestors honored by Australians are not real ancestors transformed into deities by death but mythological heroes who exercised superhuman powers even while alive (88–90).

Thus the animistic theory is refuted along with its account of the transformation of the idea of the soul into the idea of a spirit, Durkheim believed. However, he also criticized its explanation of the origin of the cult of nature because this account has been accepted by others, such as William Robertson Smith, who were not otherwise defenders of the animistic hypothesis (91). Durkheim reported that according to Tylor, the mind of the primitive considers things to be animated with souls much as the child believes its toys to be ensouled. He then mentioned that Herbert Spencer objected to Tylor's theory on the grounds that even the higher animals can distinguish animate from inanimate objects (74–76). Durkheim criticized both Tylor and Spencer on methodological grounds, accusing them of having made "doubtful analogies" from the mentality of either children or animals to that of primitives (94).

Durkheim also rejected Spencer's alternative explanation that the cult of ancestors was transformed into the cult of nature through a linguistic confusion. According to Durkheim, Spencer thought that people, led astray by the "amphibologies" of language, were unable to distinguish metaphors from reality. In Spencer's theory primitive people were originally given the names of animals, plants, stars, or other natural objects. Later generations mistakenly interpreted the names of their ancestors literally and believed themselves to be descended from tigers, lions, and so on. The cult of the ancestor was thereby transferred over to a cult of the natural object of the same name. Durkheim found it implausible that this lack of discernment assumed in Spencer's theory could have been universal (76–77).

Finally, Durkheim drew several test implications from the animistic theory. Two of these he believed to be refuted by the facts. First, he argued that if the idea of spirits derived from the idea of souls, then spirits should be conceived as residing in the things to which they are assigned in the same way that souls reside in bodies. However, he found that this was not always the case. His argument, of course, assumes that effects will resemble their causes. Durkheim refuted a second implication of the animistic theory by pointing to what he took to be the fact that sacred beings were originally conceived in the form of plants and animals and not human beings. A final implication, however, he criticized on conceptual rather than on empirical grounds. If animism were true, Durkheim argued, religious beliefs would have no basis in objective reality and rest on nothing more than confusions of dreams with waking, death with sleeping, or the inanimate with the animate. In Durkheim's opinion, however, the problem for any science of religion is precisely to discover that part of reality which gives rise to religious ideas. He asked, rhetorically, what sort of science would have for its principle discovery that its object of study does not exist? (95–99). Perhaps because he lacked a familiarity with the history of science, Durkheim failed to realize that this sort of thing happens all the time, with respect both to explanatory concepts such as phlogiston and to supposed empirical phenomena such as spontaneous generation. Be that as it may, Durkheim rejected the animistic hypothesis for its antirealist implications.

The Naturistic Hypothesis

According to Durkheim, Müller's naturistic hypothesis states that the original objects of worship were natural things and forces like the sky and fire and not souls or spirits. Müller reached this hypothesis through his study of Indo-European languages, in which he discovered that the names of the earliest gods were taken from the words for the phenomena of nature. Durkheim said that Müller justified his hypothesis through psychological considerations regarding the fear and wonder that these natural phenomena would have inspired and then generalized his thesis to include all peoples and not just Indo-Europeans. According to Durkheim, Müller theorized that the forces of nature to which religious sentiments were originally attached were subsequently transformed into living, thinking, personal agents through the distorting action of language on thought. Müller asserted that people designated the forces of nature with the words for the human actions that resembled the manifestations of these forces. The use of these words led people to conceive the things to which these actions were attributed as personal agents. Müller claimed that mythologies about these anthropomorphic agents were then invented in order to account for the phenomena of nature (103–10).

Although Durkheim cited criticisms of Müller's theory that have been raised by philologists, he left these aside as requiring a special competence in this subject. Durkheim recognized that his intellectual niche was not in the study of language.[5] However, he believed that the naturistic hypothesis could be criticized on grounds independent of the linguistic arguments made in support of it. He argued that if the purpose of religion were to provide a guiding representation of the world, its failures would have been quickly noticed and religion would not have lasted. The naturistic, like the animistic hypothesis, makes of religion a system of false, hallucinatory images (111–14). However, if the purpose of religion were *not* to put us in harmony with the material world, he argued, its failures to do so would not weaken it. If the religious sentiment were strong enough, people would find some ad hoc explanation for the failure of their religious rites to produce their desired effects. But in order for religious sentiments to be this strong, these sentiments must have their origin elsewhere than in our need to adapt to the material world (118). To find the causal origin of these religious feelings, of course, was Durkheim's explanatory goal.

Both the naturistic and the animistic hypotheses were misguided in their attempts to construct the notion of the divine from the individual's experience of natural phenomena, Durkheim believed. Where the animistic hypothesis seeks the source of this idea in our experience of what Durkheim regarded as the "biological" phenomenon of dreaming, the naturistic hypothesis seeks it in our experience of external, physical phenomena. However, he argued, it is impossible for our experience to provide us with the idea of something outside of experience (123). "Nothing comes from nothing," he said (321). That is, nothing in nature or in the play of physical forces suggests the idea of the radical heterogeneity of the sacred and the profane (118f.). The animistic and naturistic hypotheses explain the notion of the sacred in terms of error, he concluded, leaving unexplained how humanity could have persisted in these mistakes (322). Hence, Durkheim rejected these hypotheses regarding the origin of religion for their failure to explain the origin of the idea of classifying things into the distinct categories of the sacred and the profane.

In order for Durkheim to persuade the reader that totemism was the earliest form of religion, however, it does not suffice for him to show merely that the animistic and naturistic hypotheses cannot account for the origin of the idea of the radical heterogeneity of the sacred and the profane. Minimally, he needed to demonstrate that the totemistic hypothesis could explain the origin of this concept *and* that the other hypotheses could not account for this concept indirectly by explaining how totemism derived from some earlier religion. Also, in order to meet any possible objections from defenders of the animistic and naturistic hypotheses, he needed to show that the totemistic hypothesis could account for the origins of what these hypotheses assume to be the fundamental religious notions of souls and cosmic forces. In effect, Durkheim was imposing on the totemistic hypothesis the requirement that it must have at least the empirical scope of the two earlier hypotheses. These were the tasks he set for himself in book 2 of *The Elementary Forms*. Once he established to his satisfaction that totemism was the earliest religion, he believed that he could then proceed in book 3 to demonstrate through an analysis of totemic religious rites the social causes of religious experience.

Totemism and the Concept of Classification

Book 2 of *The Elementary Forms* begins with a description of the totemic system of social organization. Durkheim also brought forth evidence regarding the sacred character of the totem in order to show that totemism satisfies his preliminary definition of religion. He then offered a further refinement of this definition, stipulating that a religion is not merely a random assortment of beliefs regarding sacred objects but an organized system of such beliefs, with some sacred objects subsumed under others. According to Durkheim, a totem is endowed with religious significance not merely for the clan that venerates it. It belongs to a hierarchy of totems that mirrors the structure of the entire tribe and that constitutes the basis of the religious beliefs and practices of the entire tribe (200, 220). This hierarchical organization of religious beliefs then constitutes for Durkheim an additional problem that any theory of the origin of religion must solve but which his totemistic hypothesis is eminently suited to explain.

Durkheim proceeded to defend the well-known primitive classification hypothesis for which he and his nephew Marcel Mauss had argued nine years earlier in an article in *L'Année sociologique* (1903a(i)). According to this hypothesis, the earliest systems of classification were not only the product of society but modeled on the very structure of society. As evidence for this claim, Durkheim adduced the totemic system in which human social groupings are used to classify things in nature. According to Durkheim, the Australian considers everything in the universe to be a member of his or her tribe. When tribes are divided into two phratries, everything in the universe is divided into two phratries. The things belonging to each phratry are then distributed among the clans into which the phratry is divided. Such primitive classifications, Durkheim concluded, constitute the origin of our modern idea of classifying things according to genus and species. This result, he believed, verifies the proposition that the fundamental categories of thought are the product of social factors, as this is the case for the very category of category itself (201, 205–6).

The placement of the primitive classification thesis in a chapter following two chapters that consist largely of a catalog of totemic beliefs leaves the reader with the impression that Durkheim has derived this thesis from the facts. The procedure he actually followed, however, would be more accurately described as a method of first drawing a few low-level empirical generalizations and then explaining these generalizations in terms of a higher-level theory. This is the method that he followed in *Suicide,* also. In order to arrive at these low-level generalizations in *The Elementary Forms* he was willing even to "correct" the ethnographic record. For example, in spite of the fact that most ethnographies of Australian tribes, unlike those of Amerindian tribes, reported totems only for clans and not for phratries, Durkheim nevertheless affirmed that the use of phratries is a general characteristic of totemism. In the case of the Australians, Durkheim said, "we have been able to infer the existence of totems for phratries only from a few survivals, so little marked, for the most part, that they have escaped a number of observers" (157).

Given his assumption that the Australians represent an earlier stage in the same developmental path as the Amerindians, one might have expected him to have interpreted the system of phratries as a later development. Instead, he regarded it as an "archaic" institution that the Amerindians had preserved due to their greater social "stability." This explanation is not entirely ad hoc, as he was able to support it with linguistic evidence regarding the names of Australian phratries where they existed. Durkheim did not shun linguistic arguments when he absolutely needed them, in spite of his admitted lack of philological competence. He was forced into maintaining that the institution of phratries is a general characteristic of totemism, I believe, in order to save his primitive classification hypothesis. He could not argue that the Australians had a concept of hierarchical classification modeled on their social structure unless he could show that in fact they at one time had had a system of phratries and clans that served as the archetype.

In arguing for the thesis that the classificatory concepts of genus and species were socially constructed on social models, Durkheim did not wish to deny to the individual mind the ability to recognize similarities and differences. In fact, he believed that this mental ability is presupposed by the ability to classify (206). However, he also thought that the concepts of genus and species cannot have their source in and cannot be explained in terms of an individual's experience of resemblance.

Durkheim offered two reasons for rejecting the thesis that the concepts of genus and species have their origin in individual experience. First, he provided the argument that any individual's experience is necessarily limited and thus cannot be adequate to the full extension even of any particular genus. This argument is couched in the rather murky terms of the theory of the compounding of mental states. He began by distinguishing the perception of resemblance from the concept of a genus, characterizing the latter as the "external framework" of which the objects perceived as similar are the contents. Individual perceptions of resemblance, he said, cannot be the source of this concept of a genus, as they consist merely of "vague and fluctuating images, due to the superposition and the partial fusion of a determinate number of individual images." The framework, on the other hand, has a definite, fixed form and can be applied "to an indeterminate number of things" (208–9). "Every genus," he said,

> has a field of extension that goes infinitely beyond the circle of objects of which we have directly experienced the resemblance. That is why an entire school of thinkers has refused, not without reason, to identify the idea of a genus with that of a generic image. The generic image is only the residual representation, with uncertain frontiers, that is left in us by similar representations, when they are simultaneously present in consciousness; the genus is a logical symbol by the grace of which we think distinctly about these similarities and other analogous ones. (209)

The way in which Durkheim attempted to distinguish the concept of a genus from the mere product of a fusion of individual mental representations, however, does not help his argument that the concept of a genus is socially constructed. Although he alluded to Berkeley's argument that no mental representation can ever be adequate to the extension of a general concept, it is not clear that he grasped the full force of this argument. He seems to have interpreted Berkeley as having said merely that the extension of our concepts goes beyond the limited personal experience from which an individual's generic images could be formed. Berkeley, however, wished to deny not only that concepts (or "ideas," as Durkheim sometimes called them) could be identified with generic images, but that generic images were possible in the first place. Durkheim, on the other hand, went so far as to claim that even animals have the ability to form generic images, although they lack the capacity for conceptual thought (209). He was sufficiently pleased with his distinction between generic images and general concepts that he made the same sort of argument in the conclusion:

> The generic images that are formed in my consciousness by the fusion of similar images represent only the objects that I have directly perceived; there is nothing there

that can give me the idea of a class, that is to say of a framework that is capable of containing the *total* group of all possible objects that satisfy the same condition. (629)

However, in his conclusion Durkheim also denied that concepts can always be distinguished from sensations in terms of the greater extension of concepts. Some concepts, such as that of a god, may apply to but a single individual, he pointed out (617–18). The kind of argument he used for the social origin of the concepts of genus and species, then, could not necessarily be applied to other concepts.

One is tempted to say that Durkheim could have made his position only stronger by arguing with Berkeley that there is no such thing as a generic image and not merely that a generic image does not account for the unlimited extension of a concept. It is not clear, however, that he could abandon the notion of a generic image without giving up his theory of the compounding of mental states. Durkheim's account of the process of fusion through which generic images are formed is not unlike his description in *The Division of Labor* of the process through which collective representations are formed. The only difference appears to be that in the case of collective representations, one's representation of something fuses with one's representation of someone else expressing something similar to what one is representing (1902b:67; cf. chap. 3 above). Since Durkheim identified concepts with collective representations, he could not give up the idea that mental representations can combine with one another without giving up his account of the formation of concepts.

The second reason Durkheim offered in support of his thesis that the concept of a genus could not have its source in individual sense experience is perhaps a better argument. He added that the spectacle of nature could not have furnished the idea of a hierarchy, with species subordinated to genus and the different species of the same genus on the same level (210–11). Even if it were sound, however, this argument would establish less than Durkheim may have wanted. If the hierarchical classification of genus and species "has manifestly been constructed by men" (209), as Durkheim said, it nevertheless appears to be the *same* social construction for every society. In fact, his assumption that all societies evolved from primitive totemic societies would seem to necessitate this conclusion.

To the extent that the basic categories of thought turn out to be the same in every society, it seems, Durkheim would have to forfeit an important argument for preferring his sociological theory of the categories over the a priori theory. As we saw above, Durkheim had argued earlier that the a priori philosophy cannot account for the fact that the categories vary with time and place. In his defense, one could argue that while the system of subsuming clan or species under phratry or genus is universal for Durkheim, the numbers and names of clans or species in a phratry or genus vary from society to society. This interpretation would allow him to preserve his earlier argument.

Nevertheless, if Durkheim believed that all societies share the same concept of subsuming species under genera, his sociology of knowledge is distinctly at odds with the conceptual relativist claims advanced on his behalf by David Bloor (1982). Durkheim and Mauss's primitive classification hypothesis does not assert that each society has its own system of classification that mirrors its social structure. Durkheim and Mauss were not conceptual relativists; in fact, their very methodology of seeking the origins of our categories in primitive society would rule this possibility out. For Durkheim and Mauss, all classification systems are nothing but minor variations on a scheme modeled on a very primitive form of social organization. Whether Bloor can successfully defend a relativist interpretation of the Durkheim-Mauss primitive classification thesis is an issue that we will return to in the concluding chapter.

Totemism as the Original Form of Religious Belief

So far, Durkheim has argued that the totemic system of social organization accounts for the concepts of genus and species and that these concepts of classification do not have their origin in individual experience. In order wholly to rule out the possibility of an individualistic, psychological account of these concepts, however, he needed to show that totemism itself is not capable of such an explanation. Durkheim did this in several ways, arguing that totemism cannot be derived from either the naturist or animist religions, for which psychological accounts had been given, that totemism as a collective phenomenon cannot be derived from individual totemism, and that any account of totemism that denies its religious character will either beg the question or fail to account for the distinction between the sacred and the profane.

Animistic and Naturistic Accounts of Totemism

First, Durkheim considered the attempt by defenders of the animistic theory to derive the totemic from the animist religion. Tylor and Wilken, he said, explained totemism in terms of a belief in the transmigration of souls from dead ancestors to animals and plants, which are thereby transformed into totems. In support of this hypothesis, they adduced evidence regarding Bantu tribes as well as tribes from Java, Sumatra, the Philippines, and Melanesia, he reported. Durkheim objected on methodological grounds to the use of this evidence, however, arguing that it was drawn from societies that were more advanced than the totemic societies he was studying (239–41). He then raised the empirical objection that metempsychosis was unknown among Australian totemists.

Durkheim also accused the defenders of the animistic hypothesis of begging the question. In order to believe that one's soul could be reincarnated in an animal, Durkheim asserted, one would already have to believe that people were related to animals. The hypothesis that totemism derived from the cult of ancestors could not explain how animists could believe that their souls pass into animals if they believe that humans are distinct from animals (242). In other words, Durkheim seems to have thought that one would have to be a totemist *before* one could believe that one's soul could be reborn in an animal, as it is the totemic religion that establishes a system of relationships between people and animals. The animistic theory fails to explain the totemic religion precisely because it fails to explain the totemic system of classification in which people and animals can be members of the same clan.

In a final objection against the animistic theorists' account of totemism, Durkheim argued that to derive totemism from a belief that animals are ensouled by one's ancestors is to misconceive totemism as a form of zoolatry. Both Tylor and Wundt were accused of laboring under this misconception. Durkheim's objection that they have mistakenly defined totemism as zoolatry is both empirical and conceptual. He rejected their definition of totemism by citing evidence that would indicate that it is not the living animal but the totemic emblem that is worshiped (243–44).

Durkheim then criticized Jevons's attempt to derive totemism from naturism. According to Durkheim, Jevons explained totemism in terms of people's desire to ally themselves with powerful forces of nature through bringing these forces within their system of clan relationships. Durkheim rejected this account on the grounds that it violates his methodological prohibition against sociological explanations couched in terms of consciously pursued goals. He also drew two test implications from Jevons's theory that are refuted by the facts. First, he reasoned that if men sought the aid of supernatural beings, then they would preferably seek that of the most powerful. However, he pointed out, totemic beings are often very humble. Second, he reasoned that if a totem were an ally, a clan would want to have as many totems as possible. Unfortunately for the naturistic account of totemism, however, each clan has but one totem (244–45).

Individual and Collective Totems

Having thus dispensed with the animistic and naturistic accounts of totemism, Durkheim then turned to criticize the theory that the clan totem began as an individual totem that the founding member of a clan passed on to his descendants. This theory, he said, had been advanced by Frazer in *The Golden Bough* and by other members of the English-language anthropological tradition, including Charles Hill-Tout, Alice C. Fletcher, Franz Boas, and John R. Swanton. The problem that immediately arises for this theory is to account for the origins of the individual totem. Durkheim considered two different solutions to this problem. According to Hill-Tout, Durkheim reported, the totem began as a kind of fetish to which an individual appealed for protection. Frazer, on the other hand, claimed that the individual totem began as a subterfuge in war, that is, with the belief that the soul could seek safety by leaving the body and hiding in another object or place, Durkheim said (246–48).

Durkheim criticized both of these explanations for being nothing more than pictures drawn by the mind, lacking what he called "positive proofs" (249–50). He then raised specific objections against each of them. To reduce totemism to fetishism as Hill-Tout has done, Durkheim said, one must first establish that fetishism is earlier. However, he claimed, fetishism appeared only later and was unknown in Australia. Durkheim of course assumed the validity of his system of classifying societies in making this empirical criticism of Hill-Tout's theory. Durkheim then objected to Frazer's theory on psychological grounds, accusing him of presupposing a "fundamental absurdity" among the primitives. Assuming again that the primitive mind is not different than ours, Durkheim found it implausible that the primitive could think his soul to be more safe in the body of an animal, what with hunters about and so on. The primitive, he said, "has a logic, however strange that it may sometimes appear to us: now, unless he were totally deprived of it, he could not commit the reasoning that one imputes to him" (250). In addition, Durkheim raised an empirical criticism against Frazer's account, asserting that very often the function of the totem—conferring magical powers—is entirely different from that assigned to it by Frazer (251).

However, even if one could account for the origin of the individual totem, Durkheim argued, one would still face the problem of explaining how the individual totem gives rise to the collective totem. Alluding to Bacon's method of crucial instances, he said that there are a number of "decisive facts" that tell against the hypothesis that the collective totem has its origin in individual totems. When we analyze the arguments that follow, however, we discover that these decisive or crucial facts include not only empirical refutations of the hypothesis, but facts about which an hypothesis says nothing at all. Finally, his concept of crucial facts encompasses the unsolved problems that a hypothesis may generate through its attempts to account for the facts.

Some of the decisive facts that Durkheim raised against the hypothesis that would derive collective from individual totems do indeed turn on empirical considerations. For instance, he derived from this hypothesis the test implication that a clan would try to prevent others from killing or eating its totem much as the individual protects his personal totem. However, he said, the clan performs certain rites to ensure that the totemic species will thrive in order to guarantee a plentiful supply of food (252). Another test implication he drew from this hypothesis assumes that societies can be classified according to their degrees of primitiveness. If individual totemism were more primitive than collective totemism, he argued, individual totemism should be more widespread in more primitive societies and less so in more developed societies. But the contrary is true, he asserted, meaning that individual totems were less prevalent among Australians than among Amerindians (254–55).

Durkheim also assumed his system of classifying societies in criticizing the evidence that Hill-Tout adduced in support of the hypothesis that collective totemism derives from individual totemism. For instance, he said that the myths concerning the founding members of clans that Hill-Tout cited in support of his hypothesis were drawn from North American Indian tribes and were therefore not representative of the most primitive religions. Furthermore, Durkheim added, Hill-Tout's hypothesis is contradicted by other, earlier myths according to which the clan is descended from the totem. Durkheim then argued that the founding member myths were added later to avoid having humans descended directly from animals (253–54).

A second type of decisive fact that Durkheim cited does not so much refute the hypothesis under consideration as pose a problem that it cannot explain. In other words, a crucial fact for Durkheim can be a fact that one hypothesis explains but about which the other hypothesis says nothing at all. A crucial fact need not be a fact that actually disagrees with the test implication of one of the hypotheses in question.[6] For example, Durkheim adduced the fact that one never finds two clans of the same tribe with the same totem (252). Another fact of this variety he cited is that one's choice of individual totem is restricted to the set of subtotems for one's clan (256–57). Both these facts can be explained by the hypothesis that totemism is collective in origin but do not necessarily refute the hypothesis that totemism began with individual totems.

A third type of crucial fact for Durkheim consists of problems for one hypothesis that the other hypothesis need not explain. For example, Durkheim argued that the hypothesis that collective totems derive from individual totems overlooks important differences between individual and collective totems. The collective totem is assigned at birth, he stated, while the individual totem is usually acquired as the result of a religious rite. Some theorists have tried to solve this difficulty by postulating "as a sort of middle term" a right to transfer one's totem to whomever one pleases. However, he claimed, where this right has been observed, it has been the prerogative of only a few, such as magicians. The hypothesis that collective totems derive from individual totems, he contended, must then explain the following three things: how the prerogative of a few became the right of all, how what used to indicate a change in the religious constitution of an individual became an original element of it, and how what used to be a consequence of a rite now happens automatically (253). Durkheim's hypothesis does not have to solve these problems precisely because he did not need to explain how individual totems were transformed into collective totems. These three unsolved problems appear to have constituted a crucial fact for him in favor of his own hypothesis and against the hypothesis that collective totems originated as individual totems.

Other Accounts of the Origin of Totemism

Durkheim criticized a more recent account of totemism by Frazer as well as a theory proposed by Andrew Lang for either denying or failing to account for the religious character of totemism. Subsequent to *The Golden Bough*, Durkheim reported, Frazer published a new theory of the origins of totemism according to which it began as a way to explain conception. According to Frazer, he said, a primitive woman attributes quickening to a spirit within her that she believes to have come from either the place where she quickened or a plant or animal that she had just either seen or eaten. The animal, plant, or location then becomes the totem of the fetus (257–59).

Durkheim criticized Frazer's hypothesis on methodological and empirical grounds, arguing that it is supported by evidence concerning only the Arunta. This tribe, he said, also has a secondary totem passed on matrilineally from generation to generation. Durkheim believed this matrilineal totem to be a survival of an earlier system of clan totemism (261–62). Frazer's hypothesis also begs the question, Durkheim accused, as it fails to explain how people could believe their souls to come from plants or animals or even how they acquired the notion of the soul in the first place. Most importantly for Durkheim, however, Frazer's hypothesis fails to account for the religious or sacred character of the totem (260).

Lang's theory, according to which totemism is explained simply in terms of human groups taking their names from plants and animals, was also rejected for failing to account for the religious character of totemism. According to Durkheim, Lang anticipated this objection and denied that totemism was originally a religion, asserting that it was only subsequently brought into an independently established system of religious beliefs. For Durkheim, however, Lang's reply merely begs the question. If totemism were not originally of a religious character, Durkheim argued, why was a need felt to incorporate it with religious beliefs? Also, he pointed out, Lang recognized but failed to explain the sacred character of things such as the blood of the totemic animal (262–65).

Durkheim concluded that even those theories of the origin of totemism that do not explicitly derive it from a supposedly earlier religion implicitly assume the prior existence of religious notions such as that of sacredness or the soul. By assuming the prior existence of religious ideas, these theories in effect try to derive totemism from an earlier religion. To attempt to derive totemism from an earlier religion, however, is either to expose oneself to the objections he has raised against the animistic and naturistic hypotheses or to suppose that totemism comes from some other religion "which differs from it only by degrees." To choose the latter path, he said, "is to leave the givens of observation and enter into the domain of arbitrary and unverifiable conjectures" (266–67). This criticism, I believe, was directed against the hypothesis that the earliest form of totemism was individual totemism, a hypothesis which he said lacks "positive proofs." This hypothesis, however, is not so much unverifiable as unverified. Perhaps what Durkheim meant is not that the hypothesis of an original individual totemism is unverifiable *in principle* but only in fact or in practice. In other words, there was no empirical evidence, of either an historical or ethnographic nature, that supports this hypothesis.

Denying that totemism can be derived from an earlier religion, Durkheim was left with the task of explaining its causal origin. Before he turned to this task, however, he wanted to complete his argument that totemism is the earliest form of religion by showing that the other candidates, animism and naturism, can be derived from totemism. In order to show this, he defended the claim that the most fundamental religious notions in terms of which the proponents of the animistic and naturistic hypotheses characterize each religion actually have their origins in totemic beliefs. First he showed that naturism derives from totemism by arguing that the concept of a religious force or power derives from the notion of a totemic principle that is supposed to suffuse everything connected with the totem. His theory of the genesis of the notion of the totemic principle in turn explains the origin of the distinction between the sacred and the profane. Then he derived animism from totemism by arguing that the concept of a soul can be traced to the notion of the incarnation of this same totemic principle in individuals. The argument strategy he employed consists of showing that his totemistic hypothesis has at least the explanatory scope of both the naturistic and animistic hypotheses, accounting for more of the facts about religious belief.

Totemism and the Concept of Force

Because he believed that the naturist notion of religious forces is more basic than the animist notion of souls, Durkheim tried to explain in terms of his totemistic hypothesis the origin of the idea of force first. He believed that his theory of the origin of the idea of a religious force also explains the origin of the concept of force in general. His account of the concept of force, he thought, thereby provides a basis for his argument for the social origins of the category of causality and contributes to the explanatory scope of his sociology of knowledge.

The Concept of a Religious Force

According to Durkheim, the concept of a religious force or power has its origin in the notion of a totemic principle. He defended this hypothesis by arguing that it explains the sacred character of religious forces and grounds them in reality. The naturistic hypothesis in its misguided attempt to derive the idea of the sacred from the individual's experience of external nature, he believed, reduces religion to a collection of unfounded beliefs. For Durkheim, the idea of religious forces has a basis in an internal reality, deriving from the effect of collective representations on the mind. The difference between the effects of collective and individual representations gives rise to the distinction between the sacred and the profane.

The hypothesis that the concept of religious forces originates in the power of collective representations over the mind, Durkheim believed, can explain the ambiguity of religious forces. Religious forces are conceived as having both physical and moral aspects, he thought. He generalized that all totemic societies form a notion of a totemic principle that is ambiguous in this way. Relying heavily on evidence drawn from non-Australian societies, he asserted that totemism is not a religion of animals or their images but of some force or principle that is believed to be found in them. The totem merely provides an image by which this principle can be represented to the mind, he believed. If a person acts in a certain way towards a totem, Durkheim explained, it is not only because this person feels himself to be in the presence of a physically superior force but because he believes he has a moral obligation to act in this way. This same confusion of acting out of duty with yielding to a superior force Durkheim found to be characteristic even of so-called advanced religions such as Christianity (269–72).

Anticipating the objection that his interpretation of totemism attributes "too much" intelligence to primitives, Durkheim conceded that this notion of a totemic principle is only implicit among primitive Australian tribes. However, he said, this notion is more explicit in the Samoan religion, in the

Melanesian idea of *mana,* and in the Amerindian concepts of *wakan, orenda, manitou,* and so on (273–77). The Australians have not reached the same level of abstraction not because of any weakness of their intellectual powers, he argued, but because their social organization is less unified (280–81). In attributing to moderns and primitives alike the same confusion between physical compulsion and moral obligation, however, Durkheim could be accused of attributing too *little* intelligence to moderns rather than too *much* to primitives. As Rousseau had pointed out in *The Social Contract,* one does not have a *duty* to give in to force (Rousseau 1762: bk. 1, chap. 3). In reply, Durkheim might have argued that he was not trying to explain the genesis of the philosopher's *concept* of moral obligation but only the common man's *feeling* of moral obligation. For him to have made this reply, however, would have been to call attention to a problem with drawing analogies between primitive and modern modes of thought. The difficulty is that how moderns in fact think may very well be controversial. As I will demonstrate below, when Durkheim argued that moderns and primitives think alike, he did not insist that primitives conform to our textbook standards of rationality. To do so would be absurd. Rather, Durkheim showed that we commit the same logical fallacies that are attributed to primitives.

Be that as it may, Durkheim proceeded to account for what he saw as the ambiguity of the idea of a religious power in terms of its origin in the notion of a totemic principle. This latter notion, he argued, does not have its origin in the individual's sensations of the things that serve as totems but in his experience of the power of society over him. "The totemic principle . . . cannot be anything other than the clan itself, but hypostatized and represented to the imaginations under the sensible species of the vegetable or animal that serves as totem" (1912a:295). Society has everything necessary to give rise to the idea of a religious power, he claimed. It is something superior to oneself upon which one depends, he argued, and it imposes certain manners of acting on the individual. The physical aspect of the totemic principle, he said, is due to its association with a physical plant or animal, while the moral aspect is due to the moral authority of the clan over the individual (319).

Durkheim explained the way in which society exercises its moral authority over the individual in terms of the coercive power of collective representations. If society imposed certain manners of acting on the individual only through physical constraint, he reasoned, this constraint would give rise to the idea of only a physical force to which we must cede by necessity. Physical constraint could not give rise to the idea of a moral power. If we defer to society, he said, it is above all because society is an object of *respect.* The respect that one bestows upon a moral authority or a social institution, he believed, is induced by the superior "vivacity" or "intensity" of one's mental representations of them. The strength of these representations, which Durkheim said is the "sign" of their moral authority, is due to the combined strength of the individual representations of which they are made (296–98). In a passage that I quoted earlier in chapter three, we saw that Durkheim believed that something inspires respect "when the representation that expresses it in the consciousnesses is endowed with such a force that, automatically, it gives rise to or inhibits actions, *exclusive of any consideration relative to the useful or harmful effects of one or the other*" (296). An individual's judgment either that a person or thing is worthy of respect or that it would be prudent to respect a recognized moral authority plays no role in Durkheim's account of moral authority.

The idea of moral authority, however, cannot be derived from the experience of the power of collective representations automatically giving rise to actions. At best, only the individual's feeling of being compelled can be explained in this way. Durkheim therefore did not escape the objection that he raised against theories that trace the idea of moral authority to the experience of physical constraint. He merely substituted an internal for an external mechanical compulsion. How or why this compulsion is conceived as a *moral* compulsion, Durkheim never explained.

Whether or not he can explain the *moral* aspect of religious forces, Durkheim nevertheless believed that his theory of the origin of this concept of force could account for their *sacred* character. Individual representations of ordinary empirical objects do not inspire in us anything like the respect that collective representations do, he asserted. These two kinds of mental representation he believed form "two circles of mental states" in our mind that give rise to the impression that we are in relation with two clearly separated realms of realities: the sacred and the profane (303–4). However, I have just argued that for Durkheim the respect induced in us by collective representations is nothing but the effect of their superior energy. Thus it would appear that he grounds the distinction between the sacred and the profane in our experience of a power that is superior to our own. The only difference between Durkheim's account of the origin of the idea of religious powers and the account given by the naturistic hypothesis is that for Durkheim the superior power experienced by the individual is that of society rather than nature.

Nevertheless, Durkheim thought that his hypothesis of a social origin of the notion of a religious force escapes the objection he had raised earlier against the naturistic hypothesis. The naturistic hypothesis, he had argued, has antirealist implications, reducing religious belief to a tissue of illusion. On Durkheim's hypothesis, however, the believer is not entirely deceived when he believes in a superior moral power, as there is in fact such a power: society. He claimed that the only error made by the primitive is to associate this power with the totem rather than with the clan (319). Proponents of the naturistic hypothesis, Durkheim believed, have failed to see that the purpose of religion is to provide a person with a representation not of the physical universe but of the society of which he or she is a member and of his or her relations with it. The fact that religious rites may not produce their desired effects on the world of nature is unimportant, Durkheim reasoned, because the real function of religious rites is to strengthen the bonds of the individual to his society. Even the ecstasy experienced during religious ceremonies is no hallucination according to Durkheim but the effect of "collective effervescence" (322–23).

Durkheim then provided additional reasons for accepting his account of the origin of the notion of religious forces. These reasons take the form of additional facts that he thought his hypothesis alone can explain. Thus his reasons belong to his extended notion of a crucial or decisive fact.

The first of these crucial facts is that the totemic principle is conceived in the form of a species of plant or animal. Species of animals and plants are chosen as symbols to represent the different clans because they are familiar, near at hand, and a source of food, he said. Also, plant and animal species provide a sufficient variety of types to represent all the clans in the tribe (334–35). When the clan assembles, he thought, people have a desire to express their feelings of group membership (553). However, he reasoned, because individual consciousnesses are closed to one another, people can communicate only by means of external signs that "translate" their internal states and that everyone can perceive and understand (329). The primitive then expresses his feelings of group membership in the only way he knows how, Durkheim stated: that is, by imitating the totemic species (553). When the members of the clan all utter the same cry, make the same gesture, draw the same picture, or indicate the same object, they each recognize that the others are expressing the same feeling. Under these conditions, Durkheim believed, the individual representations of this feeling in the minds of each person fuse with their representations of others expressing the same feelings, resulting in a collective representation. A totemic image thus serves not only as a focal point for these collective sentiments, he thought, but as a "constituent element" in their formation (329–30, 598). The primitive then experiences the superior power or intensity of this collective representation and attributes the cause of this feeling to the totemic image, Durkheim asserted. This attribution gives rise to the notion that the totemic image is suffused with some sort of principle or force. Because the animals or plants of the totemic species resemble the totemic emblem, they are believed to share this totemic principle (315–17).

Durkheim's account of the role of totemic images in the formation of collective representations in primitive societies reveals a theoretical innovation for him. Only towards the end of his career did he explicitly recognize a need for arbitrary or conventional symbols in the formation of collective representations.[7] The totemic image is arbitrary in the sense that one species of animal or plant would serve as well as another for the totem of a clan as long as this species was not already being used as a totem by another clan. The totem is conventional in the sense that its use rests on an agreement among the members of the tribe. Durkheim's account not only reveals his belief that symbols are needed for the formation of collective representations, but explains that what transforms a mere physical object into a meaningful symbol for a society is a process involving collective representations. The meaning that this object has acquired and thus what makes it a symbol or totem exists in the collective representations held in the minds of the members of the society. Durkheim believed that these symbols serve to keep the social sentiments alive even when the group is not assembled. "Social life is possible only by means of a vast symbolism," he claimed (331).

Durkheim then cited what could be described as a decisive fact against both the animistic and the naturistic hypotheses. Neither the anthropomorphic instinct postulated by the animistic hypothesis nor anything in the experience of nature, he believed, can account for the fact that primitive classification schemes confuse things from the animal, plant, and mineral realms (337). Durkheim, on the other hand, believed that his theory of the origin of the idea of a religious force can account for this fact. A primitive will put an animal and a human being together in the same class when he considers them to be animated by the same totemic principle (338, 464).

Of course, defenders of the animistic and naturistic hypotheses could have tried to save their theories with the ad hoc hypothesis that the primitive thinks differently than we do. Specifically, they could have argued that the primitive does not recognize the principle of noncontradiction when he says, for example, that a man is a kangaroo. Durkheim, however, would have responded by pointing out that for a primitive to say this is no more a contradiction than for a scientist to say that heat is the motion of molecules or that light is an electromagnetic vibration (341). In other words, Durkheim believed that for a person to regard himself as related to an animal is not a matter of that person simply failing to recognize a contradiction.

Durkheim's analogy between totemists' accounts of their relations to animals and contemporary physical accounts of heat and light constitutes one of his more insightful arguments. The same argument has been used more recently by Paul Churchland to reply to the objection that the identification of the mind with the brain is a "category mistake." Both Durkheim and Churchland have argued, in essence, that what counts as a category mistake depends on what scientific theories are generally accepted (Churchland 1988:29–31). For Durkheim, a totemic system of classification functions like a scientific theory. What would count as a category mistake for the primitive depends on his or her system of totemic classification.[8]

Durkheim regarded it as a strength of his program that, unlike say Lévy-Bruhl, he had no need to postulate the existence of an "abyss" between primitive religious and modern scientific thought (1912a:342). Durkheim's attitude towards the claim that primitives think differently than we do, I think, reveals his opinion that this claim is an ad hoc hypothesis. That he had no need of this ad hoc assumption in order to account for the facts about primitive classifications constituted for him a decisive fact in favor of his own theory and against the theories of his competitors. For Durkheim, it would appear, an ad hoc explanation was no explanation at all.

The assumption that primitive mentality is no different from ours also allows greater empirical scope for Durkheim's sociology of knowledge. Without this assumption, he could not have thought that analyzing the primitive categories of thought would shed any light on our own. Specifically, he would not have been able to argue that the primitive notion of a totemic principle was the origin not only of our concept of a religious force but of our concept of force in general. Durkheim defended this thesis by adducing the explanatory roles played by the notions of *wakan* and *mana*, which he asserted are similar to the role of the concept of force in science (290–92). For Durkheim, to explain the phenomena is to establish relationships among them that are not revealed by sense experience but created by the mind. Despite the fact that scientific explanations rest on more controlled observations, he argued, they are nevertheless of the same "nature" as primitive ones: both show how one thing "participates" in another. Although we might not consider primitive explanations to be good explanations, he contended, it was essential for the mind not to remain enslaved to appearances. Once man had the sentiment that there were internal connections among things, science and philosophy became possible (339–41).

The Category of Causality

To explain the phenomena in terms of the forces that produce them is to provide a kind of causal explanation of the phenomena. For Durkheim, in fact, the concept of force or power is implicit in the concept of causality. An analysis of the concept of causality, he believed, would reveal its social origins and thereby corroborate his account of the origin of the idea of force. "Reason will confirm this result of observation," he said, and show that the idea of power or force could not have any other source (519). Specifically, he argued that the individual's experience either of external nature or of the workings of his own mind could not be the source of this idea. By elimination, he concluded that the source of this idea must be in the individual's experience of social forces. In effect, Durkheim's argument consists of showing that he could derive from individual psychology the same result that he had obtained through an analysis of the facts concerning totemic beliefs.

To say that Durkheim derived a sociological conclusion from psychological premises is not necessarily to say that he violated his injunction against psychological explanations in sociology. It could be argued in his defense that he was attempting to derive only the most primitive concept of causality and that such primitive concepts consist in the first sort of collective representations that are formed from individual representations. He was not attempting to derive more advanced concepts of causality, which consist in collective representations formed from other collective representations, from individual representations. He said, in fact, that the concept of a cause is not the same for the scientist and the lay person and may differ even between physicists and biologists (527 n. 1). He made no attempt to account for the scientist's concept of cause. Nor is it clear that his derivation of the primitive's concept of cause from individual psychology was intended as an explanation. In this instance, his appeal to psychology could be characterized instead in terms of the inverse deductive method of John Stuart Mill. According to this method, as we mentioned in the previous chapter, a low-level empirical generalization is regarded as unreliable unless it can be deduced from a higher-level theory. The inverse deductive method is therefore a method of confirmation rather than a method of explanation.

In his psychological argument for the social origins of the idea of a cause, Durkheim chose David Hume for his antagonist over any of the proponents of the animistic and naturistic hypotheses. Hume, of course, had raised skeptical arguments against our claims to have knowledge of causes. Durkheim, on the other hand, claimed that his immediate concern was merely to explicate the concept of a cause and not to establish its well-foundedness (519). However, to the extent that he wished to show that our idea of a religious power and thus of a cause is grounded in reality, Durkheim appears to have wanted to demonstrate that these ideas are well founded.

Durkheim's argument closely follows Hume's analysis of the notions of force, power, and necessary connection. Like Hume, Durkheim treated these notions as more or less synonymous with the idea of cause. Hume's analysis starts from the premise that an idea is meaningless unless it can be shown to originate in a sense impression or a combination of impressions. However, Hume argued, he cannot discover any impression of power in any single instance of the operation of causes in external bodies (1748:41–42; cf. 1739:158). Like Hume, Durkheim also began with a consideration of our experience of external objects and did not find any idea or sense impression of power there:

The senses enable us to see only the phenomena that co-exist or that follow one another, but nothing of what they perceive can give us the idea of that constraining and
determining action that is characteristic of what one calls a power or a force. . . . the
internal process that links these states escapes [the senses]. (1912:519–20)

Durkheim then followed Hume in turning to a consideration of our internal experience as a possible source for this idea of power. Hume had argued that just as we receive no impressions of power from external bodies,
we are similarly unable to observe any power in any single instance of the
operation of the will upon the body or upon the mind in calling up ideas
(1748:42–45). Like Hume, Durkheim rejected the will as the source of the
idea of power, but he did so for different reasons. Instead of arguing as
Hume had that we cannot perceive the operation of our will, Durkheim
cited two disanalogies between the concept of an individual human will and
that of power or force. Unlike human wills, he maintained, forces are "anonymous, vague, diffuse" and "impersonal" (1912a:520). Furthermore, he
said, "the forces of nature have always been conceived as capable of passing
from one object to another, of mingling, combining, of transforming themselves into one another." The self, on the other hand, he argued, "cannot
change its substratum" or "extend itself from one to another"; it is "incommunicable" (521). Underlying Durkheim's line of reasoning we again find
the assumption that the primitive thinks like we do, recognizing the same
analogies and disanalogies that we would perceive. For Durkheim, this assumption appears to be necessary for reasoning about the origin of our categories at all. Without this assumption, the fact that Durkheim saw
disanalogies between the notions of forces and human wills would be
wholly irrelevant to the issue of the origin of these ideas.

Not wishing to miss an opportunity to criticize his opponents, Durkheim added that as long as the animistic theory was regarded as true, the
hypothesis that the idea of power was conceived in the image of the human
will seemed to be confirmed by history (520). Durkheim's implied attack on
the animistic theory illustrates his methodological concerns. He revealed his
belief that empirical confirmation alone does not make a claim reliable. Even
false claims can have confirming instances. In order to be considered reliable, Durkheim thought, empirical generalizations must be explained by a
well-founded higher-level theory.

Durkheim then concluded that the source of the idea of force or power
must satisfy the following two conditions: the idea must come from our internal experience and it must be the idea of something that is impersonal or
communicable. The only things that can satisfy both of these conditions, he
believed, are social forces. Collective forces satisfy the first condition, he
said, because they are "entirely psychical; they are made exclusively of ideas

and objectified sentiments." That they satisfy the second condition is true by definition, he thought, as they are the product of cooperation (521).

It is true, Durkheim acknowledged, that physical forces also satisfy the requirement of being impersonal. Physical forces, however, do not satisfy the condition of being part of our internal experience. Durkheim argued that we can perceive only the effects of physical forces, the causes themselves lying "outside the circle of perception." "It is otherwise with social forces," he said, "they are part of our internal life, and, consequently, we know not only the products of their action; we see them acting . . . . In a word, this constraining and necessitating action that escapes us when it comes from an external thing, we here seize in the flesh because it takes place entirely within us" (522).

Earlier in his career Durkheim had believed that social forces, like physical forces, were outside of individual consciousnesses and could be known only through the "constancy of their effects" (1897a:348; cf. 1897e:250). But now, in order to maintain a distinction between physical and social forces, he seems to have reversed his position on social forces. Nor did he make clear how it is that we are supposed to be able to introspect social forces in action, especially given his attacks on introspection elsewhere. The introspection of social forces in action would seem to require an apperception of a collective representation giving rise to an action. But we no more have conscious perceptions of mental activity than we have sensations of retinal images. The mere substitution of social forces for individual wills does not allow him to escape Hume's argument that we can "only feel the event, namely the existence of an idea, consequent to a command of the will" and not "the manner, in which this operation is performed" or "the power, by which it is produced" (1748:45). Mental representations, in other words, are purely passive states of affairs. This is not to rule out the possibility of a kinetic image or series of images, but kinetic imagery is not thereby dynamic.

Another argument Durkheim raised against the hypothesis that the idea of force has its origin in the idea of the human will is that implicit in the notion of force are the concepts of domination and subordination. These concepts could have their source only in society (1912a:522–23).

Durkheim then argued that the idea of a cause contains the idea of a necessary connection as well as the idea of force and suggested that Hume had failed to account for this idea of necessity. The "empiricists," Durkheim contended, have failed to account for the sense of "authority" that the idea of necessity carries with it and that indicates that the mind is not its artisan. Under the pretext of explaining the necessary character of the causal relation, he claimed, the empiricists actually made it disappear (524, 526).

Hume, not having found the source of the idea of power or necessary connection in any single instance of the operation either of bodies or of the will, turned to a consideration of the repetition of similar such instances. Arguing that repetition alone does not give us any new idea not found in the particular instances, Hume suggested instead that the idea of a necessary connection comes from the feeling of anticipation that results from the habit we form of expecting the effect (1739:163, 165; 1748:50–52). Durkheim, on the other hand, distinguished the "state of personal expectation" from "the conception of a universal order of succession that imposes itself on the totality of minds and events" (1912a:630). The *sensation* of regular succession must be distinguished from the *category* of causality, he argued. The sensation of regular succession is individual, subjective, incommunicable, and made from our personal observations. On the other hand, he said, the category of causality is the work of the collectivity. It is a "framework" in which our experiences can be arranged and which allows us to think and to communicate with others about these experiences, he argued. The framework must be distinguished from its contents, he asserted, as it has a different origin. It is not a mere summary of individual experiences, he contended, but something that is "made to answer to the exigencies of communal life" (526).

The terms in which Durkheim drew the distinction between the category of causality and the sensation of regular succession recall the language of his account of the concept of classification by genus and species. I suspect this latter account, going back to 1903, was prior to and served as a model for his account of the category of causality. In his account of the concept of classification, as I argued above, his distinction between the framework and its contents rests upon a distinction between two kinds of compounded mental states: collective representations and individual generic images. Similarly, as I will demonstrate below, although he claimed that the concept of causality has an origin other than individual sensation, his account of this concept also appears to rely on his theory of the compounding of mental representations.

Durkheim claimed to have found in the obligatory nature of certain religious rites the source of the idea of authority which he believed is implicit in the idea of a necessary connection. The source of the idea of authority, he asserted, is no mere collective expectation that if we perform the proper ceremonies, the spirits will favor us with the desired ends. It is not sufficient merely that everyone have the same associations, he believed, for then "an imperative rule of thought would not be constituted for anyone" (524–25). Because an important social interest is at stake, he argued, society obligates us to repeat this rite as often as is necessary, and thereby requires us to be in a certain state of mind. For example,

To prescribe that one must imitate the animal or the plant in order to make it revive, is to pose as an axiom that cannot be placed in doubt that like produces like. The opinion cannot be permitted to individuals of theoretically denying this principle, without permitting them at the same time to violate it in their conduct. [Society] therefore imposes it, as well as the practices that derive from it, and thus the ritual precept is doubled by a logical precept that is only the intellectual aspect of the first. The authority of one and that of the other derive from the same source: society. (525).

The error of empiricism, Durkheim continued, has been to see in the causal link a construction of speculative thought. Speculative thought, however, can give rise only to hypothetical, provisional views, he argued, and not to an "axiom" that is placed beyond doubt (526–27). Now even if, in the spirit of charity, we agree to overlook Durkheim's somewhat questionable concept of an axiom, his account still remains highly problematic. To begin with, it would appear to be possible for a society to allow its members to have theoretical doubts about the efficacy of certain of its religious rites but still to make it absolutely obligatory for its members to perform them. Indeed, it would seem to have to allow such doubts, for how would society be able to determine what anybody's actual opinions were as long as he or she carried out the rites in question in the appropriate manner? As Durkheim himself pointed out, the mental states of individuals are private or "incommunicable." In order to coerce them into such ceremonies, it would seem that all that is required is that society possess sufficient *power* over individuals. Their intellectual assent would appear unnecessary. Durkheim's reason for believing that society requires such intellectual agreement seems to be the belief that it is only in this way that society gains power over the individual. Presumably, he thought that for an individual to have doubts about the beliefs his society would impose upon him would be for him to have individual representations that would oppose and thus weaken the collective representations. For Durkheim, in other words, the mechanism of social constraint is entirely psychological, taking place in the minds of individuals through the superior force of collective over individual representations. The superior force of collective representations for Durkheim, as I explained in chapter 3, derives from the combined strength of the individual representations from which they were compounded. Durkheim said nothing about the relative strength of generic images.

In sum, in Durkheim's account of the category of causality, the source of the idea of authority would then appear to be the superior strength or vividness of collective relative to individual representations. Since Durkheim also maintained that the origin of our idea of power is in our experience of the social forces to which collective representations give rise, the ideas of power and authority appear to have more or less the same source for him. But as I argued above, to derive the idea of authority from the experience of the power of collective representations is to make it difficult for him to maintain that the idea of authority has any moral or religious aspect. His argument against the naturistic hypothesis, however, turns on his claim to be able to account for the religious aspect of religious forces. In sum, it seems that Durkheim did not provide a better account than the naturistic hypothesis of the origin of the concept of force.

Totemism and the Concept of the Soul

After his attempt to provide a totemistic account of the notion of religious forces postulated by the naturistic hypothesis, Durkheim turned his attention next to the animistic hypothesis, according to which religion began with the animist belief in a soul. He argued that totemism is an earlier religion than animism by showing that the idea of a soul could be derived from the notion of the incarnation of the totemic principle in individuals. He was careful to make clear that he meant that the idea of the soul derives from that of the totemic principle in a purely *logical* and not a chronological sense (381). Even the most primitive societies have some notion of the soul, albeit a confused one, he asserted (343–48). However, he said, the concept of a soul logically presupposes the idea of a religious power while the idea of a religious power does not presuppose that of a soul (382).

Durkheim defended his hypothesis that the concept of the soul derives from that of a totemic principle by presenting a "well-made experiment" (352) that favors the totemistic over the animistic hypothesis. His argument consists of showing that the totemistic hypothesis provides the better explanation of conflicting ethnographic reports of Arunta beliefs about conception. According to Spencer and Gillen, Durkheim said, the Arunta believed that human souls are reincarnated generation after generation in the fetuses of women who conceive near the sacred places inhabited by the souls of the dead (353–54). Now Durkheim claimed that according to Strehlow, on the other hand, the Arunta had no concept of reincarnation, believing the souls of the dead to return for only a short while and only as guardian spirits. Among the Arunta, Durkheim interpreted Strehlow as saying, conception was attributed simply to the powers of the souls of ancestors who went into the ground near the sacred places where the women quickened. For Durkheim, however, these seemingly different accounts were but two versions of the same "mythic theme." According to both accounts, conception for the Arunta is ultimately a matter of the incarnation of the totemic principle in the fetus (357–64). This notion of conception represented a primitive notion of ensoulment for Durkheim. It is precisely because the totemistic hypothesis derives the notion of the soul from the totemic principle that it alone can reconcile these conflicting ethnographic reports, he seems to have believed.

Because the same effect always has the same cause, Durkheim argued, one can assume that what is true of the Arunta tribe will be true of the rest of Australia and elsewhere (352). He then adduced additional evidence corroborating his hypothesis that the idea of the soul derives from the notion of the incarnation of the totemic principle in individuals (367f.). However, the connections between this evidence and his hypothesis are rather loose. Durkheim's hypothesis may be able to explain the special relationships individuals have with their totems but there could be other explanations as well. He does, however, attach special probative significance to totemic beliefs in reincarnation into and from animals. Repeating an argument he had made earlier in the book, Durkheim contended that reincarnation between animals and humans cannot be explained by the hypothesis that the soul is "an essentially human principle" as it was for Tylor (375). Here, however, Durkheim used this argument as a decisive fact in favor of his theory of the origin of the idea of the soul rather than as a reason for believing that totemism cannot be derived from animism.

Durkheim then cited a number of additional facts that he believed favor the totemistic over the animistic hypothesis. The first such fact concerns the sacred character of the soul. For the soul to be nothing but a solution to the problem of dreams would not explain its religious character, he argued. The sacred character of the soul, Durkheim believed, is explained by the fact that it is a part of the collective soul. The persuasiveness of this argument then depends upon his being able to explain the sacred character of the collective soul. As we saw above, however, Durkheim tried unsuccessfully to derive its sacred and moral character from the superior strength of collective relative to individual representations (375–78).

Durkheim then anticipated the objection that the idea of the individual soul does not derive from that of the collective soul because the individual soul, unlike the collective soul, is not impersonal and cannot spread itself everywhere. In reply, he asserted that this way of conceiving the individual soul is relatively recent. For the Australian, he claimed, the soul is thought to be diffuse, able to divide and duplicate, and so on. Such beliefs about the soul survive even in supposedly advanced religions in the form of beliefs about holy relics (378–80).

Another decisive fact he cited in favor of his theory is the belief in the autonomy of the soul relative to the body. By explaining this belief in terms of the individual soul sharing in the collective soul (388–89), Durkheim was at the same time trying to show that this belief has some foundation in reality. The idea of the soul as distinct from the body is not an illusion, he said, as it is grounded in the existence of two different kinds of representations in us, one relating to our body and the other relating to something higher (377). Durkheim explained the belief in the survival of the soul upon the death of the body as something that was invented in order to explain birth and to guarantee the perpetuity of the clan, rejecting alternative hypotheses for empirical reasons (382–84).

In the final chapter of book 2 of *The Elementary Forms,* Durkheim showed how his account of the idea of the soul allowed him to explain the subsequent development of the ideas of spirits and gods. However, some of the defenders of the animistic hypothesis also derived the ideas of spirits and gods from that of the soul. By showing that he could arrive at the same explanation, Durkheim was demonstrating to the reader that beliefs about spirits and gods could not be used as decisive facts *against* his hypothesis. He did, however, cite a class of facts that does seem to favor his theory. According to Durkheim, the animal characteristics of many of the polytheistic gods are the mark of their origin as totems (418–21).

In sum, although Durkheim's totemistic hypothesis explains a number of facts more successfully than the animistic hypothesis does, both hypotheses fail to explain the sacred character of the soul. For Durkheim, the sacredness of the individual soul depends upon that of the collective soul or totemic principle of which it supposed to be a part. However, Durkheim was not able to account for the sacredness of the totemic principle. To the extent that he defined religion in terms of a distinction between the sacred and the profane, the inability to explain the sacred character of the soul and the totemic principle is a major weakness in his theory.

Totemic Rites and the Social Character of Religion

For Durkheim, the strength of the totemistic hypothesis lay not merely in its ability to account for the origin of religious beliefs and concepts but above all in its ability to account for the social function of religion. The animistic and naturistic hypotheses conceive the purpose of religion as one of providing a system of beliefs that explain the world around them. Such theories explain religious rites as nothing but the expression of these beliefs. Durkheim found this conception of religion inconsistent with that which the faithful themselves hold. To a believer, he asserted, the function of religion is to make one feel stronger and better prepared for living. It is the cult or the system of rites that has this effect on individuals, he said. Durkheim argued that he can explain this effect through his theory of the social causes of religion (594–96).

In short, Durkheim thought that his theory of the social origins of religion was confirmed by its ability to account for the function of religious rites. According to Durkheim, their function is to reaffirm the bonds among the individuals that make up a community, providing them with a feeling of personal well-being while at the same time ensuring the continued existence of the community itself. Because the function of a social phenomenon is often to preserve its cause, as Durkheim explained in *The Rules,* he seems to have thought that to show that religion serves to preserve society would corroborate his theory that society gives rise to religion in the first place.

Durkheim defended his hypothesis concerning the function of religion through the elimination of alternative accounts of the function of religious rites among totemic societies. His chief antagonist was Frazer, a defender of the animistic hypothesis, who denied that totemism is a religion and maintained that it is a form of magic that serves a purely economic function. Durkheim provided both empirical and conceptual arguments against Frazer's interpretation of totemism. Interpreting totemic rites as a kind of sacrifice without gods, Durkheim turned next to William Robertson Smith's theory of sacrifice. According to Smith, the purpose of the rite of sacrifice is to establish a bond between a people and their god. Although Durkheim criticized Smith elsewhere for accepting certain aspects of the animistic theory, he did not reject Smith's theory of sacrifice. Instead, Durkheim revised Smith's theory in such a way that it would be able to account for more of the facts. Durkheim did not seem to consider any other account of the function of religious rites that may have been proposed by the defenders of the naturistic hypothesis.

Frazer's Account of the Intichiuma Rite

Spencer and Gillen reported on a rite called the Intichiuma performed among Australian totemists. As the ostensive purpose of the Intichiuma is to make the totemic species reproduce, Spencer and Gillen assumed that the Australians performed this rite in order to assure a plentiful supply of food for the tribe. Although the members of a clan would not eat a representative of their clan's totem, each clan was responsible for the reproduction of its totemic species in order to provide food for the other clans, Spencer and Gillen said. According to their interpretation, totemism is a system of economic cooperation among the clans in a tribe (529). Accepting the interpretation of Spencer and Gillen, Frazer argued that since the Intichiuma has a purely economic function, it is not truly a religious but only a magical rite (Jones 1985:83–84).

However, Durkheim said that Strehlow has found counterevidence to the economic interpretation of the Intichiuma. According to Durkheim, Strehlow argued that some of the totemic species were not edible or useful in other ways and some were even harmful. Thus, Durkheim concluded, the natives did not always have an economic interest in ensuring the reproduction of the totemic species. Furthermore, what Spencer and Gillen claimed to be the purpose of the rite is not what the natives themselves said it is. When asked the reason they perform this rite, Strehlow asserted, the natives said they perform the rite because their ancestors did. For Durkheim, this constituted an admission on the part of the Australians that they celebrated the rite for the eminently social reason of upholding a tradition (1912a:530).

Durkheim also rejected Frazer's thesis that totemism is a form of magic rather than a religion. Citing work on magic by his close associates Hubert and Mauss, Durkheim argued that far from magic being a predecessor to religion, it presupposes religious ideas (517–18). In addition, Durkheim attacked the theory of magic upon which Frazer's interpretation of the Intichiuma is based. Frazer distinguished two kinds of magic and explained both in terms of the principles of associationist psychology. Homeopathic magic rests on the assumption that like produces like and is therefore a mistaken application of the principle of the association of ideas by similarity, he said. Contagious magic, on the other hand, he claimed, rests on the assumption that things that were once in contact continue to act upon one another after they are separated and is thus a mistaken application of the principle of the association of ideas by contiguity (Frazer 1922:12–13).

Frazer's account of homeopathic magic may apply to casting spells and charms, Durkheim granted, in which two distinct things are assimilated: the image and the original model that the image represents. However, Durkheim said, the Intichiuma rites in which Australians imitate their totemic species in order to encourage it to reproduce cannot be interpreted as a form of homeopathic magic. In such rites, only the image and not the original model is given, he argued, as the ostensive purpose of the rite is to bring about a new generation of the totemic species. That is, unlike the casting of a spell on an already existing person by sticking a pin in a waxen image of that person, the Intichiuma is directed at the creation of animals that do not yet exist (510).

Earlier in the book, Durkheim had also criticized the use of the concept of contagious magic in the interpretation of taboos, although without attributing this interpretation to Frazer. Durkheim maintained that the "contagious" character of the sacred, which is revealed through the analysis of the totemic system of interdictions, is explained better in terms of the social origins of the idea of sacredness than in terms of the association of ideas. According to the associationists, he said, the sentiments that a sacred thing or person inspires in someone spread by contagion from the representation of that thing or person to the representations of the things with which that thing or person comes into contact. Thus, according to the associationist, the faithful believe that the sacred character of sacred things actually passes into other things through contact. Taboos are instituted in order to prevent this diffusion of sacredness. Durkheim rejected this explanation of taboos, arguing that the same kind of interdictions can be found in contemporary religions such as Catholicism and that it is hard to believe that today's Catholic can be the dupe of the association of mental images (460). Not only does this argument rest on an analogy between primitives and moderns but what it says is difficult to reconcile with Durkheim's other assertions about what contemporary Catholics supposedly believe. If, as Durkheim claimed, Catholics believe that a relic contains a piece of a saint's soul and if they confuse moral obligation with physical coercion, they could easily be misled into mistaking relations among ideas for relations among the things represented in ideas.

Durkheim then offered his own account of totemic taboos. He theorized that a sacred object is expressed in the mind by a representation of high mental energy. Profane objects are not allowed to come into contact with sacred objects, he reasoned, in order to prevent the weakening of the mental representations of sacred objects through the diffusion of their energy into representations of profane objects (453–54). Sacredness is not a property of the totemic object itself, he asserted. The religious forces responsible for the sacred character of the totem he thought have their source in the feelings of comfort and dependence that society induces in the mind. These emotions are contagious and spread to other mental states. According to Durkheim, contagion is not some secondary process by which sacredness, once acquired, is propagated, but is the very procedure by which sacredness is acquired in the first place. The contagiousness of the sacred is explained by his theory of religious forces and thus serves to confirm his theory, he claimed (462–63).

Durkheim may have had valid reasons for rejecting Frazer's interpretation of the Intichiuma. However, it seems that his only basis for preferring his own account of taboos or even for thinking that it differs from the associationist account was his belief that only his theory could explain the very sacredness of the totem. His argument against associationism is then undermined by his inability to account for the religious character of religious forces.

Smith's Theory of Sacrifice

Without any knowledge of the facts about Australian totemism, Smith deduced that totemism, which he believed to be the earliest religion, must have had some form of sacrificial rite, Durkheim said. Smith's hypothesis of totemic sacrifice was now confirmed, Durkheim claimed, as the Intichiuma of the Arunta is a form of sacrifice (485). Before Smith, Durkheim reported, a sacrifice was thought to be an offering to a god analogous to the sort of tribute or homage that subjects owe to their princes. Smith, however, argued that this interpretation does not account for the fact that a sacrifice is a meal that the faithful believe themselves to take part in along with their god. For Smith, the purpose of the sacrifice is to establish a bond of kinship with the god (480–81). Finding absurd the thought that a god would depend upon human beings to be fed, Smith concluded that the idea of an offering to the god must have been a later addition that arose only when the gods began to be thought of as kings to whom subjects owed tribute (486–87).

For Durkheim, on the other hand, a sacrificial rite is both an act of communion and an offering that serves not so much to establish as to *renew* the bond between people and their god. Smith's reasons for denying that a sacrifice was originally an offering are refuted by the facts concerning the Intichiuma, Durkheim said. He reported that the Intichiuma begins with a rite directed at assuring the fecundity of the totemic species and is then followed by a ritual meal. Durkheim interpreted the first rite as a kind of offering upon which the survival of the totemic species is believed to depend (485–89).

Durkheim then explained why it is not absurd to believe that gods depend on men. The rhythms of nature are evidence to the primitive that the sacred powers on which they depend sometimes wax and sometimes wane, Durkheim claimed. As the primitive's life depends on life generally he cannot be indifferent to this spectacle and he then feels that he cannot let his gods die. Because the ostensive purpose of the Intichiuma is to encourage rebirth, he affirmed, the periodicity of these rituals bears the mark of the seasons. The primitive is not entirely deceived in believing that his gods depend upon him, as in fact the sacred beings would cease to exist as such if nobody believed in them. Religious beliefs must be constantly reinforced through the kind of social gatherings in which they are formed. By the very act of uniting, he believed, people mutually comfort one another and this emotion overcomes their personal doubts about their gods. So, Durkheim concluded, we can reject Smith's assertion that gods do not depend on men. However, it is not gifts of food but people's *thoughts* that keep their gods alive. After all, the totemic principle is nothing but the society itself "hypostatized," Durkheim said. In effect, he continued, the function of the rite of sacrifice is to periodically recreate the moral being called society. Men do not therefore deceive themselves when they think that something outside themselves is born again (491–99).

Durkheim on the Function of Religious Rites

For Durkheim, as we have seen, the function of religious rites is to renew the bonds among human beings in society rather than to establish a bond between human beings and a god. However, as the idea of god is nothing but a symbol of an underlying social reality (495), he saw his theory as deriving from and not as an alternative to Smith's theory of function of sacrifice. Having provided his reasons for modifying Smith's theory, Durkheim next attempted to corroborate this revised theory by showing that it could explain some additional facts.

The first thing that Durkheim believed his theory of the function of religious rites could explain is the fact that primitives seem to think that their rites are actually effective. Performing the rites produces feelings of well-being among the participants, he claimed, which then make them think that they have succeeded in reaching their goal. In other words, he said, the moral efficacy of the rite, which is real, leads them to believe in its physical efficacy, which is illusory. According to Durkheim, the feelings of well-being produced by a religious ceremony are "like an experimental justification" of the practices of which it consists. The primitive's capacity for fallacious reasoning of this sort then raises again the issue of the extent to which primitives and moderns think alike. Anticipating this question, Durkheim adduced what he took to be the facts regarding the justification of faith today. Preachers, he claimed, do not try to prove their faith point by point but emphasize the moral comfort that results from practicing the religion. The faithful may doubt the efficacy of each rite taken singly, he asserted, but participate in all the rites anyway. The true justification of these rites, he affirmed, lies in their effects upon the mind of the believer and not in their ability to achieve their ostensive purposes (513–15).

Faith, Durkheim went on to explain, is "impermeable to experience." The intermittent failures of the Intichiuma to make the totemic species reproduce do not dissuade the Australian from his beliefs. Nor is the attitude of the faithful all that different for Durkheim from that which lies behind the scientific method. One does not always reject a well-supported scientific law in the face of an apparent disconfirming fact, he pointed out. "It is necessary to be assured that this fact allows for only a single interpretation and that it cannot be accounted for without giving up the proposition that it seems to disconfirm" (515). The Australian, according to Durkheim, treats in a similar manner facts that would seem to be antagonistic to his or her beliefs. As the Intichiuma is celebrated just before the start of the rainy season when life is reborn in the Australian desert, the Australian's faith in this rite seems to him to be supported by a preponderance of his or her experience as well as by the good feelings it produces. The apparent failures of this rite are the exception rather than the rule and are attributed usually to sorcery, Durkheim said. When the rainy season and the renewal of life arrives early, before the tribespeople have celebrated their Intichiuma, the premature abundance of life is attributed to an otherworldly Intichiuma celebrated by spirits (515–16).

In other words, primitive thought for Durkheim involves nothing more unusual than drawing conclusions through the fallacy of post hoc ergo propter hoc and defending these conclusions with ad hoc hypotheses. Unlike some contemporary thinkers, when Durkheim defended the thesis that primitives and moderns think alike, he did not try to argue that all human beings adhere to certain "universal" (Lukes 1970:208) or "necessary" (Hollis 1970:218) criteria of rationality, such as *modus ponens* or the principle of noncontradiction. To argue in this way is to create an easy target for one's relativist opponents (for example, Barnes and Bloor 1982:35f.). The target is destroyed as soon as someone discovers a primitive committing a logical fallacy. Lucien Lévy-Bruhl, for example, according to Durkheim, opposed primitive religious thought to modern scientific thought on the grounds that primitives do not recognize contradictions and are indifferent to empirical evidence (1913a(ii)(6) and (7):34–35). Durkheim, on the other hand, recognized that even good scientists will hold on to a thesis that is opposed by contradictory evidence. Far from opposing modern scientific thought to primitive religious thought, furthermore, he saw science as having grown out of religion, depending on it for having first elaborated the basic categories of thought that make logical thought possible (ibid., 35–36).

Durkheim believed that his theory of the function of religious rites could also explain those rites for which the ostensive purpose is *not* to make the totemic species reproduce. Sometimes, he pointed out, the Intichiuma celebrates a totem, such as the laughing boy totem, that is not a plant or animal species that would reproduce (1912a:538–42). Australians also perform rites that he described as dramatic representations for the purpose of recalling the past (531). The purpose of these commemorative rites, Durkheim believed, is to revivify "the most essential elements of the collective consciousness," that is, to reaffirm the mythic beliefs of the tribe (536). These rites connect the present members of the tribe with their past and thereby strengthen their attachment with the tribe (541).

The theory that religious rites are the means by which the group periodically reaffirms itself also explains the ambiguity of the ostensive purposes of totemic rites, Durkheim believed. He pointed out that the constitutive rites of the Intichiuma can be used for purposes other than that of assuring the reproduction of the totemic species, such as serving as initiation rites. Also, he argued, the Australian believes that the same end can be brought about through different rites. For example, the reproduction of the totemic species may be assured by oblations, imitative rites, or dramatic representations. These ambiguities, he reasoned, demonstrate that the actual function of a rite is not the effect at which the faithful aim. The essential function of a rite is to bring people together to experience common sentiments and to express these feelings through common actions, Durkheim believed. The particular nature of these sentiments and actions are of only secondary importance.[9] Durkheim concluded with a hypothetical reconstruction of the history of these rites, in which he argued that their ostensive purposes are bound up with a mythology that the primitive subsequently invented in order to explain these rites to him- or herself (548–55).

Durkheim then considered what he believed to be a final decisive fact that favors his over Smith's theory of the function of religious rites. According to Durkheim, although Smith acknowledged that sacred beings can be malevolent as well as beneficent, he failed to explain this fact.[10] Durkheim believed that he could explain this fact on the basis of his account of the function of piacular or sad rites, such as rites of mourning (588–89). For Smith, the function of piacular rites is to propitiate an angry spirit. According to Durkheim, Smith thus believed these rites to date only from the time when conceptions of sacred beings were first introduced into religion.

Durkheim denied the relatively late origin of piacular rites, claiming that the expiatory sacrifices characteristic of piacular rites may be found among Australian totemists. He argued that expiatory rites do not depend upon myths about sacred beings for their existence. In primitive religions, Durkheim asserted, rites involving such things as abstinences or the letting of blood are believed to stop famines or cure illnesses directly (581–82). As the group assembles for these sad rites they comfort one another, strengthen their feelings of group solidarity, and become happier (573–74). Just as in the Intichiuma rites, these feelings of well-being then lead the primitive into believing that the rites have been effective. The notion that these rites produce their effects by propitiating a spirit or god was introduced only later in order to make these rites easier to understand, Durkheim said (582–83). Expiatory rites such as the letting of blood are a kind of offering, he reasoned, and the making of an offering led to the idea of there being someone to whom the offering is being made (575; cf. 491). Because religious forces are nothing but the expression of collective sentiments, he contended, the idea of an evil power or spirit is the product of piacular rites much as the idea of a benevolent power results from happier celebrations. Evil powers can even be transformed into benign spirits as the affective state of the collectivity changes as a result of these piacular rites (588–91; cf. 574–75). In sum, good and evil religious forces for Durkheim corresponded to two opposed states of collective life: well-being and ill-being (591).

In conclusion, Durkheim's theory of the function of religious rites differs from Smith's through substituting society for god. For Durkheim, the function of religious rites is to strengthen group solidarity rather than to establish a bond with a god. His theory of the social function of religion allows him to account for the tenacity of religion. Societies, he believed, will always have a need to reaffirm themselves through religious rites (609–10) and mythologies will continue to be invented in order to justify these rites (615–66). Religion would not persevere, he argued, if its purpose were merely to provide explanations and if religious rites were meant only to appease the sacred beings posited in these explanations. Under these conditions, religion would instead progressively give way to the superior explanations offered by science. Science and religion are compatible for Durkheim just to the extent that they fulfill different functions in society (614).

Methodological Summary

Durkheim ended *The Elementary Forms* on a conditional note. That is, he did not pretend to have *proved* his theories of the origins of religion and of the basic categories of thought. Rather, he argued that these theories provide the best explanations of the facts. Previously, he asserted, one faced the alternative of either reducing the higher faculties of man to sense and matter or explaining them in terms of a postulated reality beyond the realm of experience. Once it is granted that there exists a real entity that is society a new way of explaining man becomes possible that is subject to the control of experience, he believed. At the time, he could not determine how much of what is characteristic of human beings could be derived from society. But "what is necessary, is to try the hypothesis," he said, and that is what he claimed to have done in the present work (638).

Although society is a real entity for Durkheim, it cannot be directly observed. It consists in "collective representations" that exist only in the minds of individuals. However, hypotheses about this unobservable entity can be tested empirically through their logical consequences. We have seen that the method by which Durkheim "tries the hypothesis" derives from Bacon's method of crucial instances by way of Mill's methods of elimination. The method of crucial instances basically consists of comparing the implications of putative causal hypotheses with the facts and selecting the hypothesis in accord with the facts over the hypothesis that yielded conclusions opposed to the facts. Facts that agree with one hypothesis and disagree with the other are called crucial or decisive facts. The idea of a crucial fact as it is used in *The Elementary Forms,* however, appears to be much more comprehensive than the concept of a crucial fact in *The Rules.* Yet I believe that the methods of *The Elementary Forms* would be described more accurately as an expansion or a further development of his explicit methodology than as a deviation from it.

In *The Elementary Forms,* we have seen, Durkheim included within the scope of the concept of a crucial fact nonrefuting anomalies or those facts that one hypothesis can explain but about which the other says nothing at all. In this sense, the concept of a crucial fact is linked to the requirement that hypotheses seek the greatest possible empirical scope. Durkheim was also careful to explain all the facts that his competitors do so that they could not use the facts they explain as crucial facts against him. The concept of a crucial fact for Durkheim in addition includes facts that only one hypothesis can explain without either begging the question, drawing on ad hoc auxiliary assumptions, or giving rise to new problems that the hypothesis cannot explain. Hypotheses not only must explain the facts but explain them in an acceptable way.

The facts in comparison with which Durkheim evaluated hypotheses are very often not mere isolated facts about, say, the particular beliefs of a particular tribe, but generalizations relying on an interpretation of the facts. For example, Durkheim rejected hypotheses that fail to account for totemism *as a religion* or for the *sacred* character of the totem or for the very distinction between the sacred and the profane. The ability to explain this last "fact," which is of course a summary fact that he regarded as definitive of religion, Durkheim considered to be of the highest importance. In one case that we saw the crucial fact that required explanation was not even a fact about primitive beliefs or practices but the fact that different ethnographers had provided different interpretations of Arunta beliefs about conception. In other words, Durkheim seems to have included the ambiguity of the ethnographic record among the facts that a theory of primitive religion needed to explain.

Some of the things that Durkheim treated as crucial facts were not even facts in the usual sense but might be more accurately described as methodological or metaphysical assumptions of his program of research. For example, Durkheim rejected hypotheses that contain the implication that religious beliefs have no basis in reality. It is not, of course, a "fact" that religion *does* have any such basis. Nevertheless, Durkheim refused to accept that religion could be mere illusion and regarded it as a strength of his hypothesis that he could explain religious beliefs as obscure and confused representations of society. Perhaps he considered the assumption that all believers in religion are simply hallucinating to be ad hoc.

In addition to the use of crucial facts, there are a number of other ways in which *The Elementary Forms* follows Durkheim's explicit methodology. Most importantly, this book employs the method of historical analysis described in *The Rules,* according to which one tries to understand present conceptions by tracing their development from primitive, simple ideas. This method of analysis assumes the method of classifying societies that is also characterized in *The Rules,* according to which types of societies are classified and arranged in a linear progression according to their degree of com-

position. As I mentioned in chapter 5, Durkheim used a method of historical analysis in his papers on the incest taboo (1898a(ii)) and primitive classifications (1903a(i)) and in his published lectures on the family (1888c),
socialism (1928a), and the French educational system (1938a). Some of
these lectures date from the beginning of his career.

We also find in *The Elementary Forms* the same concern that we found in
his other works with defining one's term in order to avoid being misled by
common-sense notions. Durkheim began *The Elementary Forms* by defining
religion in terms of what he took to be its observable characteristics, much as
he defined key terms in other works. Terms are to be defined in this way to
ensure the testability of the hypotheses in which they are contained and un-
testable hypotheses are to be rejected. This concern with definition is reflected not only in Durkheim's definition of religion in general but in his
attention to how specific religions such as totemism are defined.

The methodological assumptions that lie behind his criticisms of his
competitors' hypotheses are also taken from *The Rules*. For example, Durkheim rejected hypotheses that violate his prohibition against sociological
explanations couched in terms of consciously pursued ends. Assuming a
concept of explanation that is again taken from *The Rules,* he criticized the
hypothesis that the function of religion is to explain the world on the
grounds that this hypothesis fails to explain how this function maintains
the cause of religion in existence. Although he recognized that the function
that a religious rite presently serves may not have been its original function,
he seems to have believed that the continued existence of a rite can be explained only in terms of its functional necessity. For Durkheim, the purpose
that a social phenomenon serves must be related to the preservation of the
society in which it is found.[11]

Durkheim's appeal to crucial facts in *The Elementary Forms* sometimes involved the use of the method of concomitant variation, although not as
often as in *Suicide* or even *The Division of Labor*. In all of these works, however, as Jean-Michel Berthelot (1989) points out, the method of concomitant variation is concerned only with the direction in which phenomena
vary and not with the mathematical relationship describing the way in
which one phenomenon varies as a function of the other. Durkheim's use of
concomitant variation in *The Elementary Forms* presupposes the method of
classifying societies according to their degree of composition. The degree of
composition of a society for Durkheim was a measure of the extent to which
it is either primitive or advanced. Durkheim assumed this system of classifying societies according to their degree of complexity not only for the testing
of hypotheses but for criticizing the evidence that others have adduced in
support their hypotheses. He believed that he was able to test claims that
some phenomenon is primitive against the extent to which this phenome-

non is found at different points in this scale. However, he never measured this extent of a phenomenon in a rigorous, quantitative fashion.

Robert Alun Jones (1986a:153), however, finds the way in which Durkheim would explain away apparent counterinstances to his own hypotheses according to where this evidence fits in his scale of development to consist of nothing but "ad hoc evolutionary speculations." For example, he said, Durkheim dismissed the evidence concerning the Arunta tribe that seemed to contradict his interpretation of totemism, calling the Arunta religion a "subsequent elaboration" (Jones 1985:84). In Durkheim's defense I would argue that Durkheim's assumption that the Arunta are more advanced than other Australian tribes was not ad hoc just to the extent that he maintained this assumption consistently throughout the work. That is, this assumption was supported by more than one piece of evidence. To defend Durkheim against Jones's specific charge, however, is not to say that Durkheim never resorted to ad hoc explanations. He assumed an ad hoc hypothesis, for example, when he accounted for the relative paucity of sculptures and paintings of the totemic emblem among the Australians in comparison with the Amerindians by asserting that the arts were less developed among the Australians (1912a:161).

Another way in which the arguments of *The Elementary Forms* may deviate from his explicit methodology involves Durkheim's use of the inverse deductive method, a method he had attacked in *The Rules*. In *The Elementary Forms* he made use of this method when he appealed to psychological considerations in order to confirm his hypothesis of the origin of the idea of causal forces. *The Elementary Forms* is not unique in this regard, however, Durkheim invoked psychological arguments in corroboration of his sociological results in his earlier works as well.

A final way in which Durkheim may have deviated from his methodological prescriptions concerns his use of analogies between the ways in which primitives and moderns think. He rarely had anything good to say about analogies and attacked others for relying on them. However, as I suggested above, the assumption that all human beings think alike appears to be not only presupposed by his method of historical analysis but necessary for any sociology of knowledge. In fact, as we have seen, Durkheim rejected theories of primitive religion that assume that primitives think differently than we do for making this very assumption. With this assumption, one could have little more than an ethnographic collection of curious beliefs and practices that one cannot hope to understand. In order to be able to explain these beliefs and practices, Durkheim assumed, one must suppose that at some level we all think alike. To say that the reason a primitive believes that a kangaroo is his kinsman is that the primitive has an alternative logic and thus thinks differently than we do is simply not an explanation for Durkheim.

However, as we have seen, when Durkheim said that primitives and moderns think alike, he did not mean that we all conform to some ideal principles of rationality. How we think for Durkheim is a not a matter of stipulation but a question for empirical investigation.

Jones (1981) has offered an interpretation of *The Elementary Forms* that would seem to call into question the very relevance of this discussion of Durkheim's methods, assumptions, and explanatory goals. He questions why Durkheim would have continued to hold the theory that totemism is the oldest form of religion at a time when this theory was already discredited among anthropologists. Jones attempts to answer this question in terms of Durkheim's personal feelings of rootlessness resulting from his status as an assimilated Jew. These feelings, Jones asserts, predisposed Durkheim to appreciate William Robertson Smith's account "of the stable and periodically reaffirmed bonds of the ancient Hebrew community" (1981:197). For Jones to provide this as the reason for Durkheim's adherence to the totemistic hypothesis, however, is for him to ignore Durkheim's *arguments* against the animistic and naturistic hypotheses. His explanation of Durkheim's adherence to the totemistic hypothesis is typical of externalist accounts in the history of science in that it looks plausible only if one is willing to overlook the basic assumptions about goals and methods that characterized Durkheim's research program. These assumptions constituted a set of cognitive values and norms that constrained theory choice for Durkheim. If Durkheim were to have abandoned the totemistic hypothesis, how would he then have been able to explain the social causes of religious belief, the primitive classification thesis, or the origin of the distinction between the sacred and the profane? In short, how could Durkheim have adopted one of the other theories of the origin of religion without abandoning his explanatory goals? Jones does not say.

Jones even insinuates that Durkheim's feelings of rootlessness led him to tamper with the facts: "Durkheim's aborigines often bear a closer resemblance to the ancient Hebrews than to anything described by Spencer and Gillen" (1981:200). Jones's complaint, however, overlooks the fact that Durkheim found the ethnographic record to be inconsistent. For Durkheim to have adopted what Jones (1981:195; 1986b:616) seems to take as the more acceptable hypothesis that the functions of totemism were primarily economic would be for Durkheim to have ignored Strehlow's counterevidence to the interpretation of Spencer and Gillen. Also, in order for Durkheim to have offered economic explanations of social phenomena, he would have had to abandon some of his basic assumptions about sociological explanation that characterized his program of research. Nor is the fact that Durkheim's Australians resembled ancient Hebrews all that surprising in the light of his explanatory goals and methodological assumptions. How

could Durkheim have believed that one form of religious life was the predecessor of the other unless they *did* resemble one another? Durkheim, as we
have seen, also worked very hard to make Australian totemists look like contemporary Catholics. These resemblances would be surprising only to an
unsophisticated reader who believed that Durkheim was attempting to "derive" his theories from the facts or that Durkheim would have been happy to
see his life's work refuted.

Jones's interpretation of *The Elementary Forms* resembles Selvin's account
of *Suicide* in that both are trying to find external causes for what they take to
be Durkheim's "mistakes." I would argue that just as Durkheim's decision
not to adopt new statistical methods was not a mistake in the light of his
conception of sociology, so, too, was his decision to adopt the totemistic
hypothesis not an error. Of course, to make this argument is to raise the
question as to whether his conception of sociology itself was a mistake. This
issue will be addressed in the concluding chapter.

Theoretical Considerations

In *The Elementary Forms* Durkheim continued to appeal to many of the same
explanatory mechanisms as well as to use many of the same methods of comparative theory evaluation that he employed in his earlier works. For Durkheim, the social forces that underlie religious experience are to be explained
ultimately in terms of the fusion of individual representations into collective
representations during periods of collective effervescence. The explanatory
appeal to the mechanism of the compounding of mental states can be found
in Durkheim as early as in his account of mechanical solidarity in *The Division of Labor*. In fact, *The Elementary Forms* can be read as an account of how
religion, especially through religious rites, gives rise to mechanical solidarity or to what in *Suicide* he called the forces of social integration.[12]

Of course, one does not find the terms *mechanical solidarity* or *social integration* in *The Elementary Forms*. The absence of these terms may be attributed to Durkheim's having found it unnecessary in this work to distinguish
mechanical from organic solidarity or social integration from regulation.
He had no need to express this distinction as he was concerned only with
primitive societies, which he believed to be characterized solely by mechanical solidarity or social integration. Only advanced societies with a high degree of division of labor, according to Durkheim, are characterized by
organic solidarity and social regulation.

Some of Durkheim's interpreters have perceived a shift over the course of
his career away from an emphasis on morphological factors towards an emphasis on mental representations as explanatory factors (Pickering 1984:48,
283–84; Davy 1911:43–44). The force of this objection depends upon an

interpretation of Durkheim's earlier works as having appealed to morphological factors that exist in some way independently of collective representations. But to think of the *structure* of society as having an independent existence for Durkheim would be absurd, since he believed that society itself depends upon collective representations. It would then be a mistake to interpret Durkheim as having tried to *reduce* social facts or collective representations to social structures. Furthermore, to maintain that he was no longer concerned with social morphology in *The Elementary Forms* would be to ignore both his account of primitive classifications and his insistence on comparing societies only when they are of the same social type.

One could raise the objection, however, that Durkheim's theory of the compounding of mental states violates his strictures against psychological explanations in sociology. As we have seen, Durkheim thought that the individual's experience of the social forces that result from the fusion of mental representations is the source of our concepts of force and cause. Thus his sociological theory of the categories at least in part depends upon this psychological mechanism of the compounding of mental states, just as his theory of social solidarity does. Similarly, E. E. Evans-Pritchard has argued that for Durkheim it is the emotional states of individual participants in periods of collective effervescence that is causally responsible for the beliefs and sentiments that emerge (1965:68). Pickering (1984:395–403) has attempted to defend Durkheim against this objection. Although the mechanism of collective effervescence takes place in the minds of individuals, he argues, effervescence consists in the individual losing himself as an individual and becoming part of the group. Pickering's defense is unsatisfactory, however, as he avoids the issue of precisely what this "mechanism" is and points instead to the sociological character of its effects. The latter is not what is in question: the objection concerns how these effects are to be explained. A better defense would be to return to Durkheim's 1898 distinction between two kinds of collective representations and argue that he intended the mechanism of the fusion of mental states in periods of collective effervescence to explain only the original and not the subsequent kind of collective representations.

Nevertheless, there is something going on during these periods of collective effervescence that is irreducibly social. The social forces experienced by the participants are all referred to a totemic object, which thereby acquires the status of a symbol. As we saw above, Durkheim now recognized that arbitrary and conventional symbols play a role in the formation of collective representations while these collective representations at the same time confer meaning on physical objects and transform them into symbols. Even with this theoretical emendation, however, Durkheim was still unable to explain further developments and changes in collective representations. That is, he could explain only the original type of collective representations that are formed from individual representations among members of primitive societies. Still lacking the laws of collective ideation that he had been promising, he was unable to explain the process by which these collective representations give rise to the second type of collective representations. Although the use of symbols may be necessary for this process, they do not alone suffice to explain it. Until he found these laws, the best he could do was to provide accounts of the primitive origins of our basic categories of thought.

For Durkheim, all of our basic categories of thought are but modifications of the ideas of our primitive forebears. We all have the same idea of subsuming species under genera because we all have totemists for ancestors, Durkheim thought. Also, if all of our ancestors were compelled to perform the same sort of religious rites, then it would seem that we should all have more or less the same idea of causality. Steven Collins (1985:52) and Terry Godlove (1989:166 n. 4), however, claim that Durkheim equivocated with regard to whether the categories are the same in all societies. In one sense for Durkheim, we have seen, the categories are not the same for every society: the particular *directions* of space, *divisions* of time, *number* (and names) of species in a genus, and so on, are specific to each society. However, in another sense the categories are the same in all societies: all societies arrange things spatially, mark the passage of time, classify things by subsuming species to genus, and have a notion of causal relationships. These categories, Durkheim argued, make social life possible. Society can not exist unless individuals are organized into groups and a social organization presupposes the concepts of classification. Furthermore, he said, places must be divided among these groups, requiring the concepts of divisions and directions of space. Similarly, convocations to feasts, hunts, and military expeditions necessitate a common way of fixing dates and times. Finally, he recognized that cooperation with the same end in view demands a consensus about means and ends that presupposes the concept of causal relationships (632–33).

To say that the categories are alike in some ways and different in others from society to society does not itself seem to pose any problems. The problems arise from Durkheim's attempt to account for the categories through his essentialist model of explanation, in which he collapsed logical with causal and functional relationships. Thus when he said that the categories make social life possible, he equivocated as to whether this was supposed to be an empirical claim about the functions of these concepts or about their causes or whether he meant that these concepts were logically presupposed by the concept of social life. To put the problem another way, Durkheim appears to have confounded a plausible empirical claim with a reasonable philosophical argument and drawn an invalid sociological conclusion. The plausible assertion is that the cause of a particular way of measuring space, time, and so on, is to be found in society. Systems of measurement after all are purely conventional. The argument that some way of measuring these things is necessary for social life appears to be unimpeachable. Problems arose only when Durkheim tried to put these two positions together and conclude that society is the cause of categories like space and time in general. This confusion is avoidable, I believe, once we drop the essentialist model of explanation. Once we do, we realize that from the fact that the categories are *necessary* to social life it does not follow that they *result* from it. Durkheim appears to have been misled by the ambiguity of the concept of a necessary condition.

To be sure, Durkheim did say that the categories impose themselves on us with only a "moral" necessity, analogous to that of moral obligation, and not a physical or metaphysical necessity, since the categories will vary with time and place (24–25). Unfortunately, however, Durkheim's notion of moral obligation is anything but clear. As I showed in chapter four, in his 1906 essay "The Determination of Moral Facts" he equivocated with regard to whether a society's moral rules are to be understood as analytic or synthetic statements.

The ambiguity of Durkheim's theory of the categories has led to misinterpretations. Many of the objections that have been raised against it tend to confuse his concept of a category with Kant's.[13] For instance, Terry Godlove (1986, 1989) and Steven Lukes (1973:447) suggest that Durkheim appears to have been seeking an empirical answer to what is essentially a philosophical question. That is, he seems to have wanted a causal explanation of the necessity with which the categories impose themselves on our experience. But this is to set an impossible task for a sociological or for any other empirical theory, they say. The most that experience could show us is that these categories *always* impose themselves on experience; it could never show us that they *necessarily* do so. Godlove objects that Kant's categories and forms of intuition cannot be assigned causes in the empirical world because they are the very conditions on which the possibility of experience depends (1989:32). For the categories to be contingent upon causes either within or outside the realm of experience, he argues, would be for them only to *seem* necessary and not to *be* necessary (14–15).

For Godlove and Lukes to raise this kind of objection is for them to misunderstand Durkheim's project. A sociology of knowledge is a scientific theory and as such it is obliged to account for the phenomena. The phenomena are nothing more than how things *appear* to us. To explain that the categories *seem* necessary to us is precisely what the sociology of knowledge must do. In this way, I believe, Durkheim could also try to escape the objection first raised at his dissertation defense that he had attempted to derive a moral "ought" from a physical "is." In his account of the origin of the notion of religious forces *The Elementary Forms,* one could argue, he was merely attempting to explain our *feelings* of moral obligation. His account of these feelings in terms of the experience of the compulsion of social forces, however, fails to explain the *moral* character of this feeling. It is not at all clear that submitting to moral authority even feels like submitting to force. The latter feeling may be accompanied by feelings of being treated unjustly, which itself is a moral notion. To raise this objection, however, may be only to complain that Durkheim has not accounted for our contemporary moral sentiments. Perhaps the primitive does not distinguish submitting to force and submitting to moral authority. This would appear to be an empirical question that Durkheim has not settled.

Like Godlove and Lukes, Durkheim's earliest critics also failed to appreciate the point of his theory of the categories and faulted it for not being sufficiently Kantian. Charles Elmer Gehlke and Edward Schaub accused Durkheim of treating the categories as mental contents rather than as "inherent capacities" (Gehlke 1915:52) or "active principles" (Schaub 1920:337). William Ray Dennes made this point more clearly, arguing that Durkheim's categories are more like Hume's ideas than like the categories of Kant (1924:32–39).

With some justification, Bloor argues that these early critics simply beg the question, assuming that the categories are psychological rather than social (1982:294–95). Indeed, Gehlke's *inherent capacities* does have a psychological ring to it. Another early critic, A. A. Goldenweiser, went so far as to assert that "there must be other sources in experience or the psychological constitution of a man which may engender mental categories" (1915:733). Bloor also interprets what he calls "the celebrated charge of circularity" as begging the question in favor of psychology. In order to perceive in the social model such things as spatial distribution, temporal periodicity, quantity, or class, this objection runs, we must already be in possession of certain cognitive abilities or categories. Bloor quotes Dennes as the original source of this argument:

Durkheim's theory of the origin of the categories depends upon his ambiguous conception of the mind. If he takes mind in the Kantian sense, the sense usual in epistemology, as the subject's system of cognitive faculties, it is ridiculous to say that the categories of the mind are in any sense transferences from social organization. The category of quantity would have to exist and to operate in order that an individual mind should even recognize the one, the many and the whole, of the divisions of his social group. And again, it is only by the mind's perceiving its data in the form of succession that the periodicity of religious rites could have been known at all. . . . If, on the other hand, Durkheim means by mind a mere aggregation of representations or ideas, there is sense in supposing that the first ideas of time may have been of the periodicity of primitive religious rites, the first ideas of quantity, of the divisions of the tribes, etc. But the supposition is then of merely historical importance. . . . It has no direct bearing upon either the epistemological or the psychological study of the nature or status of the categories of the mind. (Dennes 1924:39)

When I analyze this passage in the context of Dennes's entire essay, however, I come up with a different interpretation. Dennes was contrasting a Kantian conception of the mind as an integrated system of capacities with a Humean conception of the mind as a bundle of representations or ideas and arguing that *if* Durkheim had the Kantian conception, he would run into the circularity problem. However, Dennes made it quite clear that he believed that Durkheim's real difficulties stem from his Humean conception of the mind and that he was caught on the *other* horn of the dilemma. The problems that Dennes raised for Durkheim's theory of the categories are then much different from what Bloor takes them to be. As Dennes (1924:34) pointed out, if a society or a mind is nothing but a collection of representations, then what distinguishes a mind or a society from a book or an art museum, which may also be characterized as collections of representations? Today we could add that for similar reasons Durkheim would not be able to distinguish a mind from a computer program.

What Durkheim lacked was an account of that in which mental activity consists. He seems to have wanted to explain it in terms of the activity of the representations themselves, with collective representations more active than individual representations. But this will not suffice. To borrow an analogy from John Haugeland, just making the mental representations themselves active is a bit like trying to make a book more like a mind by painting the letters in the book on the backs of busy ants (1985:120). Durkheim's concept of mind is then analogous to a collection of two kinds of ants, with the stronger kind overpowering the weaker. Durkheim needed to say more— much more—before he could claim to have a theory of the mind.

One could argue that in all fairness we cannot expect a sociology of knowledge to solve long-standing problems in the philosophy of mind. Nevertheless, as Dennes argued, to the extent that Durkheim's categories are like the complex ideas of the British empiricists, his sociological theory of the categories amounts to little more than a historical thesis about how these ideas were formed over the course of generations. Durkheim's collective representations turn out to be nothing but museum pieces and his theory of the categories turns out to be nothing more than a story about how they got into the museum. Durkheim's theory of the categories, in other words, amounts to a kind of sociological history of ideas.

Bloor might want to object that it is less important whether Durkheim has solved all the problems confronting his theory of the categories than whether this theory can be mined by contemporary sociologists of scientific knowledge for fruitful suggestions. Recognizing some of the limitations of Durkheim's conception of collective representations, Bloor (1982) tries to defend a relativist interpretation of the primitive classification thesis by welding it to a more contemporary theory of semantics, drawing on the network theory of meaning of W. V. O. Quine and Mary Hesse. Whether Bloor succeeds in this endeavor is an issue that I will treat more fully in the following, concluding chapter.

IV

*An Evaluation of
Durkheim's Sociological
Research Program*

9 Conclusion

A New Niche in an Old Program

The shift in Durkheim's concerns from contemporary European to primitive Australian and Amerindian societies represents but the creation of a new niche within his program of research and not a fundamental change in his thinking. His conceptions of the subject matter of sociology and of a sociological explanation did not change. Also, his methods of comparative theory evaluation remained the same. To be sure, there were changes in his sociological theories. He abandoned explanations in terms of the social volume and density in favor of explanations in terms of the structure of society. Also, towards the end of his career he began to think about the role of symbols in the formation of collective representations. However, he always maintained that social life consisted in collective representations and that the goal of sociology was to account for the causes and functions of these collective representations.

Durkheim's turn to the sociology of knowledge can be explained in terms of his core assumptions about the goals and methods of sociology and his need to solve certain conceptual problems that arose for him due to his conception of sociology. His having taken up the study of the primitive categories, we have seen, was consistent with his method of historical analysis. Durkheim, it seems, believed that the laws of collective ideation could be discovered by tracing the development of these concepts from their simplest beginnings to what they are today. Once these laws were found, he would then be able to account for the moral and other concepts that unite contemporary societies. If he could solve these problems, he appears to have thought, his sociological research program would then be in a position to compete more effectively with other programs of research in the social sciences.

Through his turn to the study of primitive concepts, Durkheim had created an intellectual niche in which he could not only solve problems he had encountered in the sociology of morals but contribute to the growth of the

247

explanatory scope of his research program. Although Durkheim drew all of the empirical facts in *The Elementary Forms of the Religious Life* from previously published ethnographies, he had established a niche distinct from ethnography by using and explaining these facts in a new way, that is, in his sociological theory of the categories. This new niche was also distinct from philosophical theories of knowledge in so far as it concerned the causes and origins rather than the justification of our categories of thought.

It is thus unnecessary to invoke either developments in French society at large or factors in Durkheim's personal life in order to account for his turn to the study of primitives or for his theories about primitive religion and thought. Jones's (1981) appeal to Durkheim's emotional life simply ignores Durkheim's assumptions about methods and goals. Similarly, Selvin's (1976) account of Durkheim's refusal to adopt new statistical methods in terms of a lack of institutional support for such work overlooks Durkheim's explanatory goals. Alexander (1982, 1986a), we saw, accuses Durkheim of having abandoned the supposed materialism of *The Division of Labor in Society* and turned to a theory and subject matter wholly divorced from political controversy in order to avoid being taken for a Marxist and thus to advance his career. However, Durkheim never subscribed to the view that social phenomena can be explained in terms of economic, geographic, or other "material" factors. *The Division of Labor* is a deeply flawed, ambiguous book that was widely misunderstood as explaining social phenomena in terms of causes in the physical environment. If this misinterpretation is taken to be Durkheim's actual theory in this work, then there does appear to have been a major change in Durkheim's thinking. However, there is no reason to assume that Durkheim ever held the views that he was misinterpreted as holding in *The Division of Labor*. Furthermore, even Alexander concedes that Durkheim did not express such materialist views in his publications prior to *The Division of Labor*. This concession implies that Durkheim changed his concept of sociology *twice* over the course of his career and leaves the first change unexplained.

Durkheim's constant goal was to explain the causes and functions of the collective representations that constitute the collective consciousness that holds society together. He consistently held to the theory of the compounding of mental states in his accounts of collective representations. Of course, some may find nothing new or original in my reading of Durkheim and point out that this interpretation was put forth long ago by Charles Elmer Gehlke (1915). Although I agree with Gehlke that Durkheim's sociological theories all follow from this theory of mental states, I would also point out that many other things that Durkheim did *not* say could also be derived from this theory of mental states. To understand Durkheim, one must recognize that the issue is not whether but *how* sociological phenomena are to be explained in terms of mental states. After all, many of the explanations of social facts he rejected were premised on a theory of the combination of mental states according to the principles of the association of ideas. The question for an interpretation of Durkheim then becomes, why did he explain sociological phenomena in terms of mental states in just the way that he did? To raise this question is to make Durkheim's methods central to an interpretation of his works.

The method Durkheim constantly used was one of hypothesis and test. For Durkheim, the method of hypothesis begins with a provisional definition of the social fact to be explained expressed in terms of what he took to be its observable characteristics. He invariably explained these observable characteristics in terms of an underlying social reality consisting in collective representations and social forces that cannot be directly observed. His method of hypothesis allows one to postulate not only unobservable mental entities but primitive causal origins as well, as long as the hypotheses are testable. In *The Elementary Forms* he constructed a hypothetical totemism that is not identical with either Australian or Amerindian totemism. His idealized picture of a totemic society is not that of a perfectly simple society but a doubly compounded society, consisting of two phratries each divided into a small number of distinct clans. Such clans are supposed to derive from the perfectly simple horde that is nowhere observed but simply postulated in *The Rules of Sociological Method*.

Durkheim typically defended his explanations through a series of eliminative arguments. His conception of an eliminative argument derives from Mill's canons of induction. Durkheim understood these methods to capture the logic of Baconian crucial instances in which the facts decide among alternative theoretical explanations. However, Durkheim operated with a broader notion of crucial facts or experiments than Mill had explicitly defined. For Durkheim, the method of crucial facts allows a scientist to criticize a hypothesis not only for making false predictions. It also permits the criticism of an hypothesis for failing to say anything at all about a group of facts, raising more problems than it answers, begging the question, or failing in some other way to explain the facts in an acceptable manner. Thus, for example, in Durkheim's method of crucial facts one can choose between competing hypotheses even when they explain the same fact, if one of those hypotheses explains that fact in a manner that is ad hoc. Durkheim also used Mill's method of concomitant variations in a new way. The variables he compared in the method of concomitant variation were not always empirical variables. Sometimes one of the variables was nothing but a society's supposed position in his hypothetical classification of societies.

Sociology and Mental States

Durkheim identified the subject matter of sociology with a type of mental state that he called collective representations. These collective representations were supposed to contain the meanings of social facts and social actions. With this concept, Durkheim attempted to provide a role for the meanings of social actions in their causal explanations while at the same time avoiding what he took to be the problems with intentional explanations. For Durkheim, as we have seen, to explain an action by appealing to the agent's ability to reflect upon its meaning is to invoke something like a miracle. Durkheim attempted to solve the problem of how meanings can interact with the physical world by having the meanings of social actions supervene on the causal entities that are responsible for them.

In the explanation of action, one must carefully circumscribe the sort of thing that one accepts as an action. For something to count as an action, it must have some meaning for the agent. The meaning of the action also determines the type of action that it is. Thus, for example, Durkheim refused to accept self-imposed deaths as suicides if they were committed by insane people who did not understand what they were doing. It is one thing, however, to individuate an action in terms of its meaning and quite another thing to explain it in these terms. One way to explain the effects that meanings can have in the physical world without invoking magic is to appeal to the causal properties of the physical realizations of the way in which these meanings are represented in the mind. Durkheim appears to have been trying to do something like this by identifying meanings with mental representations and explaining the effects of society on the individual in terms of the power of collective over individual representations. As we saw above, individual representations for Durkheim arose from reactions among neural states and collective representations from reactions among individual representations. Ultimately, then, collective representations for him were real physical entities with causal properties.

The problem with Durkheim's attempt at a solution to the problem of the dualism of meaning and matter is not only that it relies upon a mysterious, unanalyzed notion of psychological energy, but that it identifies types of meanings with types of entities. Presumably, different meanings should have different effects on us. For Durkheim, however, it seems that what explains the differences in effects is that they result from different kinds of causal entities. There appears to be little reason, however, to assume that a class of mental representations that all have the same meaning will all be realized physically in the same way. That Durkheim makes this assumption can be explained by his essentialism. However, from the premises that types of actions are individuated by their meanings, that these meanings are represented in the minds of actors, and that these representations are physically realized in states of the nervous system, the conclusion does not follow that the same kind of neurophysiological process underlies the same kind of action in different individuals, or even in the same individual at different times.

Durkheim's mistake was first to distinguish different types of individual and collective representations as natural kinds and only second to seek the laws of sociology in the laws of "collective ideation" that govern collective representations. However, there is no more reason to assume that the laws governing social facts are the same as those governing the mental states in which they are represented than there is to believe that these psychological laws are the same as the laws of neurophysiology governing the neural states in which these mental states are realized. What Durkheim should have done is first to have sought the laws governing social phenomena and then let these laws carve up the social world into natural kinds without concerning himself about the way in which these social facts are represented in the mind.

By arguing that sociology should seek relations among social phenomena without concerning itself with how social phenomena are psychologically realized, I am proposing in effect a form of functionalism in sociology analogous to functionalism in psychology. Psychological functionalism defines types of mental states in terms of their relationships to other types of mental states and not in terms of types of brain states, although it does not deny that individual mental states are realized in brain states. For example, although neural processes may underlie one's feeling of pain, these processes are not what "pain" means. The functionalist defines it in terms of such things as the feelings of discomfort associated with it and the desire to avoid it.[1] Similarly, a sociological functionalism would define social facts in terms of other relationships to other social facts and not in terms of mental states. For example, the two types of social solidarity would be defined in terms of other sociological concepts and not in terms of the differences in the underlying mechanisms involving mental images that Durkheim invoked in *The Division of Labor.*

An objection that has been raised against functionalism in psychology is that it fails to account for the phenomenon of our conscious awareness of our mental states.[2] A cleverly designed computer or robot that preserved all of the functional relationships among the mental states in a human mind would thereby be functionally equivalent to that mind. According to the functionalist theory in psychology, for a computer to react to a pain stimulus in the same way that a human being would react to it would then be for the computer to be in pain. This conclusion seems absurd to us, as we believe a computer not to experience mental states. However, this objection cannot be raised against functionalism in sociology, as there is no problem for a functionalist sociology analogous to the problem of consciousness for a functionalist psychology.

Some might object that by defining sociology simply as the study of the relationships among social facts, sociology is removed from the realm of empirical science. For example, Steven Taylor suggests that Durkheim's realist identification of social facts with collective representations is necessary for the very conception of a subject matter for sociology. If there is nothing to social reality other than shared meanings, he argues, if the only thing that for example suicides have in common is that we call them suicides, then there would be nothing to study empirically. All that we could do, then, would be to analyze the meanings we impose on things (Taylor 1982:55). Sociology would then be, in Peter Winch's notorious phrase, "misbegotten epistemology" (1958:43). Nominalism, in other words, would wipe out sociology and leave nothing but philosophy. As Winch has argued, the connections among social phenomena are not causal as they are in the natural sciences but logical, like the links among propositions. According to Winch, the goal of sociology should be to understand social behavior through an interpretation and analysis of the *meanings* of social actions for the participants.[3] What social scientists mistakenly regard as scientific problems, Winch says, are in reality conceptual problems, requiring conceptual analysis.

I do not find the empirical status of sociology to be in such dire jeopardy. Certainly conceptual analysis is necessary for sociology. But this would still be true under a realist interpretation of its subject matter: no science can dispense with clarity. At the time Winch wrote *The Idea of a Social Science* he seemed to think that philosophy is distinguished from the sciences in that the former alone employs conceptual analysis. More recent philosophers of science, however, recognize that scientists are engaged to a considerable extent in the analysis of concepts and may even consider conceptual problems a more serious objection to a theory than empirical problems (for example, Laudan 1977). Nor does it follow from the claim that sociologists ought to be concerned with the meanings of social facts that they ought not to be concerned with their causes. As Alasdair MacIntyre (1967:20) points out, causal analysis and meaning analysis are not necessarily mutually exclusive. Finally, even if we accepted a nominalist interpretation of the subject matter of sociology it would not follow that empirical investigations have no bearing on how one concept relates to another for a given society. How the meanings we impose on things change over time and for what kinds of reasons are also empirical questions.

Far from forsaking conceptual analysis in favor of causal analysis, Durkheim, as we have seen, devoted considerable time and energy to the analysis of conceptual problems. Indeed, he departed from our ordinary notions of suicide and religion because he was concerned with the conceptual problem of ambiguity and not because he was uninterested in the meanings of social phenomena. I agree with Durkheim that to explicate the meaning of a social phenomenon like religion is not necessarily simply to report what the man in the street thinks religion is. Where Durkheim went astray was in identifying these shared meanings with types of mental representations.

Durkheim's Conception of Meaning

Durkheim's identification of meanings with types of mental representations made it difficult for him to account for the meanings of general concepts. Educated in the philosophical tradition of the way of ideas, he thought of a general concept as some sort of Lockean complex idea that is progressively compounded over time from simple ideas. Beginning with Berkeley, however, philosophers had recognized that as long as these simple ideas are conceived as some sort of mental image or picture, no compound idea could ever express the meaning of a general concept adequately. Nevertheless, the tradition of identifying the meanings of linguistic expressions with some type of mental representation persisted and Durkheim fell squarely within this tradition:

Now there is no doubt that language and, consequently, the system of concepts that it translates is the product of a collective elaboration. That which it expresses is the manner in which society as a whole represents to itself the objects of experience. The notions that correspond to the diverse elements of the language are thus indeed collective representations. (1912a:620; see also 1912a:329; 1913b:70)

In this passage, Durkheim appears to have been suggesting that we operate with two "languages." On the other hand, we have a written and spoken language that consists of words. On the other hand, he seems to have believed, we have a language of thought that consists of collective representations. According to Durkheim, "each word translates a concept" and the system of concepts with which we think is that expressed by our language (1912a:619).

To be fair, toward the end of his career Durkheim at least tried to distinguish meanings from ideas. For example, he pointed out that our concept of God does not correspond to any mental image (1912a:617–18; 1988a: 203). At least according to one set of student notes from his lectures on pragmatism given in 1913, he seems to have drawn from such examples the conclusion that concepts take their meanings from their use in certain contexts: "How do we think about them? It is not by means of general ideas. No more is it by means of images: we have never seen them. We have concepts of them: because we discuss them, and it is these concepts that serve as the point of departure for our discussions and for our reasonings" (1955a:203). However, Durkheim never seems to have drawn the further implication that the meanings of concepts can not be identified with collective representations.

In saying that Durkheim's conception of meaning led to problems for him, I am not suggesting that others in his day did much better. With the possible exception of a few pragmatists and positivists, nearly every philosopher at that time identified the meanings of words with some sort of mental entity. Of course, the pragmatists and positivists who identified the meanings of propositions with the methods for testing them encountered difficulties of their own. Furthermore, as I argued in chapter 5, Durkheim would have rejected this operationalist alternative as inconsistent with his realism.

Given Durkheim's views on meaning, the claims made by David Bloor (1983:168), Steven Collins (1985:50–51), Ernest Gellner (1962:22), and Steven Lukes (1973:437) that Durkheim's sociology of knowledge is a precursor to Wittgenstein's identification of meaning with use is overstated.[4] Granted, both thinkers said that language can exist only in a social context or "form of life." Wittgenstein, however, would have vigorously opposed the identification of meaning with any sort of mental entity, even Durkheim's *collective* representations, preferring to think of meaning in terms of a society's rules of usage. Whereas for Durkheim, "similar" individual representations combine to form collective representations or concepts, for Wittgenstein it is these very rules that establish what a society will regard as similar. Durkheim, who was a realist, assumed similarity relationships in the things themselves or at least in our representations of them; Wittgenstein, who was a nominalist, did not. For Wittgenstein, even the way in which these rules will be applied in the future is not fixed beforehand. We can attach sense to saying that a person applies these rules correctly to a new situation, he thought, only under the assumption that there are others who are in a position to point out mistakes. Wittgenstein would probably have regarded Durkheim's notion that individuals are constrained by collective representations that they have somehow internalized as being as silly as the belief that one's linguistic usage is constrained by one's checking the definitions of words in a dictionary that existed in one's imagination (Wittgenstein 1953: part 1, par. 265).

Furthermore, as Quine (1951) has argued, to think of the meanings of concepts as things, even as some sort of mental entities, to which the words of our language correspond, is to confuse meaning with reference. That is, it is to think of words referring to their meanings in the way in which names refer to persons or places. The meaning of a concept cannot be *identified* with a type of mental representation any more than with a kind of representation on paper, such as a word or symbol.

In distinguishing Durkheim's theory of linguistic meaning from that of Wittgenstein or Quine, I do not wish to be understood as thereby endorsing the nominalism of the latter thinkers. The problems with Durkheim's theory of meaning stem not so much from his realism as from his identification of meanings with types of mental representations. It is possible to regard the rules that govern meanings as real things without thereby committing ourselves to the view either that meanings are fixed and unchanging or that meanings exist independently of us or that meanings consist in mental representations. Rules may be considered real in the sense that they have real effects on us. To say that rules are real because they have real effects on us is not necessarily to identify these rules with their effects but only to say that what we *mean* when we say that something is real is that it affects us. For example, when we say that English, as opposed to some language concocted by a science fiction writer, is a real language, we do not mean that it exists independently of us. We mean only that speakers of English are constrained in their use of the language by commonly accepted rules of usage. Now one could argue, following Wittgenstein, that these rules are not fixed, that they are open to negotiation, and so on. But the fact that meanings change does not imply that they are not real any more than the fact that species evolve implies that they are not real or the fact that the temperature on a summer afternoon is changing implies that it is not really hot. Temperatures, species, and rules can be real even though they exist only as the properties of groups of things. Temperature, for example, is identified with the *average* kinetic energy of a collection of molecules. No one molecule even need have that quantity of energy at any given time in order for the temperature to be a real property of the group of molecules in question at that time. Similarly, a meaning may exist for a society at some given time without that meaning having to be represented in the mind of some individual member of that society at that time.

My criticism of Durkheim's identification of meanings with types of mental representations does not imply a rejection of the appeal to mental states in the explanation of human action. As I explained above, I agree with Durkheim that even to attribute a type of action to a person, that person must have a mental representation of the meaning of that type of action. Where I disagree with Durkheim is solely in his identification of that meaning with a mental representation of any type, even a representation that is held in the minds of all the members of a society who understand that meaning.

Durkheim and the Sociology of Scientific Knowledge

Durkheim's views about the goals, methods, and subject matter of sociology formed a coherent, defensible research program that attempted to provide a role for the meanings of social facts in causal explanations of them. As we have seen, he included among his examples of social facts having to follow certain professional practices. Presumably, he believed that scientists, like the members of any other profession, are constrained to follow specific professional practices. For scientists, these practices would include rules of method. Durkheim characterized methodological rules as social institutions governing scientific disciplines. Social institutions for Durkheim exist in collective representations that have the power to constrain individuals. Hence, according to Durkheim the social constraints on a scientist exercised by rules of method require no explanation in terms of membership in some social group or other. For Durkheim the social relations among scientists and their concepts of scientific method are on an equal explanatory footing: the meanings of both exist only in collective representations and they both constrain individuals in the same way.

Arguably, Durkheim's program is more promising than those of some of his recent champions among contemporary sociologists of scientific knowledge. Sociologists of scientific knowledge today would agree with Durkheim that the practices and even the conceptual thinking of scientists are socially constrained. Many of these contemporary sociologists, however, draw antirealist conclusions regarding the subject matter of science from the fact that science is a social product (for example, Woolgar 1983, 1988; Latour 1987). Not only do these conclusions not follow, but they leave the sociologist in the awkward position of being committed to the reality of social facts while denying the reality of what other scientists study. Durkheim avoided this conundrum. Although he believed that there are sociological explanations for the very concepts scientists use, he thought that the social origins of such concepts in no way undermine the reality of that which scientists describe with them.

A decade ago, however, Bloor (1982) attempted to defend a relativist interpretation of Durkheim's and Mauss's primitive classification thesis and to generalize it to include the systems of classification of modern science. Recognizing some of the philosophical problems that arose when Durkheim identified the meanings of concepts with a kind of mental representation, Bloor assumes along with Quine and Hesse that concepts take their meaning instead from their relationships to other concepts in our system of knowledge. Although all the elements of this network, including a system of classification, are open to negotiation, there are both physical and social or cultural constraints on what can be accepted. Included among the latter, for

Bloor, are certain social interests related to the use of scientific theories for purposes of legitimizing or justifying the social order: "And since interests derive from, and constitute, social structures it will be no surprise to find that putting nature to social use creates identities between knowledge and society of the kind predicted in *Primitive Classification*" (Bloor 1982:283).

Contrary to what Bloor says, however, Durkheim and Mauss did not "predict" any kind of "identities between knowledge and society." For Durkheim and Mauss, all classification systems are nothing but minor variations on a scheme modeled by our primitive forebears on their form of social organization. We all have the same idea of subsuming species under genera because we all have totemists for ancestors, Durkheim thought. His theory of the categories, we have seen, amounts more or less to a history of how they were formed over the course of countless generations. It provides no basis for establishing identities between scientific classifications and social classifications in contemporary society. The structure of contemporary societies, as Durkheim realized, is far more complex than the simple, hierarchical structure of his totemic societies and thus of the nested hierarchy of genus and species.

Bloor's thesis is very different in an additional way than the one proposed originally by Durkheim and Mauss in 1903. By appealing to social interests when they explain social actions, the Strong Programme violates Durkheim's strictures against explanations in terms of consciously pursued ends-in-view. According to the primitive classification thesis of Durkheim and Mauss, a particular social structure is a direct cause of a particular belief; according to Bloor, a particular social arrangement is either a goal to be attained or something to be maintained by persuading others to accept a particular belief. To explain a belief in terms of social interests is to explain it not in terms of what Durkheim regarded as social *causes* but rather in terms of the *purpose* it serves for an individual or group of individuals. Durkheim, as we have seen, believed that the fact that we desire or want something can never bring it into existence. The causes that give rise to social facts are independent of the purposes that they serve. Furthermore, the ends that social facts serve for Durkheim have to do with maintaining their existence and thus are invoked in functional rather than causal explanations of social facts. Although functional explanations of social facts may play an important role in Durkheim's sociology, they are not to be confused with explanations in terms of individual intentions.

Be that as it may, Bloor nevertheless purports to adduce evidence from the history of science in support of what he takes to be a modified version of the Durkheim-Mauss thesis. He argues that Boyle and his circle in seventeenth-century England appealed to the passivity of matter postulated in the corpuscularian philosophy in order to recommend a social structure in which the people would passively allow themselves to be governed by the Anglican Church (1982:284–91). Among other historical case studies, Bloor also cites Steven Shapin's papers on the Edinburgh phrenology debates (1975, 1979a, 1979b) and Donald MacKenzie's work in the history of statistics (1978; see also Bloor 1982:291 n. 66). Although these historical studies may exemplify Bloor's claims, to borrow an expression from Durkheim, illustration is not demonstration (see chapter 5 above). Moreover, the studies Bloor cites are equivocal. Citing the same papers by Shapin, in addition to Paul Forman's (1971) essay on causality and quantum theory in Weimar Germany, Martin Rudwick's (1974) account of Poulett Scrope's concept of geological time, as well as his own work regarding Udden's geological and social beliefs, John Law (1984) argues for a *weakened* version of the Durkheim-Mauss thesis, according to which a scientist's *beliefs* about his society may serve merely as a model for his theories.[5]

Of course, Bloor recognizes that the historical episodes he discusses are open to other interpretations. In a later paper, he considers Geoffrey Cantor's alternative to Shapin's interest account of the Edinburgh phrenology debates. According to Bloor, Cantor attributes the disagreements between the contending parties to an "incommensurability" of fundamental metaphysical, methodological, and theoretical assumptions. Shapin, on the other hand, tries to reduce the incommensurability of the ideas involved to the social interests of the relevant parties (Bloor 1984; cf. Cantor 1975a, 1975b; Shapin 1975). For the Strong Programme, social interests always take causal priority over ideas. When we investigate historical controversies, Bloor argues, "we must notice how the group takes up a body of culture and changes it in a direction consonant with their perceived interests" (1984:310).

Now even if one succeeds in explaining a scientist's beliefs in terms of his or her perceived interests, one must still explain the fact that the scientist perceives himself or herself to have these interests. "Perception" is after all a term relating to beliefs and cognition and the sociologist's of knowledge task is not complete until he or she accounts for these perceptions. Interest accounts seem to assume the existence of a stable social structure that does not depend on people's beliefs and that accounts for their perceived interests. In addition, the method of explanation of the Strong Programme assumes that social interests remain stable during the course of negotiating scientific beliefs. But as Durkheim pointed out, "nothing is more changeable than interests" (1886a:201). Cantor (1975b) replies to Shapin that the Strong Programme's approach to the sociology of knowledge cannot account for intellectual disputes just to the extent that it cannot explain how beliefs and interests change in the course of debate. At best, this approach can provide only a static view of the controversy, explaining why the parties to the dispute disagreed and held the views that they did.

The Strong Programme also faces the problem of defending the very distinction between social structure and beliefs. As John Law (1986) and Bruno Latour (1987:141–44) argue, the distinction between a stable, independent social structure and a system of beliefs is becoming less and less tenable. Even for Durkheim, as we have seen, a society's beliefs and the social relations that constitute its structure are but two aspects of the same reality, as both exist ultimately in collective representations. If we think of them as belonging to some sort of network instead of as collective representations, it is not clear why certain elements of this network should have some sort of privileged causal status with respect to the rest. Social relations may constrain beliefs about nature but beliefs about nature may also constrain social relations.

Even if the historical evidence that Bloor cites could establish the existence of identities between scientific classifications of things and social classifications of people, there remains the issue as to what sort of identities these are. What these historical case studies appear to show is that particular social classes held either favorable or unfavorable opinions about such things as phrenology or Pearson's or Yule's conception of statistical methodology. At best, Bloor has provided evidence of identities between token social classifications and token scientific classifications. A defender of the Strong Programme might object that such token-token identities provide positive, confirming instances of identities between types of social classifications and types of scientific classifications. However, whether token-token identities are in fact instances of type-type identities is equivocal. In the philosophy of mind, for example, the token-token identity that Jones's pain = Jones's C-fiber firing may be read by the reductive materialist as an instance of an identity between a type of mental state and a type of brain state. A functionalist, however, would not read it this way. For the functionalist, the same type of mental state may be physically realized in different ways in different people, much as the same software computation may be physically realized in different ways in different kinds of computers. Similarly, in the sociology of scientific knowledge token-token identities between scientific and social classification will not alone suffice to establish that the former reduce to the latter. What is called for is, first, a general theory of social classifications, what they depend upon, how they change, and so on. Second, there would have to be a general theory of scientific classifications. To complete the reduction, one would also need additional assumptions and statements of conditions, especially bridge principles of some sort—not necessarily identities—connecting the properties characterized by one theory with those characterized by the other. Finally, one would actually have to produce the derivation of one theory from the other, which is by no means a trivial task even when one is in possession of the two theories and the bridge principles connecting them. The Durkheim-Mauss hypothesis, however, could not serve as the basis for the reduction of scientific to social classifications. As we have seen, the hypothesis does not provide a general theory of types of social relations that can be related to types of cognitive relations.

Nevertheless, according to the Strong Programme, a scientists's metaphysical, methodological, and other cognitive commitments do not explain his or her beliefs or actions but are merely intervening variables that must be reduced to membership in social groups. Bloor, who as we have seen assimilates Durkheim to Wittgenstein, seems to think that a form of sociological reductionism is entailed by the theory of meaning of the later Wittgenstein. According to Bloor's interpretation, Wittgenstein says that the beliefs of social actors are to be reduced to their "form of life" much as Durkheim would explain religious belief in terms of an underlying social reality:

There can no longer be any excuse for offering the meaning of an actor's beliefs as an explanation of his behavior, or of his future beliefs. Verbalized principles, rules and values must be seen as endlessly problematic in their interpretation, and in the implications that are imputed to them. They are the phenomena to be explained. They are dependent, not independent variables. The independent variable is the substratum of conventional behavior that underlies meaning and implication. As Wittgenstein put it: "What has to be accepted, the given, is—so one could say—*forms of life*." (Bloor 1983:137)

The notion that scientific controversies can be reduced to conflicts between "forms of life" or "language games" has also been adopted by Steven Shapin and Simon Schaffer to account for the Hobbes-Boyle controversy (1985:15). For Wittgenstein, however, these concepts had to do with the analysis of linguistic meaning in terms of the practices of a group of language users. As Richard Hadden (1988:442) has argued, Wittgenstein intends that language games make our actions *intelligible;* they are not supposed to provide *reasons* or *causes* for our actions.

Bloor defends his reductionist interpretation of Wittgenstein through an analysis of the latter's discussion of following a rule. Arguing that a scientist's beliefs cannot be explained in terms of such "epistemic" factors as methodological principles, Bloor says, "Wittgenstein's point is that 'no course of action could be determined by a rule, because every course of action can be made to accord with the rule'" (1984:304). Bloor explains that whether an utterance or an act of reasoning is an instance of or is in accordance with a rule is a matter for conventional decision. The inference from the premises to the conclusion of an argument is always socially constructed. Why certain implications are socially constructed by particular social groups and why other groups constructed different implications is to be explained in terms of differences among their forms of life (1984:298–305).

Criticizing sociologists of scientific knowledge like Bloor who try to ground their reductionism in Wittgenstein, Michael Lynch distinguishes two readings of this philosopher. The first, which he calls "rule skepticism, takes Wittgenstein to be arguing that the relation between rules and conduct is indeterminate" (Lynch 1992:217). Skeptics then draw the implication that one must invoke social conventions, institutions, interests, socialization, and other such "extrinsic factors" to explain human actions and "the relation between rules and their interpretations" (1992:226). The "antiskepticist" position, on the other hand, follows Baker's and Hacker's interpretation of Wittgenstein, according to which rules are "internally related" to their applications or instances (Baker and Hacker 1985:172). According to Lynch, the antiskepticist "holds that Wittgenstein treats rules inseparably from practical conduct, so that there is no basis for explaining the relation between rules and conduct by invoking extrinsic factors" (1992:217).

In his reply to Lynch, Bloor begins by asking whether consensus and other social processes "really do lie outside" Baker's and Hacker's "internal relations" or whether they "are actually constitutive of it" (1992:271–72). Bloor provides an example from Anscombe about teaching the rules of a game to children. In teaching the game to the child, the child cannot continue to ask why, Bloor explains, because "there are no reasons for the rules outside the game. The game itself is cited as the reason for the rules, and yet the rules are constitutive of the game." Anscombe thus shows us "in quite general terms just how internal relations are constructed, and how they are grounded in social institutions, e.g., the institution of the game" (1992:272). He then proceeds to draw sociological implications from Anscombe's example, saying that her "implicitly sociological explanation of the internal relation between rules and their application is not designed to replace or displace our more ordinary talk of rules. It is the usual story identified in a new way" (1992:272–73).

Bloor compares Anscombe's account of rule following to a reductive explanation of lightning in terms of electrical discharges, which he contrasts with the replacement or elimination of a "magical" account of thunderstorms by a "meteorological" account. To complete the analogy with lightning, Bloor says that he does not want to *eliminate* explanations that appeal to the internal relationships between rules and practices. Rather, he wants merely to *reduce* such accounts to explanations that appeal to social relationships. "The internal relations between rule and applications just is a social relationship," he says (1992:273). Bloor's argument, however, purports to establish more than it really shows. On the one hand, the analogy with light-ning and electricity suggests that the identities between internal and social relationships that he has in mind are type-type identities. On the other hand, it is not clear that the example he takes from Anscombe—or another example he draws from Wittgenstein—illustrate anything more than token-token identities.

Wittgenstein's example involves a teacher trying unsuccessfully to teach a rule of arithmetic to a student who systematically misunderstands, having constructed his or her own set of definitions and internal relationships among rules and practices. What "breaks the deadlock" between the teacher's and the recalcitrant student's sets of rules, Bloor argues, is a social consensus. The rules alone do not suffice (1992:273–74). What Bloor fails to establish with such examples, however, is that there is a general theory of social relations that can always explain what breaks the deadlock between competing sets of rules.

Bloor contrasts his position that a rule by itself never dictates a particular course of action with what he calls "normative determinism," according to which norms are rules or principles that are first internalized and then guide action (1984:296, 306). Ironically, however, "normative determinism" is an apt characterization of Durkheim's explanatory appeals to collective repre-sentations and the social forces to which they give rise. Parsons certainly reads Durkheim as a normative determinist (1937:385–86).[6] Once we take rules and norms out of the head and put them into a social context where their interpretations can be negotiated, we have come a long way from Durkheim. It is surprising then that Bloor not only chooses Durkheim as a patron saint but believes that Durkheim developed "a social theory of knowledge so close to Wittgenstein" (1983:168).

Even if we grant Bloor his rejection of normative determinism and his argument that inferences are socially constructed, however, it does not fol-low that in the explanation of belief epistemic factors must always be re-duced to non-cognitive social factors. According to Bloor, "the meaning, and hence the implications, of the epistemic factors in question can only be ascertained when we know the goals and interests that inform their use. We need to know what their users were trying to achieve with them, and the context in which they saw themselves pursuing these goals" (1984:302).

The issue, as I see it, is precisely whether a scientist's *cognitive* goals play a role in the explanation of his or her actions and beliefs. Interest theorists simply refuse to recognize intellectual desires as being on a par with desires for power, prestige, money, or sex. Although cognitive goals, like any other goals, may not determine a course of action, they do constrain actions. Durkheim, as I explained above, could not have adopted either Tarde's or Quetelet's accounts of suicide rates without violating his conception of a sociological explanation. To maintain that subscribing to a cognitive goal always reduces to belonging to a social group is to make an empirically problematic assertion. For example, not all of Durkheim's collaborators on *L'Année sociologique* agreed with his explanatory goals. As I showed in chapter 3, for instance, in his review of *Suicide,* Simiand (1898) rejected Durkheim's realist interpretation of social forces. A problem for a sociological reductionist such as Bloor would then be to subdivide the collaborators on this journal into social groups in a way that would explain Simiand's opinions of Durkheim without being ad hoc. That is, a reductionist would have to show that there is something that unites those who rejected Durkheim's realism other than the fact that they all shared in this rejection.

Science, like any other human social institution, is defined in terms of the norms and values that govern it. To the extent that science aims at the growth of *knowledge,* it is characterized by *cognitive* norms and values. Cognitive values specify the aims of science, while cognitive norms specify the means to achieve these goals. Both cognitive values and norms range widely. Cognitive values may include everything from a scientist's position regarding the ontological status of unobservable entities to the desire to solve a specific set of problems or to explain a particular set of facts. Cognitive norms may range from rules governing the forms of persuasive argument that can be brought forth in defense of one's theory in a journal article to procedures for manipulating "inscription devices" in the laboratory. To say that such cognitive factors should play a role in the sociology of scientific knowledge is not to say that all scientific activity must be explained exclusively in terms of cognitive factors. There is no question that scientists can be and have been influenced by many noncognitive interests. However, it does not follow from this fact that cognitive goals must always be reduced to noncognitive goals and interests.

When I say that science as a social institution is characterized by cognitive norms and values, I do not mean that all of science is characterized by the same set of cognitive commitments. As Durkheim recognized, different scientific communities or "disciplines" may subscribe to different and even opposing sets of intellectual goals and methods. Differences of opinion regarding empirical claims among scientists can be explained at least in part in terms of differences in their assumptions about methods and values. Similarly, differences in cognitive values and norms can be explained at least in part in terms of differences in beliefs about the empirical world.

When explaining a scientist's beliefs in terms of his or her cognitive commitments, of course, one must take care to distinguish his or her professed methodological norms and those that are actually operative. In the case of Durkheim, as I have argued, although he did not violate his methodology, he nevertheless used several forms of argument that go beyond what he explicitly defended. For example, he criticized his opponents' theories for their conceptual problems and nonrefuting anomalies. Some of the operative norms in the sciences are less narrowly technical and reflect beliefs shared with the wider society. An instance of such a norm would be Durkheim's ethnocentric assumption that Australian tribes represent an early developmental stage of society that is a precursor to modern European society. At different times in history and in different societies various types of scientific arguments have been considered persuasive. The sociologist of scientific knowledge should try to find out the norms in use in both historical cases and contemporary science in order to assess just why something seems persuasive to scientists at a particular point in time.

Principles of rationality are included among the cognitive norms that characterize science. As long as these principles are understood in a descriptive rather than a normative sense,[7] explanations of belief that invoke such principles are no less sociological than explanations premised upon social, economic, and political interests. Indeed, explanations of belief premised upon social interests must themselves presuppose principles of rationality linking interests with beliefs that promote these interests. Furthermore, interest theorists seem to assume that the principles of rationality linking beliefs and desires with actions will pick from a field of competing theories the one that is unambiguously the best for realizing the scientist's social interests. That is, as John Nicholas (1984) argues, they overlook the possibility that principles of rationality in conjunction with social interests no less than principles of rationality in conjunction with statements of evidence may underdetermine theory choice. Finally, to the extent that sociologists of scientific knowledge take for granted principles of rationality linking interests with theory choice, they fail to investigate how such principles change with time and circumstance. Understanding the reasons and causes for variation in such cognitive norms, however, would seem to be an important explanatory goal for the sociology of scientific knowledge.

In sum, although I agree with David Bloor that there are fruitful insights in Durkheim that contemporary sociologists of scientific knowledge should follow up (1976:2), I disagree about what those insights are. What I like about Durkheim is that he did not maintain the absurdity of being a realist about the explanatory concepts of sociology while denying any reality to the concepts of the natural sciences. He recognized that the sociology of knowledge needs to keep epistemological and metaphysical issues distinct. I also approve of the way that he avoided relativism. He defended the idea that all human beings think alike without committing himself to the view that we are all perfect logicians. Also, although he recognized that concepts may differ from social group to social group, he nevertheless believed that these differences amount to nothing more than variations of the same set of primitive or original concepts and their extensions. In particular, I find his hypothesis that our concepts of genus and species have their origin in subsuming social groups under one another to be far more plausible than Bloor's hypothesis that all societies use systems of natural classification that mirror their social organization.

To say that Durkheim avoided relativism while maintaining the variability of concepts is to raise the issue of just what I mean by relativism. I take it that relativism asserts that different societies differ radically in their thinking and not merely with respect to variations in the meaning and extension of their concepts. Evidence that relativism is true might consist of societies that had no system of classification or that did not differentiate directions in space or measure the passage of time. A society with no system of classification, however, would be a society without language. As Donald Davidson (1974) argues, if a society has a language at all, then they cannot be all that different from us.

Some of Durkheim's insights into the sociology of scientific knowledge are to be found not in works like *The Elementary Forms* and "Primitive Classification" but in *The Division of Labor*. It is in the latter work that he explained how specialization in science is a solution to the struggle to continue to exist as a scientist in a given social environment. Scientific specialities for Durkheim are distinguished not only by their empirical subject matter but by the cognitive values and norms that they bring to bear on their subject. As the cognitive commitments that characterize research programs are brought to bear on new empirical domains, we have seen, new specialized niches are created that permit the research program to grow and thrive. A sociology of scientific knowledge that ignores cognitive values and norms then may be unable to provide an adequate account of specialization in science. After all, it is Durkheim's cognitive commitments that distinguish his sociology of knowledge in *The Elementary Forms* from the ethnographic reports of people like Spencer and Gillen.

One may object that Durkheim's account of specialization belongs to the sociology of science rather than to the sociology of scientific knowledge. That is, it concerns only the organization of the scientific community and not the actual content of scientific beliefs. In reply, I would argue that the same cognitive goals and norms that distinguish his work from ethnography also help to explain the very contents of some of his beliefs. For example, I appealed to Durkheim's explanatory goals in order to explain his decision to accept the hypothesis that totemism rather than animism is the oldest form of religion. For an externalist like Jones, on the other hand, Durkheim's decision ended up looking like a mistake.

I have tried to explicate Durkheim's cognitive goals and the cognitive norms to which he subscribed. I have also tried to explain his creation of a niche in the sociology of knowledge in terms of his cognitive commitments and his need to solve certain conceptual problems to which they gave rise. In order to defeat my claims about the explanatory role of cognitive goals and norms, one could do the following: take the group of scholars who founded and edited *L'Année sociologique* and without appealing to any cognitive factors explain how they divided the labor among themselves and how the structure of the journal changed over time.

Unlike Durkheim, I do not believe that intellectual biographies such as the present study necessarily fall outside the concerns of sociology. As Mac-Intyre argues, all thinking "has a social character" (1986:92). Thinking involves the application of rules and the use of a rule presupposes a social context. A society's shared system of rules includes standards for what counts as a good reason for doing something. What counts as a good reason depends on the social role that the individual in question is playing. For example, a person may have a good reason *as a scholar* but not as a parent for working late into the evening. To say that what counts as a good reason is determined by shared standards, however, is not to imply that there may not be disagreement in a society over what constitutes a good rule. Such disagreement, in fact, makes it possible for these rules to change over time. Nevertheless, because thinking has this social character, one can make one's reasoning persuasive to others if and only if one finds it persuasive oneself. By analyzing the reasoning of a thinker like Durkheim, therefore, we are not investigating his personal psychology so much as gaining some insight into what he believed to be the standards accepted in his society that governed what counted as a good reason. The extent to which Durkheim remains persuasive reveals the extent to which we share some of the same standards.

Notes

Chapter One

1. The article he cited is 1903a(i).

2. On the role of conceptual problems in science, see Larry Laudan (1977: chap. 2).

3. 1938a:180–81. Throughout this book I have cited only French editions of Durkheim, providing my own English translations of quotations from his work. Durkheim's translators have not always been sensitive to the meanings that certain of his terms held within the philosophical tradition in which he was educated. At the end of each Durkheim reference in the bibliography, I have indicated where translations may be found.

4. 1925:375 n. 2. The article to which he referred is 1898a(ii).

5. Stjepan Meštrović (1988a:19–20) reports that Durkheim's papers were left behind when his surviving relatives left Paris for the south of France during the Nazi occupation. The family that subsequently moved into the Durkheims' Paris home simply threw his papers away.

6. 1918b:121–22. Condillac did not so much deny the possibility of abstract general ideas, as Berkeley did, as argue that they are the result of an analysis that requires language. See Condillac 1746:24a; 1775:410a; 1780:402a.

7. Durkheim said that language is necessary for general ideas or concepts—he seems to treat these as equivalent expressions—also at 1906b:74; 1911b:119; 1911c(1):46–47; 1914a:317; 1955a:204.

8. See Jones 1977, 1978, 1984; Johnson 1978; Gerstein 1983; Kelly 1990.

9. The critical reader should be cautioned that social scientists attach different concepts to terms like *positivism* and *idealism* than philosophers do. For Parsons, these are terms for theories of social action and order rather than for semantic, epistemological, or metaphysical theories. It is not at all clear that there is any logical relationship between a positivist philosophical theory and a positivist theory of action. Similar remarks could be made regarding Jeffrey Alexander's use of terms like *instrumentalism*.

10. These interpreters include Alexander (1982, 1986a), Giddens (1971b), Gieryn (1982); Hamilton (1974:103), Llobera (1980:398), Lukes (1982:6); Göran Therborn (1976:254), Kenneth Thompson (1982:67, 83; 1985:15, 17), and Jonathan Turner (1990:1094).

11. Mauss 1928:vii–viii, v. According to Terry Clark, Comtean sociology had

come to be associated with the followers of Le Play who were organized around the journals *La Réforme sociale* and *La Science sociale* (1972:152, 158). The Le Playists tended to be conservative, upper class, Catholic, and outside the universities (Lukes 1973:394), which tended toward anticlericalism and socialism (Clark 1972:156–57). Clark's account of the Le Playists, however, is somewhat called into question by Walter Goldfrank (1972), who points out that Le Play himself was not a Comtean.

12. Alexander 1982: vol. 2, 483 n. 168. Here Alexander cites a passage from page xli of the 1938b translation of *The Rules* (Durkheim 1901c:ix). He cites no specific page reference to the preface to the second edition of *The Division of Labor*, however.

13. Bloor 1976:7. The Strong Programme consists of the historians and sociologists associated with the Science Studies Unit at Edinburgh University, especially Barry Barnes, David Bloor, and (formerly) Steven Shapin.

14. 1987:258, 247. More recently—in fact the following year—Latour (1988) seems to reject the idea that the sociology of scientific knowledge should even be in the business of providing explanations.

15. Jeffrey Alexander criticizes Bloor for using Durkheim in this way: "Though Bloor identifies his approach as Durkheimian, I will try to demonstrate . . . that such a materialistic sociologism cannot be viewed as an accurate application of Durkheim's mature work" (1982: vol. 1, 166–67 n. 27).

Chapter Two

1. Darwin 1859:112. Silvan Schweber (1985) suggests that Darwin may have come to this explanation of the divergence of characteristics through his reading of the political economists and thinking about their claims that the division of labor allows more people to be supported in each country. Jonathan Turner (1984) believes that Durkheim may have learned of this explanation from Spencer and not from Darwin, but he is not interested in building a strong historical case.

2. Works by mainly academic social scientists were reviewed (Karady 1983:83). Marxist works were slighted (Llobera 1981).

3. On this division of intellectual labor, see Paul Lapie's letter to Célestin Bouglé of 14 March 1897 (Lapie 1979:37) and Durkheim's letters to Bouglé of 13 June 1900 and 6 July 1900 (Durkheim 1976).

4. Letter dated 9 July 1897 (Lapie 1979:40). A generation later, Gaston Richard reported that *Suicide* was the "least known" of Durkheim's works (1923:129).

5. Similar methods were used in his first publication concerning suicide. In this paper he compared the population growth rates with the suicide rates for the departments of France by dividing these departments into five groups according to suicide rates and comparing the percentages of departments in each group with growth rates above and below the average. Then, in what he called the "inverse experiment," he divided the departments into four groups according to growth rates, and compared the average suicide rate for each group (1888d:451–57).

6. In the preface to the first edition of *Suicide*, Durkheim acknowledged Tarde for his assistance (1897a:xii). This preface was omitted in later editions.

7. More recently, Theodore Porter would seem to disagree with this assessment

of *Suicide.* He compares it favorably with similar works by Adolf Wagner, M. W. Drobisch, Georg Mayr, Alexander von Oettingen, and even Quetelet, saying that Durkheim's "use of numbers to draw sociological conclusions . . . was more sophisticated than virtually any of his predecessors." At the same time, however, Porter believes that Durkheim's "critique of statistical knowledge . . . was among the most confused of all such efforts" (1986:68–69). One of the anonymous reviewers of this book suggests that what impresses Porter about Durkheim's use of statistics may not be so much his mathematical as his dialectical skills.

8. The data in this article were drawn entirely from table 22 in *Suicide* (1897a:209).

9. See Durkheim 1899a(iii); 1899a(iv):184, 191; 1900c:20, 33; 1903c:493–94. Ratzel was a contributing editor to volume 3 of *L'Année sociologique.*

10. See Durkheim 1887c:49–58, 275–82; 1888a:98–100; 1900c:33; 1903c: 490–91, 495–96. Steinmetz was also a contributing editor to volume 3 of *L'Année sociologique.*

11. In a review of Lucien Lévy-Bruhl's *La Morale et la science des moeurs,* Durkheim proposed that a practical morality could be established on the basis of a scientific study of moral phenomena and their causes (1904a(5)). See also Durkheim's lectures on moral education (1925a), given in the academic year 1902–3.

12. Durkheim appears to have used the terms *crucial facts* (1895d:608), *well-established facts* (ibid.), *well-studied facts* (1901a(iii)17:275), and *well-made experiments* (1912a:132–35) all in the same way when criticizing ethnographers. I believe that all these terms refer to the method of crucial facts he described in *The Rules of Sociological Method* (1895a:98). This concept will be discussed in chapters 5 and 8. For additional criticisms of the methods of ethnographers, see 1907a10:586.

13. 1895e:76–77; 1900b:133. Until he took ill in 1898, Letourneau held the first chair in sociology in France at Broca's École d'Anthropologie. This chair was originally called "History of Civilizations." His lectures each year, which were attended by some of Durkheim's associates, would focus on a single social institution and trace its development through the ethnographic literature on animal as well as preliterate societies, and then proceed through ancient Egypt, Japan and China, ancient Greece and Rome, and contemporary Europe (Clark 1973:118–20; Hammond).

14. The letter in question is dated 7 April 1900. Lévy-Bruhl's *La Philosophie d'Auguste Comte* had appeared earlier that year. The letter was first published in 1969b.

15. Mauss also considered the history of science to be a serious lacuna in their sociology (Fournier 1982:65).

16. The article referred to is Hubert 1905. Durkheim reviewed Lévy-Bruhl's *Les Fonctions mentales dans les sociétés inférieures* (1910) in 1913a(ii)(6) and (7).

17. See Durkheim's letter to Bouglé of 14 December 1895, in which Durkheim tries to persuade Bouglé that the two are in closer agreement than Bouglé thinks (Durkheim 1976:166–67). This letter, of course, belongs to the period in which the journal was being organized.

18. Condominas 1972:491; Lévi-Strauss 1945:527. As Mauss explained to Lévi-Strauss, "it is easier to study the digestive process in the oyster than in man; but

this does not mean that the higher vertebrates were formerly shellfishes" (Lévi-Strauss 1945:527).

Chapter Three

1. Elsewhere in the same work he characterized moral facts as "rules of acting" (1902b:xxxvii) or "conduct" (1893b:23) or as duties that are generally admitted within a society (1893b:5).

2. 1901c:ix; 1897a:352; 1900b:128. Cf.: "Everything that is social consists in representations, consequently is a product of representations" (1898a(ii):100); "Social life is a system of representations, of mental states" (1901d:704); "In social life, all is representations, all is ideas, sentiments, and nowhere does one observe better the efficacious force of representations" (1908f:133).

3. Although Gurvitch (1950:354–55) interpreted Durkheim as having asserted that there are several "levels" of social reality, he regarded the level of collective representations as most basic, being causally responsible for the rest. Filloux (1970) and Thompson (1982) have adopted Gurvitch's interpretation of Durkheim as their own.

4. He discussed the sociological use of the study of law in terms of a sign of an underlying reality in one of his first sociology courses given at Bordeaux (1888c:268).

5. Durkheim's reasons for rejecting introspection will be discussed at greater length in the chapter on method.

6. The passage to which he referred is Fauconnet and Mauss (1901:171).

7. See also Cresswell (1972:143); S. Turner (1986:134–35). Elsewhere Durkheim cited Quetelet as having been the first to argue for an underlying social reality from the fact that each society has it own characteristic birth rate, marriage rate, crime rate, and so on (1903c:492).

8. See Takla and Pope (1985) for another defense of a realist interpretation of Durkheim's concept of social forces.

9. The issue of operational definitions will be discussed at greater length in chapter 5.

10. Contrary to the interpretation of Robert Strikwerda (1982:79), who sees Durkheim as having made a much more radical break from Comte than I do.

11. He provided similar accounts of the hypotheses of light waves and rays of heat on pp. 460 and 487, respectively, of volume 1.

12. For a scholarly introduction to these developments in nineteenth-century methodology, see Laudan (1981). For cogent summaries of the arguments of this work, see Laudan (1984:55–61; 81–84).

13. Of all the commentators on Durkheim I have read, only Georges Sorel seems to have been aware of the scientific and technical connotations of this term, having discussed the problems posed by the concept in meteorology (Sorel 1895:15).

14. George Simpson chose the former term (Durkheim 1933b:97), and W. D. Halls the latter (Durkheim 1984:53). Halls's translation of this passage is even worse than Simpson's, having rendered *autour d'elle* as "around *us*" instead of "around *itself.*" Halls fails to distinguish the third person singular from the first person plural.

15. 1912a:209, 329, 339, 629; 1913b:84, 99; 1919b:311. This notion of "fusion" can be found in Espinas, who also wrote of the energy of mental representations increasing through some sort of "repercussion" or "echo" effect (Espinas 1878:532–34).

16. Note, however, that the very term *coercion* [*coercition*] was used earlier in his definition of social facts in *The Rules* (1895a:8). I think what he meant to say in the later work is that the earlier definition should not be construed as necessarily implying *physical* coercion.

17. Parsons credited Durkheim, along with Freud, as the source of his inspiration for the concept of "internalization" (1956; 1964:18–19).

18. Cf.: "Collective representations are of an entirely different nature than those of the individual" (1897a:352). See also 1898b:16; 1912a:22.

19. Schaub (1920:335) also objected to the analogy.

20. This charge has been widely reported in the secondary literature but never adequately analyzed. See Catlin (1938:xxviii), Jones (1986a:78), Keat and Urry (1975:83, 187), Lukes (1973:11; 1982:4), Ossowska (1977:36), and Parsons (1937:361).

21. He objected to the charge of hypostatizing the collective consciousness also at 1895a:127 n. 1, and in a letter to Bouglé dated 14 December 1895 (Durkheim 1976:166).

22. The most recent critic of Durkheim who is subject to this confusion is Mike Gane, who treats "reification" and "hypostatization" as equivalent expressions (1988:70).

23. Letter to Bouglé, Dec. 1896, in Durkheim 1975a, vol. 2, 393. Besnard disagrees with Karady's dating of this letter, assigning the date April 1897 (Besnard et al. 1983:44).

24. He also wrote about association as a "cause" of new phenomena at 1902b:342 n. 3.

25. Durkheim used this analogy repeatedly, in 1897a:350, 1898b:40, 1900b:128, 1900c:364, 1901c:xv, 1909e:144, 1911b:114, 1912a:605.

26. This position has been referred to as Durkheim's "associational" or "relational social realism" by Harry Alpert (1939:155–57) and Ernest Wallwork (1972:9).

27. According to Marshall Sahlins, Lévi-Strauss also read Durkheim as believing that social relations exist in collective representations (Sahlins 1976:121).

28. Durkheim explains this distinction in several places, including 1902b:74, 1903b:92, 1911c(1):41, 1912a:23, 1913b:64, 73, 1914a:316–20, 330.

29. Georges Davy made a similar accusation a decade later (1911:43–44), which, as Pickering points out, Durkheim found unworthy of a reply (Pickering 1984:288).

30. Letters to Bouglé, 14 Dec. 1895, 6 July 1900 (1976:166, 176). Durkheim expressed similar views at 1897a:352; 1898b:49 n. 1; 1908f:133. Bouglé appears to have been convinced (1938:19).

31. He compared social psychology to the German *Völkerpsychologie* of Wundt, Lazarus, and Steinthal (1888a:102; 1903c:492 and n. 27).

32. One of the anonymous reviewers of this book suggests that Durkheim may

have moved away from using the term *social psychology* because it made sociology seem less autonomous.

33. The term *collective consciousness* appears on p. 86. See also 1888a:97; 1900b:124.

34. Meštrović seems to think that Wundt as well as Renouvier played an influential role in this regard (ibid., 204). So does Wallwork (1972:27), who claims to find even the concept of *collective* representations in Wundt. Durkheim claimed to have first read Wundt only in 1887 (1907b:613). More recently, Meštrović has stressed the importance of Schopenhauer's philosophy in Durkheim's thinking about mental representation (Meštrović 1988a, 1988b, 1989a, 1989b, 1989c).

Chapter Four

1. The same ambiguity that is found in "expliquer" is carried over into "rendre compte," which I translate as "to give an account."

2. Pickering (1984:187) also sees Durkheim as having offered empirical theories through his definitions. Although Durkheim talked about definitions explaining, he also said that tautologies do not explain (1902b:38).

3. Durkheim's attack on prenotions will be discussed in the following chapter on method.

4. Durkheim's conception of provisional definitions will be discussed in greater detail in the following chapter on method.

5. Similarly, Durkheim faulted what he perceived to be the nominalist view of society held by historians, according to which every society and every event is unique and generalization is impossible (1895a:94–95).

6. See, for example, Putnam (1975:73). Laudan criticizes this form of argument (1984: chap. 5).

7. Starting with the "lowest," the six sciences in Comte's hierarchy are mathematics, astronomy, physics, chemistry, biology, and sociology.

8. As Durkheim's teacher Émile Boutroux and colleague Lucien Lévy-Bruhl have argued, in spite of Comte's antimetaphysical attitude there is nevertheless a "latent metaphysics" in Comte's willingness to accept the existence of natural classes of phenomena and of laws that govern them (Benrubi 1933:16–17).

9. He had Comte in mind here; cf. 1895a:96, 121–22.

10. Similarly, when Durkheim said that "nothing exists except through [*par*] representation" (1912a:493 n. 1), he was not endorsing a form of idealism but asserting only that nothing exists *for us* except in a representation.

11. Stephen Turner thinks that the word "logic" here refers to Mill's methods of eliminative induction and that Durkheim was arguing that the principle of the plurality of causes would render these methods useless in the social sciences (1986:130). However, if Durkheim thought that the use of these methods conflicts with the principle of the plurality of causes, it is not clear why he would say that this principle is "linked" to these methods.

12. See, for example, Hempel (1965).

13. The point abut a possible plurality of causes being merely apparent is also made in *The Division of Labor* and in *The Elementary Forms of the Religious Life:* "The unity of the effect reveals the unity of the cause" (1902b:35); "The same fact cannot

have two causes, unless this duality is only apparent and at bottom they are but one" (1902b:50); "It is not conceivable, in fact, that, according to the circumstances, the same effect may be due sometimes to one cause, sometimes to another, unless, at bottom, the two causes are really only one" (1912a:594).

14. I have drawn on the French translation of Durkheim's Latin dissertation for this passage, rather than the English, which seems less clear; cf. 1960b:47.

15. Robert Hall, however, has argued that Durkheim's distinction between normal and pathological "is essential to sociology only because it is the link through which Durkheim claimed that the sociology of morals can have any significance for ethical judgments" (1987:23). This is in fact the gloss on this distinction that Durkheim provided in *The Rules,* but his *original* reasons for drawing this distinction in *The Division of Labor,* as I explain below, were different.

16. Durkheim also believed the comparison of normal with pathological cases to be already a part of the methods of psychology (1893c:227).

17. On Durkheim's debt to Quetelet, see S. Turner (1986:111) and Porter (1986:68–69). Durkheim, however, rejected Quetelet's concept of the "average man" (Turner 1986:152).

18. For Durkheim, the nature of species is neither stable nor fixed but evolves over time. Also, what is normal for an individual, whether an organism or a society, varies with its age (1893b:34–35).

19. My analysis of Durkheim's concept of causality, then, shows this concept to be identical to Mackie's (1974) analysis of the concept of causality of Hume and Mill.

20. Durkheim also rejected accounts of law that attribute them to one or another lawgiver at 1960b:11–12.

21. "Sociological facts can be truly explained only by other social facts" (1888a:92).

22. A social phenomenon, he said, "can be produced only by another social phenomenon" (1895e:101).

23. Letter to Bouglé, 14 Dec. 1895 (1976:166). Durkheim's page reference is to *The Rules of Sociological Method.* Similar passages about society or social facts being a reality sui generis may be found at 1895a:12 and 1912a:22.

24. On customs, moral rules, the family, and the law, see 1887c:40; 1888c:267–68; 1960b:11; 1925a:31–32. On values and social functions and their relations, see 1902b:48–49, 357. On social structure, see 1885c:632–33; 1888a:105. On religion, see 1886a:195. This last reference contains a discussion of Spencer, who was probably the source of this explanatory concept for Durkheim. The last work in which Durkheim seems to have explained social facts in terms of habit was his series of lectures on moral education, presented in 1902–3 and published in 1925. However, he still described the social being in each of us as a collection of "ideas, sentiments, and habits" as late as 1911c(1):41.

Chapter Five

1. The passage cited in *Suicide* occurs within the context of a criticism of Gabriel Tarde for relying on introspection (1897a:350–51).

2. This is also the reason he provided in the preface to the second edition of *The Rules of Sociological Method* for rejecting introspective methods (1901c:xi).

3. These views have been attributed to Durkheim by Alexander (1982:211), Catlin (1938:xiv), Douglas (1967:153), Giddens (1976:132; 1977:38), Hirst (1975:97); Meštrović (1989c), Nye and Ashworth (1971), and Taylor (1982:34, 203 n. 26, 204 n. 27).

4. Hirst seems to have based his interpretation almost entirely on *The Rules of Sociological Method* alone and on a faulty English translation of this work at that (1938b).

5. The so-called theory-ladenness of observations is not just a twentieth-century discovery in the philosophy of science but can be found at least as early as in the writings of Auguste Comte: "In order to apply itself to the making of observations, our mind has need of some sort of theory. If, in contemplating the phenomena, we do not immediately attach them to some principles, . . . most often, the facts would remain unperceived before our eyes" (1830–42: vol. 1, 21). For a detailed discussion of Comte's philosophy of science, see Larry Laudan (1971).

6. Alexander (1982:211) mistakenly attributes this sentence to Durkheim's 1888a, citing p. 57 of Traugott's translation (Durkheim 1978). Alexander also claims to quote the following passage from p. 58 of the same translation: "There is only one manner to arrive at the general, and it is to observe the particulars, not only superficially and in general, but minutely and by detail." Although nothing like this sentence appears in 1888a, a similar sentence did appear in 1887c; it reads: "But the only way ever to obtain an adequate definition . . . is to study by detail all the particular forms, all the nuances and varieties" (1887c:280). Note, however, that this sentence is about arriving at adequate definitions, not generalizations.

7. "An illustration does not constitute a demonstration" (1897a:vi).

8. Sorel (1895:1) reads the attack on economists as directed against Marx's *Capital*.

9. Cf.: "Logicians recognize that one can establish a law upon one well-made experiment" (1908a(3):233). I will discuss Durkheim's concept of a well-made experiment at greater length in the chapter on *The Elementary Forms*.

10. In a review of Westermarck published around the same time as *The Rules,* Durkheim identified what he called well-established and demonstrative facts with Bacon's crucial facts (1895d:608).

11. Of course, there is a sense in which all experiments are crucial experiments. That is, any experiment designed to test the hypothesis H that A is the cause of a may be thought of as a crucial test between H and the null hypothesis H^0 that A is *not* the cause of a.

12. Ironically, Tiryakian (1978:201) cites this same source in support of his induction by enumeration interpretation of Durkheim's methodology.

13. As I will discuss in chapter 8, Durkheim also argued that the primitive who refuses to abandon his religion when a rite does not produce its desired outcome is acting no differently, no less logically, than a contemporary scientist who refuses to abandon a theory in the face of anomalies.

14. Laudan makes this distinction in his criticism of Kuhn (1984:88).

15. W. D. Halls translates the French *notion* with the English word "notion" (Durkheim 1982:60).

16. He made similar injunctions against "prenotions," "prejudices," and "preconceived ideas" at 1888c:272; 1899a(ii):142; 1912a:32.

17. Fauconnet and Mauss also described methodical induction as making use of hypotheses: "If the work of induction has been done with method, it is not possible that the results at which sociology arrives will be stripped of all reality. Hypotheses express some of the facts, and consequently they always possess at least a portion of the truth: science may complete them, rectify them, transform them, but it never fails to utilize them" (1901:175).

18. Durkheim actually used the term *extension* in complaining of the ambiguity of word *society:* "There are few words that may be taken up with an acceptation so little defined. Sometimes one understands by it the entire society; sometimes only a part of this society. Even when one understands the word in this last acceptation, the limits within which its extension varies differs according to the case" (1900e:433–34).

19. The problem that the extension of ordinary terms may include too much is also mentioned at 1895a:47–48 and 1895d:612; the problem that it may include too little is also cited at 1912a:37.

20. Douglas 1967:178; Gibbs 1961:227; Giddens 1976:133; 1978:124; Halbwachs 1930:451–80; LaCapra 1972:150; Lacombe 1926:105; P.-Lévy 1981:108; Lukes 1973:200; MacIntyre 1967:26; Pope 1976:10–11; Simiand 1898:650; Thompson 1982:117; Winch 1958:110. For a sympathetic defense of Durkheim against the charges of MacIntyre and Winch, see Rosenberg (1980:28).

21. The distinction between imputing knowledge and imputing intent has long been recognized in the common law, although not in criminal statutes. That is, even when it cannot be established that one had an intent to deceive, one can be held liable in a suit at law for a negligent misrepresentation on the grounds that a reasonable and prudent person should have known the representation in question to be false.

22. Douglas, either utterly failing to grasp this point or deliberately misrepresenting Durkheim, interprets Durkheim as maintaining that suicide is never intended (1967:381).

23. Madge (1962:18), at least, also interprets Durkheim in this way.

24. A generation later, Durkheim's former collaborator Halbwachs attempted to draw the distinction between suicide and self-sacrifice in terms of societal approval (1930:475–80).

25. He made similar methodological prescriptions at 1897a:vii, 2.

26. That provisional definitions are not truly explanatory is also discussed at 1895a:44 and 1912a:31. He explained the purpose of his preliminary definition of religious facts in similar terms at 1899a(ii):140 and 1908a2:200. Fauconnet and Mauss provided a similar account of the function of provisional definitions in science (1901:172–73).

27. Harry Alpert did not seem to understand the hypothetical nature of these provisional definitions, for he said that "his misconceptions of the role and nature of definitions can be traced . . . to his failure to appreciate the vital part played by initial hypotheses and a priori assumptions . . . His theory of definition ignores the definer, and seems to imply that somehow or other things define themselves" (1939:114).

28. The recent translation of "emprunter" (1895a:54) as "derive" (1982:81) rather than "borrow" can be very misleading. The English word *derive* has a technical

meaning for philosophers, connoting the presence of some sort of logical argument, that Durkheim did not intend in this context.

29. On the method of defining facts according to their resemblances and differences, see also 1895a:44–45, 97; 1899a(ii):140–42. A few years earlier he had said, failing to maintain a clear distinction between essences and the signs by which we know them, that "the essential properties of a thing are those that one observes everywhere where this thing exists and that belong only to it" (1893c:227).

30. Similarly, in describing the monograph series of *L'Année sociologique*, Durkheim said that one aspect of method these monographs will share will be the attempt "to show by which bias [the facts] can be treated sociologically" (1908b:vi).

31. Gibbs and Martin 1958:140; 1964:6; Johnson 1965:883–86; Martin 1968:77–80; Pope 1976:30–44, 56; Thompson 1982:119–20. The failure to provide operational definitions of these terms has led some of these critics to question the very distinction between social integration and regulation. To depend on society for aid and support, however, is clearly not the same thing as to live according to its rules. For a defense of Durkheim's distinction, see Besnard (1984).

32. Also, as he explained, what is normal for a type of society varies with the stage of development it has reached (1893b:34–35). Finally, he warned, we must be on guard against mistaking what is merely a passing crisis for what is normal (1893b:36).

33. See especially 1897a:142, where he explained that he was interested in only those types of suicide as a function of which the suicide rate varies.

34. The corresponding argument in *Suicide* is expressed from a practical point of view, in terms of a lack of data, rather than from a purely theoretical point of view, in terms of the logical possibility of an infinite number of individual characteristics (1897a:139–40). The way in which it is expressed in *Suicide* is appropriate for a work that is more empirical than philosophical in character.

35. Cf.: "Every classification implies an hierarchical order" (1903a(i):399).

36. In *The Elementary Forms of the Religious Life*, religions are similarly arranged in a hierarchy in accordance with their degree of complexity (1912a:3).

37. The history of other intellectual disciplines using the concept of a mathematical series as a model actually predates even Descartes. In the sixteenth century, Jean Bodin had already proposed a method modeled on mathematical series for the classification of laws (Desan 134–35).

38. This interpretation is suggested by Mill's characterization of the method of crucial experiments in an essay on the method of political economy (Mill 1844:328). See also Mill (1843: bk. 3, chap. 8, sec. 3), where he said that it has been understood since the time of Bacon that these methods capture the essence of experimentation.

39. In his earliest publication on suicide, for example, he refers to statistical comparisons as "experiments" (1888d:447–59). In another early paper, he said that history provides "ready-made experiments" (1886a:202).

40. He gave voice to this last concern also at 1888c:262.

41. Durkheim seems to have been describing what Mill referred to as the "joint method of agreement and difference" (Mill 1843: bk. 3, chap. 8, sec. 4).

42. The recent translation of *The Rules of Sociological Method* translates the second "deduction" of the first sentence as "induction" (1982:152).

43. The articles I have in mind are 1888d and 1906d.

44. Pickering (1984:102) reads Durkheim's criticisms of Frazer as a wholesale rejection of the comparative method. Durkheim, however, was opposed only to indiscriminate comparisons that are not controlled for social types.

45. Among the historians responsible for the shift from simple narrative to institutional history, Durkheim cited Fustel de Coulanges, Sumner Maine, Maurer, and Wilda.

46. According to Christopher Prendergast, Durkheim's methodological assumption that the essence of modern social institutions is to be found in simple, primitive societies reflects the influence of his teacher Fustel de Coulanges. For Fustel, this assumption represents the application of Descartes's rules of method to the study of history (Prendergast 1983–84:57 and n. 5).

47. 1895a:145; 1900b:119; 1903c:470; 1904b:83; 1909e:147; 1914b:35; 1915a:7; 1928a:327–28. Durkheim's account of Comte's three-state law is not accurate and for this reason his criticisms of Comte are less than fair. Comte claimed merely that each one of the sciences and not society as a whole passes through the theological, metaphysical, and positive states (1830–42:1.21). Thus, for example, mathematics and astronomy had already reached the positive stage at a time when biological and sociological thinking were still metaphysical or theological (see Schmaus 1982:251–52). Of course, Comte's three-state law is not usually read in this way. His claim that social thought proceeds through these three stages is usually understood as the claim that society itself passes through these stages. This is an understandable confusion, given that for Comte, it is the history of ideas that governs the history of mankind.

Chapter Six

1. Turner relies entirely on Simpson's 1933 English translation of *The Division of Labor* and thus fails to recognize that such factors are only signs or symptoms and not causes of what Durkheim called the moral density of a society. See note 12 below.

2. Durkheim conflated the a priori with introspective methods, as for him the question as to whether a principle is a priori is a question of its *source* and not a question of the way in which it is justified.

3. The term *diffuse* was meant to signify that moral sanctions, unlike legal sanctions, are not applied by specific social organizations constituted for the purpose of enforcing the rules by applying sanctions (1893b:27).

4. 1893b:43/1902b:8. From this point onward, the text of the first edition was preserved intact in the second and all subsequent editions. I will make all subsequent citations using the pagination of the second edition, which is far more readily available.

5. Even towards the end of his career, Durkheim described the "consciousnesses" of those participating in relationships of exchange as "remaining outside one another" (1913b:84).

6. This is the account of mental representations that alluded to Cartesian vortices that I discussed in chapter 3. At 1902b:65 Durkheim cited Maudsley's *Physiology of the Mind* (1876) as his source for the effects that a mental representations can have.

7. Durkheim cited Espinas's *Sociétés animales* (1877) in a footnote (1902b:67 n. 1).

8. As I mentioned in chapter 3, he also used this concept of the fusion of mental representations in his later works.

9. In *Suicide* (1897a), four years later, Durkheim recognized that the causes of suicide are more complex than mere unhappiness and broadened his concept of suicide to include acts of self-sacrifice in primitive societies that result not from unhappiness but from something like a sense of duty.

10. Durkheim criticized both von Jhering (1887c:52) and Montesquieu (1960b:59–60) for having attempted to explain law and society in terms of the physical environment.

11. Johnson et al. (1984:159) also argue that moral density is a property of the collective consciousness.

12. "La condensation progressive des sociétés au cours du développement historique se produit de trois manières principales" (1902b:238). The French verb *produire* is not limited in meaning to the sense in which the cause produces its effect, as the first English translator, George Simpson, seems to have thought (1933b:257). It can have such varied meanings as to show or exhibit evidence or even to introduce a person. In its reflexive form it can mean to put oneself forward. Halls has left this sentence vague, translating *se produit* as "occur" (1984:201).

13. In a review of a work by Antonio Labriola a few years later, Durkheim said that he shared with the Marxists the belief that the causes of social phenomena lie outside the conscious awareness of individuals. However, he denied that these causes are to be found in the state of industrial or economic development of a society, rather than in "the manner in which associated individuals are grouped" (1897e:250–52).

14. It would seem, however, that given what Durkheim said elsewhere, the number of social relations would not only have to change but to increase and that this increase in turn would be reflected in developments in the means of transportation and communication.

15. See chapter 2 above, where I discussed Durkheim's theory of the division of intellectual labor in science.

16. I would like to think one of the anonymous reviewers of this book for suggesting this interpretation to me.

17. See Berkeley 1710, esp. pars. 1–13 of the introduction. This argument can also be found in Hume (1748: chap. 12, part 1) and Rousseau (1755: part 1). Durkheim alluded to this argument in the account of the category of genus he provided in *The Elementary Forms,* as we shall see in chapter 8.

18. Toward the end of his career Durkheim characterized both the individual and the collective consciousness as "realities of an experimental order" (1913b:86).

Chapter Seven

1. One cannot say that A is necessarily true without committing the fallacy of affirming the consequent.

2. Mill differed from Durkheim, however, in so far as Mill regarded a sociological theory itself as unreliable unless it could be explained by an "ethological" theory and ultimately by a psychological theory (Mill 1843: bk. 6, chap. 10).

3. He also suggested that the English may conceal suicides, reporting them as accidental deaths (1897a:160 n. 1).

4. That the level of education is an unreliable indicator of the strength of the collective consciousness is what Durkheim should have said about Norway and Sweden. Of course, in Scandinavia the Protestants were not a religious minority and he would have needed some other explanation of their combination of a high rate of school attendance with a low rate of suicide.

5. Morselli pointed out that the suicide rate for Jews was higher than that for Catholics not only in Bavaria but in Bohemia, parts of Prussia, Westphalia, and elsewhere (1882:122–23). Durkheim ignored this additional data and said that Bavaria was the sole exception (1897a:154 n. 1).

6. Halbwachs (1930:38) first raised the possibility of Catholics concealing suicides, much as Durkheim suggested that the English did (note 3 above). Pope (1976:11) suggests that suicides will be concealed in Catholic countries where religious rather than secular authorities are responsible for recording deaths. The issue of Catholics concealing suicides has been raised more recently by Day (1987), who also cites evidence that there is little significant difference between the suicide rates of Catholics and Protestants.

7. For example, Pope and Danigelis (1981); Stark, Doyle, and Rushing (1983); and Bainbridge and Stark (1981). Bankston, Allen, and Cunningham (1983) found a higher suicide rate among Catholics. Breault and Barkey (1982, 1983) argue that it is religious integration generally rather than specific denominations that preserve one from suicide, "thus contradicting Durkheim's operationalization of his theory" (Breault and Barkey 1982:630). More recently, however, Breault (1986:652) finds support for "Durkheim's proposition that Catholics have lower suicide rates than non-Catholics."

8. Taylor (1982:35) also argues that Durkheim cannot be refuted by a finding of a high suicide rate among Catholics.

9. In the preface to *Suicide* he thanked Tarde for making available to him the 26,000 dossiers on suicides held at the Ministry of Justice and thanked his nephew Marcel Mauss for having analyzed this data (1897a:xi; 1975a:1.43–49). This preface was removed subsequently to the first edition but appears in the English translation (1951a:35–39).

10. See Gibbs and Martin (1958:140; 1964:6), Johnson (1965:883–86), Martin (1968:77–80), Pope (1976:30–44, 56), and Thompson (1982:119–20). For a defense of Durkheim that replies to these arguments, see Besnard (1984). At least some of the confusion in the English-language literature, perhaps, can be traced to Spaulding's and Simpson's decision to translate both the definiendum of integration, which is *tenir su sa dépendance,* and *régler* as "control" (1897a:223, 264, 428; cf. 1951a:209, 241, 373).

11. Differences between the suicide rates of heterosexuals and homosexuals would also seem to be relevant to a theory of the regulation of sexual desires, but Durkheim no doubt lacked the appropriate data to test the implications of his theory for homosexuals.

12. Durkheim also characterized the social suicidal tendency as "one of the essential elements of the social coenaesthesis" (1897a:343). His use of this last term

represents one of his more awkward attempts to avoid the unwanted metaphysical connotations of the term "collective consciousness." The term *coenaesthesis* was used by nineteenth-century philosophical psychologists such as Théodule Ribot to designate that which combines perceptions from the various senses.

13. The only previously unpublished data in *Suicide,* the statistics that he acquired from Tarde and that are summarized in tables 21 and 22, date from 1889–91. In fact, none of the statistics he cited are more recent than 1893. Pickering (1984:51, 228) reports that, according to Gaston Richard, Durkheim may have actually completed *Suicide* well before 1897.

14. These objections have been raised by Giddens 1978:120; Jones 1986:113; Lukes 1973:32, 202; Pope 1976:185–86; and Tosti 1898:474–75.

15. Similarly, he said in his lectures on pragmatism, "representations begin to have . . . an appearance of a life of their own" already at the level of mental images formed in the minds of individuals (1955a:169).

16. In *The Elementary Forms of the Religious Life* he appealed to this analogy in order to distinguish his views from the crude determinism of historical materialism (1912a:605–6, and 606 n. 1).

17. The distinction under discussion was already implicit in his review of Labriola published the preceding year: "If the different forms of collective activity also have their substratum, and if in the last instance they derive from it, once they exist, they become, in their turn, original sources of action, they have an efficacy that is proper to them, they react upon the very causes on which they depend" (1897e: 254).

18. This interpretation of this essay seems to derive from the work of Benoît-Smullyan (1948).

19. Gisbert (1959:356) also considered this essay a "Wundtian" piece, and Joas (1984:572) reads it as a defense of this theory against William James.

Chapter Eight

1. Durkheim had expressed this analogy as early as May 1899 in a letter to Gaston Richard, in which he said that religion "stands in the same relationship to the representative life of society as does sensation to the life of the individual. Just as we experience things through their colors, etc., so society represents its life and the life of the objects related to it as sacred objects. It colors them with religiosity" (1928b:299 n. 1; Pickering 1984:192).

2. Although Steven Collins (1985) dismisses Mauss's claim, he provides no persuasive reason for believing that Durkheim's concept of a category resembles Kant's.

3. Empirical refutations of Durkheim's sociology of knowledge have been offered by Jean Cazeneuve (1958), Rodney Needham (1963), and Peter Worsley (1956:57–60), and summarized by Steven Lukes (1973:446).

4. More recently, this objection has been raised by Pickering (1984:238).

5. Durkheim's nephew Marcel Mauss, on the other hand, did have a facility with languages, including Sanskrit and ancient Iranian (Condominas 1972:122). Henri Hubert, Mauss's sometime collaborator, was a philologist with a knowledge of Celtic languages (128). Also, as Lukes reports, Durkheim's son André had been

training to be a linguist under Antoine Meillet, but was killed in World War I (1973:554–55). Meillet was also a contributing editor of *L'Année sociologique* (655).

6. More recently, Laudan has called this kind of crucial fact a "non-refuting anomaly" (1977:29).

7. Durkheim discussed the role of symbols in the formation of collective representations also in several papers from the same period as *The Elementary Forms* (see 1911b:118; 1913b:84; 1914a:328).

8. Durkheim's self-professed follower David Bloor made a similar move in his interpretation of Zande witchcraft, arguing that they think much like we do but have a very different way of classifying things. However, choosing to identify logic with classification rather than with reasoning, Bloor nevertheless concluded that the Azande have a different logic than ours (1976:129–30). A few years later, however, he adopted a different tactic, arguing that all inferences are socially constructed (Bloor 1984). I will discuss the latter argument in chapter 9.

9. Durkheim did not seem to recognize that this same ambiguity of religious rites could serve as the source of a possible objection against his hypothesis that the idea of a necessary connection is grounded in the obligatory nature of these rites.

10. Durkheim also seems to have overlooked the problem that the notion of malign religious forces would appear to raise for his assertion that religious forces are conceived as having moral authority.

11. Takla and Pope (1985:80) try to defend Durkheim's use of functional explanation in *The Elementary Forms* by appealing to what they regard as the "thinking, understanding, self-conscious" nature of society in Durkheim. For Durkheim to have thought of society in this way, however, would be for him to have violated his strictures against sociological explanations in terms of the conscious pursuit of ends-in-view.

12. Prendergast (1983–84) reaches a somewhat similar conclusion about the theoretical continuity of Durkheim's work, except that he does not discuss the theory of mental representations that underlies Durkheim's conception of mechanical solidarity. Prendergast also finds the germ of the idea of mechanical solidarity in Fustel de Coulanges's *The Ancient City* and thus regards this work as standing in the same theoretical tradition as Durkheim's *The Elementary Forms*.

13. In addition to Godlove and Lukes, Benoît-Smullyan (1948:518 n. 67), Bloor (1982:268), Collins (1985:46f.), Mary Douglas (1975:xv), Giddens (1978:111), R. A. Jones (1984:74), R. A. Jones and Vogt (1984:54), and Meštrović (1989a), also interpret Durkheim as having tried to provide a sociological account of the Kantian categories.

Chapter Nine

1. The classic statement of this functionalist position is Putnam (1967).

2. See, for example, Block (1978).

3. This concept of the goal of sociology was subsequently adopted by proponents of a phenomenological research program such as Jack Douglas (1967), whom Taylor criticizes.

4. Hund (1990) also criticizes Bloor for assimilating Durkheim to Wittgenstein.

5. Lugg (1984) and Smith (1984) point out additional ambiguities surrounding the interpretation of these historical case studies. Not only is the nature of the relationship between putative knowledge claims and social interests ambiguous but so too is the very notion of "interests." As Buchdahl (1982:303) and Yearley (1982:382) argue, due to the ambiguity of "interests," this notion ends up being little more than an ad hoc redescription of the event to be explained, allowing one always to make one's historical case study fit the interest model. Furthermore, as Woolgar (1981) cautions, one could always argue that social interests are socially constructed by historians and sociologists of science just as constructivists say that scientific facts are socially constructed in the laboratory.

6. As I mentioned in chapter 3, note 19, Durkheim is one of the sources of Parsons's notion of the internalization of social norms.

7. For the distinction between descriptive and normative appeals to principles of rationality, see Gutting (1984:99).

References

For Durkheim's writings, I have adopted the numbering system invented by Steven Lukes (1973). It is the most complete in existence and is beginning to be adopted by Durkheim scholars as a standard numbering system for his works (see for example Durkheim 1975b and 1979). I have also extended this system to newer editions and translations of Durkheim's works that are not included in Lukes's bibliography.

Lukes's numbering system takes as its starting point the standard academic practice of adding letters after the date when there is more than one publication by the author in the same year (for example, 1895a, 1895b). Books and articles in French are always listed before works in English or Italian. When Durkheim has more than one publication in the same book or the same journal in the same year, however, Lukes gives all of them the same year and letter designation and distinguishes them by numbering them consecutively in parentheses following the letter. For example, 1908a(2) and 1908a(3) indicate Durkheim's second and third contributions, respectively, to volume 8 of the *Bulletin de la société française de philosophie*. Lukes adopts this system because the number of Durkheim's publications in some years is greater than the number of letters in the alphabet. This is true above all for his publications in *L'Année sociologique*. In numbering these publications Lukes has added another layer of complexity, distinguishing among editorial prefaces, research articles, notes, and reviews. When all four types of contributions are present in the same volume, they are indicated with the small roman numerals i–iv, respectively, in parentheses. When there is more than one contribution in the same category, they are numbered consecutively with arabic numerals in additional parentheses. For example, 1899a(iv)(33) is the thirty-third review published by Durkheim in the second volume of this journal. When there are no contributions by Durkheim in a given category in a given volume, however, each subsequent category moves up in rank. For example, a note in the second volume is indicated as 1899a(iii), while a note in the fifth volume, which contains no editorial preface by Durkheim, is indicated as 1902a(ii). Because I follow Lukes's system even when I do not cite everything Durkheim published in a given year, the numbering is not always consecutive in my bibliography, which contains only those works I have actually cited.

At the end of each Durkheim reference, I have indicated where English translations can be found. Where complete translations exist, partial translations are not mentioned. Because Durkheim has not always been carefully translated in a way that is sensitive to the meanings that his words held in the philosophical tradition in which he was educated, however, it is advisable to read his works in the original French.

For those who do not read French, I would recommend the newer translation of *The Rules of Sociological Method* (1982) over the older one (1938b), which as I explain in chapter 5 seriously distorts Durkheim's meaning. The newer edition also contains translations of some of Durkheim's subsequent writings on method. However, the two translations of *The Division of Labor in Society* are of about the same quality. The older translation (1933b) is more complete than the newer (1984), as the older one provides a translation of Durkheim's analytical table of contents and the entire introduction to the first French edition (1893b), which was omitted in the second edition (1902b). Multiple translations do not exist for Durkheim's other book-length works. For his shorter works, in addition to the appendix to 1982 there are several useful anthologies that provide translations of the complete texts. These include 1953b, 1956a, 1960c, 1973, 1975b, 1978, 1979, 1980, 1983b, and 1986.

Works by Émile Durkheim

1885a. Review of *Bau und Leben des sozialen Körpers: Erster Band,* by Albert Schaeffle. *Revue philosophique* 19:84–101; rpt. 1975a:1.355–77; trans. 1978:93–114.

1885c. Review of *Grundriss der Soziologie,* by Ludwig Gumplowicz. *Revue philosophique* 20:627–34; rpt. 1975a:1.344–54.

1886a. "Les Études de science sociale." *Revue philosophique* 22:61–80; rpt. 1970a:184–214; 61–69 trans. *Sociological Inquiry* 44 (1974):205–14 and 1975b:13–23; 78–79 trans. 1972:55–57.

1887b. Review of *L'Irréligion de l'avenir,* by Marie Jean Guyau. *Revue philosophique* 23:299–311; rpt. 1975a:2.149–65; trans. 1975b:24–38.

1887c. "La Science de la morale en Allemagne." *Revue philosophique* 24:33–58, 113–42, 275–84; rpt. 1975a:1.267–343; excerpts trans. 1972:67–68, 90–95.

1888a. "Cours de science sociale: leçon d'ouverture." *Revue internationale de l'enseignement* 15:23–48; rpt. 1970a:77–110; trans. 1978:43–70.

1888c. "Introduction à la sociologie de la famille." *Annales de la faculté des lettres de Bordeaux* 10:257–81; rpt. 1975a:3.9–34; trans. 1978:205–28.

1888d. "Suicide et Natalité: Étude de Statistique Morale." *Revue philosophique* 26:446–63; rpt. 1975a:2.216–36.

1889b. Review of *Gemeinschaft und Gesellschaft,* by Ferdinand Tönnies. *Revue philosophique* 27:416–22; rpt. 1975a:1.383–90; trans. 1978:115–22.

1890a. "Les Principes de 1789 et la sociologie." Review of *Les Principes de 1789 et la science sociale,* by Th. Ferneuil. *Revue internationale de l'enseignement* 19:450–56; rpt. 1970a:215–25; trans. 1973:34–42.

1892a. *Quid secundatus politicae scientiae instituendae contulerit.* Bordeaux: Gounouilhou. Trans. French 1953a; trans. English 1960b.

1893a. Review of *Essai sur l'origine de l'Idée de Droit,* by Gaston Richard. *Revue philosophique* 35:290–96; rpt. 1975a:233–41; trans. 1983b:146–54.

1893b. *De la division du travail social: Étude sur l'organisation des sociétés supérieures.* Paris: Alcan. Trans. 1933b, 1984. The pages from the introduction to the first edition that were removed in subsequent editions are reprinted in 1975a:2.257–88 and are translated in 1933b:411–35.

1893c. "Note sur la définition du socialisme." *Revue philosophique* 36:506–12; rpt. 1970a:226–35; trans. 1986:113–20.

1895a. *Les Règles de la méthode sociologique.* Paris: Alcan. Trans. 1938b, 1982.

1895d. "L'Origine du mariage dans l'espèce humaine d'après Westermarck." *Revue philosophique* 40:606–23; rpt. 1975a:3.70–92.

1895e. "Lo Stato attuale degli studi sociologici in Francia." *La Riforma sociale* 3:607–22, 691–707; trans. French 1975a:1.73–108.

1897a. *Le Suicide: Étude de sociologie.* Paris: Alcan. The preface to the first edition that was removed in subsequent editions is reprinted in 1975a:1.43–48 and translated in 1951a.

1897e. Review of *Essais sur la conception materialiste de l'histoire,* by Antonio Labriola. *Revue philosophique* 44:645–51; rpt. 1970a: chap. 9; trans. 1978:123–30; 1982:167–74; 1986:128–36.

1897f. Contribution to "Enquête sur l'oeuvre d'Hippolyte-Adolphe Taine." *Revue blanche* 13:287–91; rpt. 1975a:1.171–77.

1898a(i). Preface to *L'Année sociologique* 1:i–vii; rpt. 1969c:31–36; trans. 1960c:341–47, 1980:47–51.

1898a(ii). "La Prohibition de l'inceste et ses origines." *L'Année sociologique* 1:1–70; rpt. 1969c:37–100; trans. 1963a.

1898b. "Représentations individuelles et représentations collectives." *Revue de métaphysique et de morale* 6:273–302; rpt. 1924a:13–50; trans. 1953b:1–34.

1899a(i). Preface to *L'Année sociologique* 2:i–vi; rpt. 1969c:135–39; trans. 1960c:347–53, 1980:52–55.

1899a(ii). "De la définition des phénomènes religieux." *L'Année sociologique* 2:1–28; rpt. 1969c:140–65; trans. 1975b:74–99.

1899a(iii). "Morphologie sociale." *L'Année sociologique* 2:520–21; rpt. 1969c:181–82; trans. 1978:88–90, 1982:241–42.

1899a(iv)(33). Review of *Politische Geographie,* by Friedrich Ratzel. *L'Année sociologique* 2:522–32; rpt. 1969c:183–92; trans. 1986:62–72.

1900b. "La Sociologie en France au XIXᵉ siècle." *Revue bleue* (4th ser.) 12:609–13, 647–52; rpt. 1970a:111–36; trans. 1973:3–22.

1900c. "La Sociologia ed il suo dominio scientifico." *Rivista italiana di sociologia* 4:127–48; trans. French 1975a:1.13–36; trans. English 1960c:354–75.

1900e. "L'État." *Revue philosophique* 148 (1958):433–37; rpt. 1975a:3.172–78; trans. 1986:45–50.

1901a(i). "Deux lois de l'évolution penale." *L'Année sociologique* 4:65–96; rpt. 1969c:245–73; trans. 1978:153–80, 1983b:102–32.

1901a(ii)(2). "Technologie." *L'Année sociologique* 4:593–94; rpt. 1975a:1.246.

1901a.(iii)(17). Review of "Die neuren Forschungen zur Geschichte der menschlichen Familie," by Sebald-Rudolf Steinmetz. *L'Année sociologique* 4:340–42; rpt. 1969c:274–75; trans. 1980:184–85.

1901c. *Les Règles de la méthode sociologique.* 2d. ed. Paris: Alcan. Trans. 1938b, 1982.

1901d. "Lettre au Directeur." *Revue philosophique* 52:704; rpt. 1975a:1.52–53; 253–54; trans. 1982:253–54.

1901h. "Compte-rendu d'une conférence sur 'Religion et libre pensée,' devant les membres de la Fédération de Jeunesse Laïque." *La Petite Gironde,* 24 May 1901; rpt. 1975a:1.425–27.

1902a(i). "Sur le totemisme," *L'Année sociologique* 5:82–121; rpt. 1969c:315–52.

1902a(ii). "Civilisation en général et types de civilisation." *L'Année sociologique* 5:167–68; rpt. 1975a:1.53–54; trans. 1982:243–44.

1902b. *De la division du travail social.* 2d. ed. Paris: Alcan. Trans. 1933b, 1984.

1903a(i). With Marcel Mauss. "De quelque formes primitives de classification: Contribution à l'étude des représentations collectives." *L'Année sociologique* 6:1–72; rpt. 1969c:395–461; trans. 1963b.

1903b. "Pédagogie et sociologie." *Revue de métaphysique et de morale* 11:37–54; rpt. 1922a:81f.; trans. 1956a.

1903c. With Paul Fauconnet. "Sociologie et sciences sociales." *Revue philosophique* 55:465–97; rpt. 1975a:1.122–53; abridged trans. 1905d; trans. 1982:175–208.

1904a(5). Review of *La Morale et la science des moeurs,* by Lucien Lévy-Bruhl. *L'Année sociologique* 7:380–84; rpt. 1969c:467–70; trans. 1979:29–33, 1980:127–30.

1904b. "La Sociologie et les sciences sociales." *Revue internationale de sociologie* 12:83–84, 86–87; rpt. 1975a:1.160–65.

1905a(i). "Sur l'organisation matrimoniale des sociétés australiennes." *L'Année sociologique* 8:118–47; rpt. 1969c:483–510.

1905b. Contribution to "La Morale sans Dieu: Essai de solution collective." *La Revue* 59:306–8; rpt. 1975a:2.334–37; trans. 1979:34–36.

1905c. "On the Relation of Sociology to the Social Sciences and to Philosophy." *Sociological Papers* 1:197–200, 257 (rejoinders); trans. French 1975a:1.166–70; rejoinders rpt. 1982:255–56.

1905d. "Sociology and the Social Sciences." *Sociological Papers* 1:258–80 (partial trans. of 1903c).

1906a(2). Review of *Sociologia e storia,* by Alexandru Dimitrie Xenopol. *L'Année sociologique* 9:139–40; rpt. 1975a:1.197–99; trans. 1980:73–74.

1906b. "La Détermination du fait moral." *Bulletin de la société française de philosophie* 6:169–212; rpt. 1924a:51–101; trans. 1953b:35–79.

1906d. "Le Divorce par consentement mutuel." *Revue bleue* 5:549–54; rpt. 1975a:2.181–94; trans. 1978:240–52.

1907a(10). Review of *The Origin and Development of the Moral Ideas,* by Edward Westermarck. *L'Année sociologique* 10:383–95; rpt. 1969c:584–95; trans. 1979:40–51, 1980:150–59.

1907b. "Lettres au Directeur." *Revue néo-scolastique* 14:606–7, 612–14; 2d letter rpt. 1975a:1.401–5; trans. 1982:257–60.

1907e. "Nécrologie d'Octave Hamelin." *Le Temps,* 18 Sept. 1907; rpt. 1975a:1.428–29.

1907g. "Débat sur les rapports de l'ethnologie et la sociologie." *Bulletin de Comité des travaux historiques et scientifiques: Section des sciences economiques et sociales,* 199–200; rpt. 1975a:1.256–58; trans. 1982:209–10.

1908a(2). Contribution to "La Morale positive: Examen de quelques difficultés."

Bulletin de la société française de philosophie 8:189–200; rpt. 1975a:2.341–54; trans. 1979:52–64.

1908a(3). Contribution to "L'Inconnue et l'inconscient en histoire." *Bulletin de la société française de philosophie* 8:229–45, 247; excerpts rpt. 1975a:1.199–217; trans. 1982:211–28.

1908b. "Aux lecteurs de *L'Année sociologique.*" Preface to Célestin Bouglé, *Essais sur le régime des castes.* Paris: Alcan. Rpt. 1975a:1.430–32.

1908f. Contribution to "Enquête sur la sociologie." *Les Documents du progrès* 2 (Feb.):131–33; rpt. 1975a:1.58–61; trans. 1982:245–47.

1909d. "Sociologie religieuse et théorie de la connaissance." *Revue de métaphysique et de morale* 17:733–58; parts 1 and 2 rpt. 1912a:1–28; trans. 1915d:13–33; part 3 rpt. 1975a:1.185–88; trans. 1982:236–40.

1909e. "Sociologie et science sociale," *De la méthode dans les sciences.* Paris: Alcan, pp. 259–85. Rpt. 1970a:137–58; trans. 1978:71–87.

1910a(ii)(1). With Célestin Bouglé. "Les Conditions sociologique de la connaissance." *L'Année sociologique* 11:41–42; rpt. 1975a:1.189; trans. 1980:106–7.

1910a(iii)(2). Review of *Soziologie des Erkennens,* by Wilhelm Jerusalem. *L'Année sociologique* 11:42–45; rpt. 1975a:1.190–94; trans. 1980:107–10.

1910b. Contribution to "La Notion d'égalité sociale." *Bulletin de la société française de philosophie* 10:59–70; rpt. 1975a:2.373–85; trans. 1979:65–76, 1986:186–93.

1911b. "Jugements de valeur et jugements de realité." *Atti del IV Congresso Internazionale di Filosofia—Bologna* 1:99–114; *Revue de métaphysique et de morale* 19:437–53; rpt. 1924a:102–21; trans. 1953b:80–97.

1911c(1). "L'Éducation, sa nature et son role." *Nouveau Dictionnaire de pédagogie et d'instruction primaire.* Paris: Hachette, pp. 529–36. Rpt. 1922a:31–58; trans. 1956a.

1911c(3). "Nature et méthode de la pédagogie." *Nouveau Dictionnaire de pédagogie et d'instruction primaire.* Paris: Hachette, pp. 1538–43. Rpt. 1922a:59–80; trans. 1956a.

1912a. *Les Formes élémentaires de la vie religieuse.* Paris: Alcan. Trans. 1915d.

1913a(i)(1). With Marcel Mauss. "Note sur la notion de civilisation." *L'Année sociologique* 12:46–50; rpt. 1969c:681–85; trans. *Social Research* 38 (1971):808–13.

1913a(ii)(6) and (7). Review of *Les Fonctions mentales dans les sociétés inférieures,* by Lucien Lévy-Bruhl, and *Les Formes élémentaires de la vie religieuse,* by Durkheim. *L'Année sociologique* 12:33–37; rpt. 1975a:1.405–7; trans. 1972:246–49, 1975b:169–73, 1978:145–49.

1913a(ii)(8). Review of *Elemente de Völkerpsychologie,* by Wilhelm Wundt. *L'Année sociologique* 12:50–61; rpt. 1969c:685–96.

1913a(ii)(15). Review of *Le Conflit de la morale et de la sociologie,* by Simon Deploige. *L'Année sociologique* 12:326–28; rpt. 1975a:1.405–7; trans. 1980:159–61.

1913b. Contribution to "Le Problème religieux et la dualité de la nature humaine." *Bulletin de la société française de philosophie* 13:63–75, 80–87, 90–100, 108–11; rpt. 1975a:2.23–64; trans. *Knowledge and Society* 5 (1984):1–44.

1914e. "Le Dualisme de la nature humaine et ses conditions sociales," *Scientia* 15:206–221; rpt. 1970a:314–22; trans. 1960c:325–40, 1973:149–63.

1914b. "Une Nouvelle Position du problème moral." *Bulletin de la société française de philosophie* 14:26–29, 34–36; rpt. 1975a:1.64–70.

1915a. "La Sociologie." *La Science française*. Paris: Larousse, vol. 1, pp. 5–15. Rpt. 1975a:1.109–18; trans. 1960c:376–85.

1915d. *The Elementary Forms of the Religious Life.* Trans. Joseph Ward Swain. New York: Free Press, 1965.

1917b(2). Contribution to "Vocabulaire technique et critique de la philosophie (société)." *Bulletin de la société française de philosophie* 15:57; rpt. 1975a:1.71; trans. 1982:248.

1918b. "Le 'Contrat social' de Rousseau." *Revue de métaphysique et de morale* 25:1–23, 129–61; rpt. 1953a; trans. 1960b.

1919b. "La Conception sociale de la religion." *Le Sentiment religieux à l'heure actuelle*. Ed. Frank Abauzit et al. Paris: Vrin, pp. 97–105, 142–43. Rpt. *Archives de sociologie des religions* 27 (1969):73–77 and 30 (1971):89–90; Durkheim (1970a):305–13; 1975a:2.146–48; trans. 1975b:181–89.

1920a. "Introduction à la morale." *Revue philosophique* 89:79–87; rpt. 1975a: 2.313–31; trans. 1978:191–202, 1979:77–96.

1922a. *Éducation et sociologie.* Paris: Alcan. Trans. 1956a.

1924a. *Sociologie et philosophie.* 1951; 1974; Paris: PUF. Trans. 1953b.

1925a. *L'Éducation morale.* Paris: Alcan. Trans. 1961a.

1928a. *Le Socialisme.* Paris: Alcan. Trans. 1958b.

1928b. Letter to Gaston Richard, 11 May 1899. Gaston Richard, "L'Enseignement de la sociologie à l'école normale primaire," *L'Éducateur protestant* 7:298–99, no. 1; trans. Pickering 1984:192.

1933b. *The Division of Labor in Society.* Trans. George Simpson. New York: Macmillan.

1938a. *L'Évolution pédagogique en France.* 1969; Paris: PUF. Trans. 1977.

1938b. *The Rules of Sociological Method.* Trans. Sarah A. Solovay and John H. Mueller. Ed. George E. G. Catlin. New York: Free Press.

1951a. *Suicide: A Study in Sociology.* Trans. John A. Spaulding and George Simpson. New York: Free Press.

1953a. *Montesquieu et Rousseau, precurseurs de la sociologie.* Trans. A. Cuvillier. Paris: Marcel Rivière.

1953b. *Sociology and Philosophy.* Trans. D. F. Pocock. London: Cohen and West; rpt. New York: Free Press, 1974.

1955a. *Pragmatisme et Sociologie.* Paris: Vrin. Partial trans. 1960c:386–486; trans. 1983a.

1956a. *Education and Sociology.* Glencoe, Ill.: Free Press.

1958b. *Socialism and Saint-Simon.* Yellow Springs, Ohio: Antioch Press; rpt. London: Routledge and Kegan Paul, 1959; rpt. under the title *Socialism*, New York: Collier Books, 1962.

1960b. *Montesquieu and Rousseau: Forerunners of Sociology.* Trans. R. Manheim. Ann Arbor, Mich.: University of Michigan Press.

1960c. *Émile Durkheim, 1858–1917: A Collection of Essays, with Translations and a*

Bibliography. Ed. Kurt H. Wolff. Columbus, Ohio: Ohio State University Press; rpt. under the title *Essays on Sociology and Philosophy,* New York: Harper and Row, 1964.

1961a. *Moral Education.* New York: Free Press.

1963a. *Incest: The Nature and Origin of the Taboo.* With Albert Ellis, *The Origins and Development of the Incest Taboo.* New York: Lyle Stuart.

1963b. *Primitive Classification.* Trans. Rodney Needham. Chicago: University of Chicago Press.

1963c. *Émile Durkheim: Selections from His Work.* Trans. and ed. George Simpson. New York: Thomas Y. Crowell.

1969b. "Une Lettre inédite d'Émile Durkheim." *L'Année sociologique* (3d ser., 1967):14; rpt. *Revue philosophique* 94:381, 1975a:1.177.

1969c. *Journal sociologique.* Ed. Jean Duvignaud. Paris: PUF.

1970a. *La Science sociale et l'action.* Ed. Jean-Claude Filloux. 1987; Paris: PUF.

1972. *Émile Durkheim: Selected Writings.* Ed. Anthony Giddens. Cambridge: Cambridge University Press.

1973. *Émile Durkheim on Morality and Society.* Ed. Robert Bellah. Chicago: University of Chicago Press.

1975a. *Textes.* Ed. Victor Karady. Paris: Les Éditions de Minuit.

1975b. *Durkheim on Religion.* Ed. William S. F. Pickering. London: Routledge and Kegan Paul.

1976: "Lettres à Célestin Bouglé." *Revue française de sociologie* 17:165–80; letter of 14 Dec. 1895 trans. 1982:249–50.

1977. *The Evolution of Educational Thought.* London: Routledge and Kegan Paul.

1978. *Émile Durkheim on Institutional Analysis.* Trans. and ed. Mark Traugott. Chicago: University of Chicago Press.

1979. *Durkheim: Essays on Morals and Education.* Ed. William S. F. Pickering. London: Routledge and Kegan Paul.

1980. *Émile Durkheim: Contributions to L'Année Sociologique.* Ed. Yash Nandan. New York: Free Press.

1982. *The Rules of Sociological Method and Selected Texts on Sociology and Its Method.* Trans. W. D. Halls. New York: Free Press.

1983a. *Pragmatism and Sociology.* Trans. J. C. Whitehouse. Cambridge: Cambridge University Press.

1983b. *Durkheim and the Law.* Ed. Steven Lukes and Andrew Scull. New York: St. Martin's Press.

1984. *The Division of Labor in Society.* Trans. W. D. Halls. New York: Free Press.

1986. *Durkheim on Politics and the State.* Trans. W. D. Halls. Ed. Anthony Giddens. Stanford, Calif.: Stanford University Press.

General References

Alexander, Jeffrey. 1982. *Theoretical Logic in Sociology.* Berkeley: University of California Press.

———. 1986a. "Rethinking Durkheim's Intellectual Development I: On 'Marxism' and the Anxiety of Being Misunderstood." *International Sociology* 1:91–107.

———. 1986b. "Rethinking Durkheim's Intellectual Development II: Working out a Religious Sociology." *International Sociology* 1:189–201.

Allcock, John B. 1983. Introduction to Durkheim 1983.

Alpert, Harry. 1939. *Émile Durkheim and his Sociology.* New York: Russell and Russell, 1961.

Andler, Charles, 1896. "Sociologie et democratie." *Revue de métaphysique et de morale* 4:243–56.

Bacon, Francis. 1620. *Novum Organum; or True Suggestions for the Interpretation of Nature.* In *Advancement of Learning and Novum Organum.* Ed. James Edward Creighton. New York: Colonial Press, 1900.

Bainbridge, William S., and Rodney Stark. 1981. "Suicide, Homicide, and Religion: Durkheim Reassessed." *Annual Review of the Social Sciences of Religion* 5:33–56.

Baker, G. P., and P. M. S. Hacker. 1985. *Wittgenstein, Rules, Grammar, and Necessity,* vol. 2 of *Analytical Commentary on the Philosophical Investigations.* Oxford: Basil Blackwell.

Bankston, William B., H. David Allen, and Daniel S. Cunningham. 1983. "Religion and Suicide: A Research Note on Sociology's 'One Law.'" *Social Forces* 62:521–28.

Barnes, Barry, and David Bloor. 1982. "Relativism, Rationalism, and the Sociology of Knowledge." In *Rationality and Relativism.* Ed. Martin Hollis and Steven Lukes. Cambridge, Mass.: MIT Press.

Barth, Paul. 1897. *Die Philosophie der Geschichte als Soziologie.* Leipzig: O. R. Reisland.

Benoît-Smullyan, Émile. 1948. "The Sociologism of Émile Durkheim and his School." In *An Introduction to the History of Sociology.* Ed. Harry Elmer Barnes. Chicago: University of Chicago Press.

Benrubi, Isaak. 1933. *Les Sources et les courants de la philosophie contemporaine en France.* Paris: Alcan.

Berkeley, George. 1710. *A Treatise Concerning the Principles of Human Knowledge.* 2d ed., 1734. Indianapolis, Ind.: Bobbs-Merrill, 1957.

Berthelot, Jean-Michel. 1989. "Principe de causalité et raisonnement expérimental chez Durkheim." *Revue philosophique* 179:25–50.

Besnard, Philippe. 1983. "The *Année sociologique* team." In Besnard et al. 1983.

———. 1984. "Modes d'emploi du *Suicide.* Intégration et régulation dans la théorie durkheimienne." *L'Année Sociologique* 34:127–63.

Besnard, Philippe et al. 1983. *The Sociological Domain: The Durkheimians and the Founding of French Sociology.* Cambridge: Cambridge University Press.

Block, Ned. 1978. "Troubles with Functionalism." In *Perception and Cognition: Issues in the Foundations of Psychology.* Ed. C. Wade Savage. Minneapolis, Minn.: University of Minnesota Press. Rpt. in vol. 1 of *Readings in the Philosophy of Psychology.* Ed. N. Block. Cambridge, Mass.: Harvard University Press, 1980.

Bloor, David. 1976. *Knowledge and Social Imagery.* London: Routledge and Kegan Paul.

———. 1982. "Durkheim and Mauss Revisited: Classification and the Sociology of Knowledge." *Studies in History and Philosophy of Science* 13:267–97.

————. 1983. *Wittgenstein: A Social Theory of Knowledge*. New York: Columbia University Press.

————. 1984. "The Sociology of Reasons; or, Why 'Epistemic Factors' are Really 'Social Factors.'" In Brown 1984.

————. 1992. "Left and Right Wittgensteinians." In *Science as Practice and Culture*. Ed. Andrew Pickering. Chicago: University of Chicago Press.

Bouglé, Célestin. 1938. *The French Conception of "Culture Générale" and Its Influences upon Instruction*. New York: Columbia University.

Breault, Kevin D. 1986. "Suicide in America: A Test of Durkheim's Theory of Religious and Family Integration, 1933–80." *American Journal of Sociology* 92:628–56.

Breault, Kevin D., and Karen Barkey. 1982. "A Comparative Analysis of Durkheim's Theory of Egoistic Suicide." *Sociological Quarterly* 23:321–31.

————. 1983. "Reply to Stack: Durkheim Scholarship and Suicidology: Different Ways of Doing Research in History of Social Thought, and Different Interpretations of Durkheim's *Suicide*." *Sociological Quarterly* 24:629–32.

Brown, James Robert, ed. 1984. *Scientific Rationality: The Sociological Turn*. Dordrecht: D. Reidel.

Brunschvicg, Léon, and Élie Halévy. 1894. "L'Année philosophique 1893." *Revue de métaphysique et de morale* 2:564–90.

Buchdahl, Gerd. 1982. "Editorial Response to David Bloor." *Studies in History and Philosophy of Science* 13:299–304.

Bureau, Paul. 1924. *Introduction à la méthode sociologique*. Paris: Librairie Bloud et Gay.

Camic, Charles. 1986. "The Matter of Habit." *American Journal of Sociology* 91:1039–87.

Cantor, Geoffrey N. 1975a. "The Edinburgh Phrenology Debate." *Annals of Science* 32:195–218.

————. 1975b. "A Critique of Shapin's Social Interpretation of the Edinburgh Phrenology Debate." *Annals of Science* 32:245–56.

Carrithers, Michael, Steven Collins, and Steven Lukes, eds. 1985. *The Category of the Person: Anthropology, Philosophy, History*. New York: Cambridge University Press.

Catlin, George E. G. 1938. "Introduction to the Translation." In Durkheim 1938b.

Cazeneuve, Jean. 1958. "Les Zuñis dan l'oeuvre de Durkheim et de Mauss." *Revue philosophique* 83:452–61.

Churchland, Paul. 1988. *Matter and Consciousness*. Rev. ed. Cambridge, Mass.: Bradford/MIT Press.

Clark, Terry N. 1972. "Émile Durkheim and the French University." In *The Establishment of Empirical Sociology*. Ed. Anthony Oberschall. New York: Harper and Row.

————. 1973. *Prophets and Patrons*. Chicago: University of Chicago Press.

Collins, Steven. 1985. "Categories, Concepts, or Predicaments? Remarks on Mauss's use of Philosophical Terminology." In Carrithers, Collins, and Lukes 1985.

Comte, August. 1830–42. *Cours de philosophie positive*. Ed. Michel Serres et al. 1975; Paris: Hermann.

Condillac, Étienne Bonnot de. 1746. *Essai sur l'origine des connaissances humaines*. In vol. 1 of *Oeuvres philosophiques*. Ed. Georges Le Roy. 1951; Paris: PUF.

———. 1775. *Cours d'études pour l'instruction du Prince de Parme*. In vol. 1 of *Oeuvres philosophiques*.

———. 1780. *La Logique*. In vol. 2 of *Oeuvres philosophiques*.

Condominas, Georges. 1972. "Marcel Mauss, père de l'ethnographie française." *Critique: Revue générale des publications françaises et etrangères* 297:118–39: 301:487–504.

Coser, Lewis A. 1960. "Durkheim's Conservatism and Its Implications for his Sociological Theory." In *Émile Durkheim, 1958–1917: A Collection of Essays, with Translations and a Bibliography*. Ed. Kurt H. Wolff. Columbus, Ohio: Ohio State University.

Cresswell, Peter. 1972. "Interpretations of *Suicide*." *British Journal of Sociology* 23:133–45.

Darwin, Charles. 1859. *On the Origin of Species*. 1964, 1972; New York: Atheneum.

Davidson, Donald. 1963. "Actions, Reasons, and Causes." *Journal of Philosophy* 60:685–700; rpt. in *Essays on Actions and Events*. Oxford: Clarendon Press, 1980.

Davidson, Donald. 1974. "On the Very Idea of a Conceptual Scheme." In *Inquiries into Truth and Interpretation*. Oxford: Clarendon Press, 1984.

Davy, Georges. 1911. *Durkheim*. Paris: L. Michaud.

Day, Lincoln H. 1987. "Durkheim on Religion and Suicide: A Demographic Critique." *Sociology* 21:449–61.

Delvolvé, Jean. 1932. *Réflexions sur la pensée comtienne* Paris: Alcan.

Dennes, William Ray. 1924. "The Methods and Presuppositions of Group Psychology," *University of California Publications in Philosophy*. Berkeley: University of California Press.

Desan, Philippe. "Jean Bodin et la mathématisation de l'histoire" (unpublished, undated typescript).

Descartes, René. 1628. *Rules for the Direction of the Mind*. Trans. Elizabeth S. Haldane and George Robert Thomson Ross. In *The Philosophical Works of Descartes*. London: Cambridge University Press, 1931.

———. 1647. *Principes de la philosophie*. In vol 9, part 2 of Charles Adam and Paul Tannery, eds. *Oeuvres de Descartes*. Paris: Vrin, 1971.

Douglas, Jack. 1967. *The Social Meanings of Suicide*. Princeton, N.J.: Princeton University Press.

Douglas, Mary. 1975. *Implicit Meanings*. London: Routledge and Kegan Paul.

Duhem, Pierre. 1906. *The Aim and Structure of Physical Theory*. Trans. Philip P. Wiener. Princeton, N.J.: Princeton University Press, 1954.

Espinas, Alfred. 1878. *Les Sociétés animales*. 2d ed. Paris: Librairie Germer Baillière et Cie.

Evans-Pritchard, Edward Evan. 1965. *Theories of Primitive Religion*. Oxford: Clarendon Press.

Fauconnet, Paul, and Marcel Mauss. 1901. "Sociologie," *La Grande Encylopédie* 30:154–76.

Filloux, Jean-Claude. 1970. Introduction to Durkheim 1970a.

Forman, Paul. 1971. "Weimar Culture, Causality, and Quantum Theory, 1918–1927." In vol. 3 of *Historical Studies in the Physical Sciences*. Ed. Russell McCormmach. Philadelphia: University of Pennsylvania Press.

Fournier, Marcel. 1982. "Durkheim et la sociologie de la connaissance scientifique," *Sociologie et sociétés* 14.2:55–66.

Frazer, James. 1922. *The Golden Bough: A Study in Magic and Religion*. New York: MacMillan, 1951.

Gane, Mike. 1988. *On Durkheim's Rules of Sociological Method*. London: Routledge.

Gehlke, Charles Elmer. 1915. *Émile Durkheim's Contribution to Sociological Theory*. New York: AMS Press, 1969.

Gellner, Ernest. 1962. "Concepts and Society." In Wilson 1970.

Gerstein, Dean R. 1983. "Durkheim's Paradigm: Reconstructing a Social Theory." In *Sociological Theory*. Ed. Randall Collins. San Francisco: Jossey-Bass.

Gibbs, Jack. 1961. "Suicide." In *Contemporary Social Problems*. Ed. Robert K. Merton and Robert Nisbet. New York: Harcourt, Brace and World.

Gibbs, Jack, and Walter T. Martin. 1958. "A Theory of Status Integration and Its Relationship to Suicide." *American Sociological Review* 23:140–47.

———. 1964. *Status Integration and Suicide*. Eugene, Ore: University of Oregon Press.

Giddens, Anthony. 1972b. "Four Myths in the History of Social Thought." *Economy and Society* 1:357–85.

———. 1976. *New Rules of Sociological Method*. New York: Basic Books.

———. 1977. "Positivism and its Critics." In *Studies in Social and Political Theory*. New York: Basic Books.

———. 1978. *Émile Durkheim*. New York: Viking.

Gieryn, Thomas F. 1982. "Durkheim's Sociology of Scientific Knowledge." *Journal of the History of the Behavioral Sciences* 28:107–29.

Gisbert, Pascual. 1959. "Social Facts in Durkheim's System," *Anthropos* 54:353–69.

Godlove, Terry F. 1986. "Epistemology in Durkheim's *Elementary Forms of the Religious Life*." *Journal of the History of Philosophy* 24:385–401.

———. 1989. *Religion, Interpretation, and Diversity of Belief: the Framework Model from Kant to Durkheim and Davidson*. New York: Cambridge University Press.

Goldenweiser, Alexander A. 1915. Review of Durkheim, *Les Formes élémentaires de la vie religieuse*. *American Anthropologist* 17:719–35; rpt. Durkheim 1975b:209–27.

Goldfrank, Walter I. 1972. "Reappraising Le Play." In *The Establishment of Empirical Sociology*. Ed. Anthony Oberschall. New York: Harper and Row.

Gruner, Rolf. 1967. "Plurality of Causes." *Philosophy* 42:367–74.

Gurvitch, Georges. 1950. *La Vocation actuelle de la sociologie*. Paris: PUF.

Gutting, Gary. 1984. "The Strong Programme: A Dialogue." In Brown 1984.

Hacking, Ian. 1991. "A Tradition of Natural Kinds." *Philosophical Studies* 61:109–26.

Hadden, Richard W. 1988. "Mathematics, Relativism, and David Bloor." *Philosophy of the Social Sciences* 18:433–45.

Halbwachs, Maurice. 1930. *Les Causes du suicide*. New York: Arno Press, 1975.

Hall, Robert T. 1987. *Émile Durkheim: Ethics and the Sociology of Morals*. New York: Greenwood Press.

Hamilton, Peter. 1974. *Knowledge and Social Structure: An Introduction to the Classical Argument in the Sociology of Knowledge*. London: Routledge and Kegan Paul.

Hammond, Michael. "The Case of Charles Letourneau's Obscurity in the History of French Sociology" (unpublished, undated typescript).

Haugeland, John. 1985. *Artificial Intelligence: The Very Idea*. Cambridge, Mass.: Bradford Books/MIT Press.

Hempel, Carl G. 1965. *Aspects of Scientific Explanation and Other Essays in the Philosophy of Science*. New York: Free Press.

Hirst, Paul Q. 1975. *Durkheim, Bernard, and Epistemology*. London: Routledge and Kegan Paul.

Hollis, Martin. 1970. "The Limits of Irrationality." In Wilson 1970.

———. 1985. "Of Masks and Men." In Carrithers, Collins, and Lukes 1985.

Hubert, Henri. 1905. "Étude sommaire de la représentation du temps dans la religion et la magie." In Hubert and Marcel Mauss, *Mélanges d'histoire des religions*. Paris: Alcan, 1909.

Hubert, Henri, and Marcel Mauss. 1899. "Essai sur la nature et la fonction du sacrifice." *L'Année sociologique* 2:29–138.

Huff, Toby E. 1975. "Discovery and Explanation in Sociology: Durkheim on Suicide." *Philosophy of the Social Sciences* 5:241–57.

Hull, David L. 1988. *Science as a Process: An Evolutionary Account of the Social and Conceptual Development of Science*. Chicago: University of Chicago Press.

Hume, David. 1739. *A Treatise of Human Nature*. Ed. L. A. Selby-Bigge. Oxford: Clarendon Press, 1888.

———. 1748. *An Enquiry Concerning Human Understanding*. Indianapolis: Hackett, 1977.

Hund, John. 1990. "Sociologism and Philosophy." *British Journal of Sociology* 41:197–224.

Isambert, François-André. 1979–80. "Les Avatars du 'fait moral'." *L'Année sociologique* (3d ser.) 30:17–55.

———. 1982. "De la définition: Refléxions sur la stratégie Durkheimienne de détermination de l'objet." *L'Année sociologique* (3d ser.) 32:163–92.

Joas, Hans. 1984. "Durkheim et le pragmatisme: La Psychologie de la conscience et la constitution sociale des catégories." *Revue française de sociologie* 25:560–81; trans. in *Pragmatism and Social Theory*. Chicago: University of Chicago Press, 1993.

Johnson, Barclay. 1965. "Durkheim's One Cause of Suicide." *American Sociological Review* 30:875–86.

Johnson, Harry. 1978. Comment on Jones 1977. *American Journal of Sociology* 84:171–74.

Johnson, Terry, Christopher Dandeker, and Clive Ashworth. 1984. *The Structure of Social Theory*. Basingstoke: Macmillan.

Jones, Robert Alun. 1977. "On Understanding a Sociological Classic." *American Journal of Sociology* 82:279–319.

———. 1978. "Subjectivity, Objectivity, and Historicity: A Response to Johnson," *American Journal of Sociology* 84:175–81.

———. 1981. "Robertson Smith, Durkheim, and Sacrifice: An Historical Context for *The Elementary Forms of the Religious Life.*" *Journal of the History of the Behavioral Sciences* 17:184–205.

———. 1984. "Demythologizing Durkheim: A Reply to Gerstein." *Knowledge and Society: Studies in the Sociology of Culture Past and Present* 5:63–83.

———. 1985. "Durkheim, Totemism, and the *Intichiuma.*" *History of Sociology* 5:79–121.

———. 1986a. *Émile Durkheim: An Introduction to Four Major Works.* Beverly Hills, Calif.: Sage Publications.

———. 1986b. "Durkheim, Frazer, and Smith: The Role of Analogies and Exemplars in the Development of Durkheim's Sociology of Religion." *American Journal of Sociology* 92:596–627.

Jones, Robert Alun, and W. Paul Vogt. 1984. "Durkheim's Defense of *Les Formes élémentaires de la vie religieuse.*" *Knowledge and Society: Studies in the Sociology of Culture Past and Present* 5:45–62.

Kant, Immanuel. 1787. *Critique of Pure Reason;* trans. New York: St. Martin's Press, 1929.

Karady, Victor. 1983. "The Durkheimians in Academe: A Reconsideration." In Besnard et al. 1983.

Keat, Russell, and John Urry. 1975. *Social Theory as Science.* London: Routledge and Kegan Paul, 1982.

Kelly, Colm. 1990. "Methods of Reading and the Discipline of Sociology: The Case of Durkheim Studies." *Canadian Journal of Sociology* 15:301–24.

Kuhn, Thomas S. 1962. *The Structure of Scientific Revolutions.* Chicago: University of Chicago Press, 1970.

La Capra, Dominick. 1972. *Émile Durkheim: Sociologist and Philosopher.* Ithaca, N.Y.: Cornell University Press.

Lacombe, Roger. 1925. "L'Interprétation des faits matériels dans la méthode de Durkheim." *Revue philosophique* 99:369–88.

———. 1926. *La Méthode sociologique de Durkheim.* Paris: Alcan.

Lapie, Paul. 1979. Letters to Célestin Bouglé. *Revue française de sociologie* 20:33–42.

Latour, Bruno. 1987. *Science in Action.* Cambridge, Mass.: Harvard University Press.

———. 1988. "The Politics of Explanation: An Alternative." In Woolgar 1988.

Laudan, Larry. 1971. "Towards a Reassessment of Comte's *Méthode positive.*" *Philosophy of Science* 38:35–53; rpt. Laudan 1981.

———. 1977. *Progress and Its Problems.* Berkeley: University of California Press.

———. 1981. *Science and Hypothesis.* Dordrecht: Reidel.

———. 1984. *Science and Values.* Berkeley: University of California Press.

Law, John. 1984. "Durkheimian Analysis of Scientific Knowledge: The Case of J. A. Udden's Particle Size Analysis." *Knowledge and Society: Studies in the Sociology of Culture Past and Present* 5:85–112.

———. 1986. "Power/Knowledge and the Dissolution of the Sociology of Knowledge." *Power, Action, and Belief: A New Sociology of Knowledge?* Ed. Law. London: Routledge and Kegan Paul.

Levine, Donald N. 1985. *The Flight from Ambiguity: Essays in Cultural and Social Theory.* Chicago: University of Chicago Press.

Lévi-Strauss, Claude. 1945. "French Sociology." In *Twentieth Century Sociology.* Ed. Georges Gurvitch and Wilbert Ellis Moore. New York: Philosophical Library.

P.-Lévy, Françoise. 1981. "Le Suicide chez Durkheim: Un Problème de définition." *Cahiers Internationaux de Sociologie* 70:101–10.

Lévy-Bruhl, Lucien. 1900. *The Philosophy of Auguste Comte.* New York: G. P. Putnam's Sons, 1903.

Llobera, Joseph R. 1980. "Durkheim, the Durkheimians, and Their Collective Misrepresentation of Marx." *Social Science Information/Information sur les sciences sociales* 19:385–411.

———. 1981. "Marx's Social Theory and the Durkheimian School. A Note of the Reception of Marxism in the *Année Sociologique.*" *Études durkheimiennes. Bulletin d'information,* no. 6, pp. 7–16.

Locke, John. 1690. *An Essay Concerning Human Understanding;* London: Dent, 1961.

Lugg, Andrew. 1984. "Two Historiographical Strategies: Ideas and Social Conditions in the History of Science." In Brown 1984.

Lukes, Steven. 1970. "Some Problems about Rationality." In Wilson 1970.

———. 1973. *Émile Durkheim: His Life and Work.* Stanford, Calif.: Stanford University Press, 1985.

———. 1982. Introduction to Durkheim 1982.

Lynch, Michael. 1992. "Extending Wittgenstein: The Pivotal Move from Epistemology to the Sociology of Science." In Andrew Pickering 1992.

MacIntyre, Alasdair C. 1967. "The Idea of a Social Science." *Proceedings of the Aristotelian Society,* suppl. vol. 41, pp. 95–114; rpt. *The Philosophy of Social Explanation,* ed. Alan Ryan (Oxford: Oxford University Press, 1973).

———. 1986. "Positivism, Sociology, and Practical Reasoning: Notes on Durkheim's *Suicide.*" In *Human Nature and Natural Knowledge,* ed. Alan Donagan, Anthony N. Perovich Jr., and Michael V. Wedin. Boston: Reidel.

MacKenzie, Donald A. 1978. "Statistical Theory and Social Interests: A Case Study." *Social Studies of Science* 8:35–83.

———. 1981. *Statistics in Britain 1865–1930.* Edinburgh: Edinburgh University Press.

Mackie, John Leslie. 1974. *The Cement of the Universe: A Study of Causation.* Oxford: Oxford University Press.

Madge, John. 1962. *The Origins of Scientific Sociology.* New York: Free Press.

Manicas, Peter T. 1987. *A History and Philosophy of the Social Sciences.* Oxford: Basil Blackwell.

Mannheim, Karl. 1936. *Ideology and Utopia: An Introduction to the Sociology of Knowledge.* Trans. Louis Wirth and Edward Shils. New York: Harvest Books.

Martin, Walter T. 1968. "Theories of Variation in the Suicide Rate." In *Suicide,* ed. Jack Gibbs. New York: Harper and Row.

Mauss, Marcel. 1928. Introduction to Durkheim 1928a.

———. 1930. "L'Oeuvre de Mauss par lui-même." *Revue française de sociologie* 20 (1979):209–20; trans. in Besnard et al. 1983.

———. 1938. "Une Catégorie de l'esprit humaine: La Notion de personne, celle de 'moi'." *Journal of the Royal Anthropological Institute* 68; trans. W. D. Halls in Carrithers, Collins, and Lukes 1985.

Merton, Robert K. 1945. "Paradigm for the Sociology of Knowledge." In *The Sociology of Science*. Chicago: University of Chicago Press, 1973.

Meštrović, Stjepan G. 1985. "Durkheim's Renovated Rationalism and the Idea that 'Collective Life Is Only Made of Representations.'" *Current Perspectives in Social Theory* 6:199–218.

———. 1988a. *Émile Durkheim and the Reformation of Sociology*. Totowa, N.J.: Rowman and Littlefield.

———. 1988b. "Durkheim, Schopenhauer and the Relationship between Goals and Means: Reversing the Assumptions in the Parsonian Theory of Rational Action." *Sociological Inquiry* 58:163–81

———. 1989a. "Reappraising Durkheim's *Elementary Forms of the Religious Life* in the Context of Schopenhauer's Philosophy." *Journal for the Scientific Study of Religion* 28:255–72.

———. 1989b. "Rethinking the Will and Idea of Sociology in the Light of Schopenhauer's Philosophy." *British Journal of Sociology* 40:271–93.

———. 1989c. "Searching for the Starting Points of Scientific Inquiry: Durkheim's Rules of Sociological Method and Schopenhauer's Philosophy." *Sociological Inquiry* 59:267–86.

Mill, John Stuart. 1843. *A System of Logic*. Vols 7 and 8 of *Collected Works*. Toronto: University of Toronto Press, 1974.

———. 1844. *Essays on Some Unsettled Questions of Political Economy*. Vol 4 of *Collected Works*, 1967.

Morselli, Enrico. 1882. *Suicide: An Essay on Comparative Moral Statistics*. New York: D. Appleton and Company.

Muhlfeld, L. 1893. "Débat sur les types de solidarité et la division social du travail." *Revue universitaire* 2:440–43; rpt. in Durkheim 1975a:2.288–91.

Needham, Rodney. 1963. Introduction to Durkheim 1963b.

Newton, Isaac. 1730. *Opticks, or A Treatise of the Reflections, Refractions, Inflections and Colours of Light*. 4th ed. New York: Dover Publications, 1952.

Nicholas, John M. 1984. "Scientific and Other Interests." In Brown 1984.

Nye, D. A., and C. E. Ashworth. 1971. "Émile Durkheim: Was He a Nominalist or a Realist?" *British Journal of Sociology* 22:133–48.

Ossowska, Maria. 1977. "Morality as a 'Social Fact.'" *Dialectics and Humanism* 4:35–45.

Parsons, Talcott. 1937. *The Structure of Social Action*. New York: Free Press.

———. 1956. Foreword to Durkheim 1956a.

———. 1964. *Social Structure and Personality*. New York: Free Press.

———. 1975. Comment on "Parsons' Interpretation of Durkheim" and "Moral Freedom through Understanding in Durkheim." *American Sociological Review* 40:106–11.

Peirce, Charles Sanders. 1940. *Philosophical Writings of Peirce*. Ed. Justus Buchler. New York: Dover Publications, 1955.

Pickering, Andrew, ed. 1992. *Science as Practice and Culture*. Chicago: University of Chicago Press.

Pickering, William S. F. 1984. *Durkheim's Sociology of Religion: Themes and Theories*. London: Routledge and Kegan Paul.

Pope, Whitney. 1973. "Classic on Classic: Parsons's Interpretation of Durkheim." *American Sociological Review* 38:399–415.

———. 1975. "Parsons on Durkheim, Revisited." *American Sociological Review* 40:111–15.

———. 1976. *Durkheim's "Suicide": A Classic Analyzed*. Chicago: University of Chicago Press.

Pope, Whitney, and Nick Danigelis. 1981. "Sociology's One Law." *Social Forces* 60:495–516.

Porter, Theodore M. 1986. *The Rise of Statistical Thinking 1820–1900*. Princeton, N.J.: Princeton University Press.

Prendergast, Christopher. 1983–84. "The Impact of Fustel de Coulanges' *La Cité antique* on Durkheim's Theories of Social Morphology and Social Solidarity." *Humboldt Journal of Social Relations* 11:53–73.

Putnam, Hilary. 1967. "The Nature of Mental States." In *Mind, Language, and Reality. Philosophical Papers,* vol. 2. Cambridge: Cambridge University Press, 1975.

———. 1975. *Mathematics, Matter, and Method. Philosophical Papers*, vol. 1. 2d ed. Cambridge: Cambridge University Press, 1979.

Quine, Willard V. O. 1951. "Two Dogmas of Empiricism." In *From a Logical Point of View*. New York: Harper Torchbooks, 1961.

Renouvier, Charles Bernard. 1854–64. *Essais de critique générale*. Paris: Ladrange.

Richard, Gaston. 1923. "L'Athéisme dogmatique en sociologie religieuse." *Revue d'histoire et de philosophie religieuse* 7:125–37, 229–61; trans. in Durkheim 1975b.

Rosenberg, Alexander. 1980. *Sociobiology and the Preemption of Social Science*. Baltimore, Md.: Johns Hopkins University Press.

Rousseau, Jean-Jacques. 1755. *Discourse on the Origin and Foundation of Inequality*. In *Basic Political Writings of Jean-Jacques Rousseau*. Trans. and ed. Donald A. Cress. Indianapolis, Ind.: Hackett, 1987.

———. 1762. *On the Social Contract, or Principles of Political Right*. In *Basic Political Writings of Jean-Jacques Rousseau*. Trans. and ed. Donald A. Cress. Indianapolis, Ind.: Hackett, 1987.

Rudwick, Martin. 1974. "Poulett Scrope and the Volcanoes of the Auvergne: Lyellian Time and Political Economy." *British Journal for the History of Science* 7:205–42.

Rueschemeyer, Dietrich. 1982. "On Durkheim's Explanation of Division of Labor." *American Journal of Sociology* 88:579–89.

Runciman, Walter Garrison. 1974. "Relativism: Cognitive and Moral." *Proceedings of the Aristotelian Society,* suppl. vol. 48, pp. 191–208.

Sahlins, Marshall. 1976. *Culture and Practical Reason*. Chicago: University of Chicago Press.

Schaub, Edward L. 1920. "A Sociological Theory of Knowledge." *Philosophical Review* 29:319–39.

Schmaus, Warren. 1982. "A Reappraisal of Comte's Three-State Law." *History and Theory* 21:248–66.

———. 1985. "Hypotheses and Historical Analysis in Durkheim's Sociological Methodology: A Comtean Tradition." *Studies in History and Philosophy of Science* 16:1–30.

———. 1992. "Research Programs as Intellectual Niches." *Social Epistemology* 6:13–22.

Schweber, Silvan S. 1985. "The Wider British Context in Darwin's Theorizing." In *The Darwinian Heritage*. Ed. David Kohn. Princeton, N.J.: Princeton University Press.

Seigel, Jerrold. 1987. "Autonomy and Personality in Durkheim: An Essay on Content and Method." *Journal of the History of Ideas* 58:483–507.

Selvin, Hanan. 1976. "Durkheim, Booth, and Yule: The Non-diffusion of an Intellectual Innovation." *Archives européennes de sociologie* 17:39–51.

Shapin, Steven. 1975. "Phrenological Knowledge and the Social Structure of Early Nineteenth-Century Edinburgh." *Annals of Science* 32:219–43.

———. 1979a. "The Politics of Observation: Cerebral Anatomy and Social Interests in the Edingurgh Phrenology Debates." In *On the Margins of Science: The Social Construction of Rejected Knowledge*. Ed. Roy Wallis. Sociological Review Monograph 27:139–78. Keele: University of Keele Press.

———. 1979b. "Homo Phrenologicus: Anthropological Perspectives on an Historical Problem." In *Natural Order: Historical Studies of Scientific Culture*. Ed. Barry Barnes and Shapin. London: Sage Publications.

Shapin, Steven, and Simon Schaffer. 1985. *Leviathan and the Air-Pump: Hobbes, Boyle, and the Experimental Life*. Princeton, N.J.: Princeton University Press.

Simiand, François. 1898. "L'Année Sociologique 1897." *Revue de métaphysique et de morale* 6:608–53.

Smith, Joseph Wayne. 1984. "Primitive Classification and the Sociology of Knowledge: A Response to Bloor." *Studies in History and Philosophy of Science* 15:237–43.

Sorel, Georges. 1895. "Les Théories de M. Durkheim." *Le Devenir Social* 1:1–26, 149–80.

Stigler, Stephen M. 1986. *The History of Statistics: The Measurement of Uncertainty before 1900*. Cambridge, Mass.: Harvard University Press.

Stark, Rodney, Daniel F. Doyle, and Jesse Rushing. 1983. "Beyond Durkheim: Religion and Suicide." *Journal for the Scientific Study of Religion* 22:120–31.

Strikwerda, Robert. 1982. *Émile Durkheim's Philosophy of Science: Framework for a New Social Science*. Ph. D. diss. University of Notre Dame.

Takla, Tendzin N., and Whitney Pope. 1985. "The Force Imagery in Durkheim: The Integration of Theory, Metatheory, and Method." *Sociological Theory* 3:74–88.

Tarde, Gabriel. 1898. *Études de psychologie sociale*. Paris: V. Giard & E. Briere.

———. 1901. "La realité sociale." *Revue philosophique* 52:457–77.

Taylor, Steve. 1982. *Durkheim and the Study of Suicide*. New York: St. Martin's Press.

Therborn, Göran. 1976. *Science, Class and Society: On the Formation of Sociology and Historical Materialism*. London: NLB.

Thompson, Kenneth. 1982. *Émile Durkheim*. Chichester: Ellis Horwood.

Tiryakian, Edward A. 1978. "Émile Durkheim." In *A History of Sociological Analysis.* Ed. Tom Bottomore and Robert Nisbet. New York: Basic Books.

Tosti, Gustavo. 1898. "Suicide in the Light of Recent Studies." *American Journal of Sociology* 3:464–78.

Turner, Jonathan H. 1984. "Durkheim's and Spencer's Principles of Social Organization: A Theoretical Note." *Sociological Perspectives* 27:21–32.

———. 1990. "Émile Durkheim's Theory of Social Organization." *Social Forces* 68:1089–1103.

Turner, Stephen P. 1984. "Durkheim as a Methodologist, Part 2: Collective Forces, Causation, and Probability." *Philosophy of the Social Sciences* 14:51–72.

Turner, Stephen P. 1986. *The Search for a Methodology of Social Science: Durkheim, Weber, and the Nineteenth-Century Problem of Cause, Probability, and Action.* Dordrecht: Reidel.

Vogt, W. Paul. 1976. "The Use of Studying Primitives: A Note on the Durkheimians, 1890–1940." *History and Theory* 15:33–44.

———. 1979. "Early French Contributions to the Sociology of Knowledge." *Research in Sociology of Knowledge, Sciences, and Art* 2:101–21.

———. 1983. "Obligation and Right: The Durkheimians and the Sociology of Law." In Besnard et al. 1983.

Wallwork, Ernest E. 1972. *Durkheim: Morality and Milieu.* Cambridge, Mass.: Harvard University Press.

Wilson, Bryan, ed. 1970. *Rationality.* Oxford: Basil Blackwell.

Winch, Peter. 1958. *The Idea of a Social Science and Its Relation to Philosophy.* London: Routledge and Kegan Paul.

Wittgenstein, Ludwig. 1953. *Philosophical Investigations.* Trans. Gertrude Elizabeth Margaret Anscombe. New York: Macmillan.

Woolgar, Steve. 1981. "Interests and Explanation in the Social Study of Science." *Social Studies of Science* 11:365–94.

———. 1983. "Irony in the Social Study of Science." In *Science Observed: Perspectives on the Social Study of Science.* Ed. Karin Knorr-Cetina and Michael Mulkay. Beverly Hills, Calif.: Sage Publications.

———. 1988. *Science: The Very Idea.* Chichester: Ellis Horwood.

Woolgar, Steve, ed. 1988. *Knowledge and Reflexivity: New Frontiers in the Sociology of Knowledge.* Beverly Hills, Calif.: Sage Publications.

Worsley, Peter M. 1956. "Émile Durkheim's Theory of Knowledge." *Sociological Review* (n.s.) 4:47–62.

Yearley, Steven. 1982. "The Relationship between Epistemological and Sociological Cognitive Interests: Some Ambiguities Underlying the Use of Interest Theory in the Study of Scientific Knowledge." *Studies in History and Philosophy of Science* 13:353–88.

Index

actions: individual, 38; meanings of, 250, 255; social, 38–40; unanticipated results of, 80–81. *See also* explanation; social facts

Alexander, Jeffrey, 15–18, 56, 89, 121, 146–47, 248, 267, 268, 274

Allen, H. David, 279

Alpert, Harry, 101, 271, 275

altruism, 43, 98–99

Amerindians, 33, 104, 106, 190, 192, 202, 208, 212, 235, 247, 249

analytic, 73–74

Ancient City, The (Fustel de Coulanges), 27, 281

Andler, Charles, 16, 50, 147

animal series, 105

animal societies, 66. *See also* Espinas, Alfred

animism, 196, 205, 210–11

Année Sociologique (journal), 9, 11, 24, 29–30, 38, 146, 269; Durkheim's collaborators on, 9, 11, 15, 21, 24–26, 32, 54, 177, 263, 266; Durkheim's papers in, 180, 201, 234; monograph series of, 276; policies of, 90

anomaly. *See* method

anomie, 43, 146. *See also* suicide

Anscombe, G. E. M., 261–62

anthropology, 18, 30, 32, 236, 269; Durkheim's criticisms of, 112; and race, 154. *See also* Amerindians; Australians; ethnography and ethnology

Aristotle, 190–91

Arunta. *See* Australians

Ashworth, C. E., 92, 274

association: effects of, 51–52, 271; of individuals in society, 180–81. *See also* ideas; psychology

Australians, 33, 104, 106, 115, 180, 190, 192–93, 198, 201–2, 206–8, 211–12, 223, 225–26, 229–31, 236–37, 247, 249, 264; Arunta tribe, 192, 209, 222–23, 227, 233, 235

Bacon, Francis: method of crucial tests, 3, 7, 90; and preconceptions, 86, 93, 95. *See also* crucial experiments; induction

Bainbridge, William S., 279

Baker, G. P., 261

Bankston, William B., 279

Barkey, Karen, 279

Barnes, Barry, 230, 268

Barth, Paul, 16, 147–48

Bau und Leben des sozialen Körpers (Schaeffle), 55

Belot, Gustave, 89, 94

Benoît-Smullyan, E., 56, 280, 281

Berkeley, George, 10, 52, 145, 203–4, 252, 267

Bernard, Claude, 107

Bernhöft, France, 27

Berthelot, Jean-Michel, 109, 234

Bertillon, Jacques, 26, 167, 171

Besnard, Philippe, 276, 279

Blainville, Henri de, 105

Block, Ned, 281

Bloor, David, 18–20, 66–68, 81, 205, 230, 241–43, 254, 256–65, 268, 281

Boas, Franz, 207

Bodin, Jean, 276

Bouglé, Célestin, 9, 24, 32, 51, 54, 86, 268, 269, 271, 273

Boutroux, Émile, 82, 101, 272

Boyle, Robert, 257, 260

Breault, K. D., 279

Broca, Paul, 83, 133, 269

Brunschvicg, Léon, 16, 27, 71, 121, 147

Buchdahl, Gerd, 282